T0213747

Springer Monographs in Mathematics

For further volumes:
http://www.springer.com/series/3733

Mark V. Sapir

Combinatorial Algebra: Syntax and Semantics

With Contributions by
Victor S. Guba
Mikhail V. Volkov

Springer

Mark V. Sapir
Department of Mathematics
Vanderbilt University
Nashville, TN, USA

ISSN 1439-7382   ISSN 2196-9922 (electronic)
ISBN 978-3-319-37590-8   ISBN 978-3-319-08031-4 (eBook)
DOI 10.1007/978-3-319-08031-4
Springer Cham Heidelberg New York Dordrecht London

Mathematics Subject Classification (2010): 05Exx, 05E15

© Springer International Publishing Switzerland 2014
Softcover reprint of the hardcover 1st edition 2014
This work is subject to copyright. All rights are reserved by the Publisher, whether the whole or part of
the material is concerned, specifically the rights of translation, reprinting, reuse of illustrations, recitation,
broadcasting, reproduction on microfilms or in any other physical way, and transmission or information
storage and retrieval, electronic adaptation, computer software, or by similar or dissimilar methodology
now known or hereafter developed. Exempted from this legal reservation are brief excerpts in connection
with reviews or scholarly analysis or material supplied specifically for the purpose of being entered
and executed on a computer system, for exclusive use by the purchaser of the work. Duplication of
this publication or parts thereof is permitted only under the provisions of the Copyright Law of the
Publisher's location, in its current version, and permission for use must always be obtained from Springer.
Permissions for use may be obtained through RightsLink at the Copyright Clearance Center. Violations
are liable to prosecution under the respective Copyright Law.
The use of general descriptive names, registered names, trademarks, service marks, etc. in this publication
does not imply, even in the absence of a specific statement, that such names are exempt from the relevant
protective laws and regulations and therefore free for general use.
While the advice and information in this book are believed to be true and accurate at the date of
publication, neither the authors nor the editors nor the publisher can accept any legal responsibility for
any errors or omissions that may be made. The publisher makes no warranty, express or implied, with
respect to the material contained herein.

Printed on acid-free paper

Springer is part of Springer Science+Business Media (www.springer.com)

Оле, Жене, Яше и Рэчел-Ханне

Contents

Introduction

What Is This Book About

In analyzing proofs of results about various algebraic objects (groups, semi-groups, rings), it is easy to notice two types of results: syntactic results involving words, automata, languages, and semantic results involving algebraic properties of subalgebras and homomorphic images, geometric properties of certain associated objects (graphs, manifolds, and more complicated metric spaces), dynamical properties of associated actions, etc. One of the goals of this book is to demonstrate deep connections between syntax and semantics and to show how syntax and semantics interact when we study fundamental questions concerning algebras. These include the Burnside-type questions (what makes an algebra finite?), the questions about growth (how large is an infinite algebra?), and the finite basis question (can this class of algebras be nicely described?).

The interaction between syntax and semantics is mutually beneficial. Sometimes syntax "helps" semantics. For instance, combinatorics of words and their 2-dimensional images—diagrams—helps solve Burnside-type problems and show that certain semigroups, groups and rings are infinite or finite. On the other hand semantic information about algebraic structures helps to prove whether they have or do not have finite bases of identities. Sometimes in order to prove syntactic results one needs to study semantic properties of associated objects. For example, in order to describe "avoidable identities" of semigroups, one needs semantic properties of subshifts associated with certain infinite sets of words.

What Is in the Book

The book has five chapters. The first chapter contains basic general definitions from algebra, language theory and symbolic dynamics that are used throughout the text. The main recurrent topics of this book—Burnside problems, growth of algebras, and the finite basis property—are also introduced.

The second and third chapters contain results about avoidable words and identities, including the description of semigroup varieties where the Burnside problem has a positive solution. Although the results are about words, the methods are semantic: with every infinite semigroup, one can associate a certain subshift, and recurrence properties of that subshift are used to establish algebraic properties of the semigroup. The third chapter contains Trahtman's recent (but already famous) proof of the old road coloring conjecture by Adler, Goodwyn and Weiss. The conjecture has its origin in dynamical systems but the proof is basically syntactic and belongs essentially to semigroup theory. The third chapter also contains applications of road coloring to the classification of subshifts of finite type.

The fourth chapter is about the Burnside and growth properties for associative rings, and a big part of the fifth chapter is devoted to the same properties for groups.

In particular, Chapter 4 includes short proofs of the celebrated Shirshov height theorem, the classical result by Dubnov, Ivanov, Nagata and Higman about local finiteness of rings satisfying nil identities, the Kruse–L'vov theorem about finite bases of identities of finite rings, and Belov-Kanel's counterexample for the Specht problem (that problem was one of the main problems that inspired the whole theory of varieties of rings). Gelfand–Kirillov dimension of associative algebras with polynomial growth is also considered there.

Chapter 5 begins with showing how to convert words that are equal to 1 in a group into 2-dimensional pictures—van Kampen diagram and back. Greendlinger's theorem about Dehn–Greendlinger's algorithm for small cancellation groups is proved next, which is followed by a discussion of various syntactic properties of hyperbolic groups. Grigorchuk's example of a group of intermediate growth (a solution of Milnor's problem) and the Bass–Guivarc'h computation of the growth functions of finitely generated nilpotent groups are also in that chapter. The chapter also includes, for the first time in the literature, a "road map" of a proof (due to Olshanskii) of one of the most important results in group theory: the Novikov–Adian theorem that there are infinite finitely generated groups of finite exponent. The main goal of the road map is to give the reader a relatively short and gentle introduction (the main ideas, methods, and "points of interest") to the very difficult original proof. Chapter 5 ends with a section about amenable groups. In particular, a solution (due to Adian) of the von Neuman–Day problem is explained there.

One of our goals in the book is to show that different algebraic objects have quite similar syntactic features and are strongly related. In order to do that we use several tools and objects as "recurring characters" through-out the book. For example, full binary trees are used in describing terms of free non-associative algebras, in one of the definitions of the R. Thompson group F, and in the proof that Grigorchuk's group has intermediate growth. Zimin words play a crucial role in studying Burnside properties of semigroups, in the definition of Baer radical in rings, etc. Zimin words even appeared, at least in spirit, in Olshanskii's proof of the Novikov–Adian theorem (Section 5.2.3) and as Zel'manov's words in one of the important applications of Zel'manov's solution of the restricted Burnside problem. Uniformly recurrent words (which have origin in symbolic dynamics) are used to study Burnside-type and finite basis properties of semigroups and inverse semigroups. Finite automata of several kinds as well as Church–Rosser rewriting systems also appear throughout the book.

   This book is not only about results but also about methods. There are several "universal" methods used in many proofs in the book. For example, rewriting systems are often used to find canonical forms of words and other objects. The diversity of interconnected subjects discussed in the book is manifested in the fact that this is one of a very few books where both the ergodic theorem of George Birkhoff and the HSP theorem of Garrett Birkhoff are used (Theorems 3.9.14 and 1.4.27).

What Is Not in the Book

The main purpose of the book is to cover the foundations of combinatorial algebra and several important applications. The largest area that is barely touched on in the book is "Algorithmic problems in algebra", in particular the word problem. The reason is that the area is so large that it requires a separate book (the survey by Kharlampovich and Sapir [177] is 250 pages long and does not even mention the most important results in the area obtained during the last 15 years).

For Whom Is the Book Written

I have tried to make the book self-contained. Basically any undergraduate student who took standard linear algebra and abstract algebra classes can read this book. I taught courses based on this book for undergraduate as well as graduate students, and even a 6-week course on avoidable words (Chap-ters 1–3) for high school students in one of the Canada/USA Mathcamps. In short, no significant knowledge of mathematics is required. Nevertheless prob-lem solving experience and certain mathematical maturity would definitely help in reading the book.

Exercises

The book contains more than 350 exercises. Some of them are easy, others are relatively hard. Although I do not provide solutions, harder exercises contain hints, which should help find solutions. In some cases I decided that it would be instructive to let the reader prove a theorem on his/her own, so the proof is divided into a series of exercises, each of which is not too difficult. Also quite often I formulate a statement that is almost, but not quite, obvious. After such a statement I write "why?" or "prove it!". These are little exercises, which help, I hope, understand the proofs better. These usually replace the phrases like "It is easy to see that..." and "By a straightforward computation we obtain...", which are often used in mathematical texts and which I find intimidating. The exercises "embedded" in the proofs in the book contain technical statements whose proofs can be skipped if the reader wants to learn the main ideas of the proofs as opposed to all the details of it. If I gave detailed proofs of all the exercises, the book would become twice or three times as long while containing essentially the same information.

Further Reading and Open Problems

Chapters 2–5 end with sections "Further reading and open problems". These are surveys of main recent results and open problems in the relevant areas. Each of the open problem formulated there can be a topic of a PhD thesis.

Acknowledgement. I would like to thank Miklos Abért, Laurent Bartholdi, Alexei Belov-Kanel, George Bergman, Mikhail Ershov, Rostislav Grigorchuk, Yves Guivarc'h, Susan Hermiller, Tony Huynh, David Galvin, Alexei Krasilnikov, Gerry Myerson, Ralph McKenzie, George McNulty, Sasho Nikolov, Jan Okniński, Alexander Olshanskii, Olga Sapir, Lev Shevrin, Lev Shneerson, Arseny Shur, Vladimir Trofimov, and Benjamin Weiss for their help.

Several people, including Samuel Corson, Arman Darbinyan, Gili Golan, Mikhail Kharitonov, Inna Mikhailova, Jordan Nikkel, Olga Sapir, Lev Shevrin, and Mohammad Shahryari found many misprints and errors in the original text of the book and gave many useful suggestions, which improved the book.

I am especially grateful to Victor Guba for his notes on Olshanskii's proof of the Novikov–Adian theorem, and to Mikhail Volkov for writing about the road coloring problem, the Baer radical, the Kruse–L'vov theorem, and his important contributions to several other sections of the book. The work on the book was partially supported by NSF and BSF.

Chapter 1
Main Definitions and Basic Facts

This chapter introduces the main characters that will appear in this book: sets, words, graphs, automata, rewriting systems, various kinds of (universal) algebras, varieties, free algebras (including free semigroups and groups) and subshifts. We also introduce the main properties of algebras that we are interested in: the Burnside property, the finite basis property, properties of the growth function and the growth series, etc.

Theorems proved in this chapter include the following:

- Theorem 1.2.16 of Fine and Wilf showing that a periodic word knows its period,
- Theorem 1.4.30 of Jónsson and Tarski showing that most relatively free algebras know their ranks,
- Theorem 1.4.40 of Baker, McNulty and Werner giving many examples of inherently non-finitely based finite algebras,
- George Birkhoff's recurrence Theorem 1.6.2,
- Hedlund's Theorem 1.6.12 characterizing homomorphisms of subshifts,
- Newman's diamond Theorem 1.7.10 about confluent rewriting systems,
- Levi's Theorem 1.8.6 characterizing free semigroups,
- Klein's ping-pong Theorem 1.8.32 about free groups,
- Magnus' Theorem 1.8.37 about the orderability of free groups,
- Theorem 1.8.39 of Chomsky and Schützenberger about the growth series of a rational language.

1.1 Sets

We will use standard notions related to sets such as subsets, their intersection, union and complements, permutations of sets, and maps between subsets: injective (one-to-one), surjective (onto) and bijective. We will also

© Springer International Publishing Switzerland 2014
M.V. Sapir, *Combinatorial Algebra: Syntax and Semantics*,
Springer Monographs in Mathematics, DOI 10.1007/978-3-319-08031-4_1

use standard notions from basic arithmetic such as prime numbers, facto-
rials, modular arithmetic. These notions are studied in any undergraduate
Abstract Algebra course and should be known to the reader. To refresh the
memory of the reader, we included some exercises below. We will refer to
these exercises later.

Exercise 1.1.1. Prove that

- the number of subsets of an n-element set is 2^n,
- the number of permutations of an n-element set is $n!$,
- the number of maps from an n-element set to an m-element set m^n,
- the number of k-element subsets of an n-element set (i.e., the *binomial
  coefficient* $\binom{n}{k}$) is $\frac{n!}{k!(n-k)!}$.

Exercise 1.1.2. Find a formula for the number of *partial maps* from an n-element set A to itself (i.e., the number of maps from subsets of A into A).

Exercise 1.1.3. Prove that if p, n are integers, p is a prime and $0 < n < p$, then $\binom{p}{n}$ is divisible by p.

Exercise 1.1.4. Prove the Pascal triangle formula $\binom{n-1}{k} + \binom{n-1}{k-1} = \binom{n}{k}$.

Let A_i, $i \in I$, be sets. The *Cartesian product* (named after René Descartes) $\prod_{i \in I} A_i$ of A_i (I is an arbitrary set of indices) is the set functions $f \colon I \to \bigcup A_i$ such that $f(i) \in A_i$ for every $i \in I$. For every $i \in I$, the value $f(i)$ is called the ith *coordinate* of f. If I is almost countable then one can view $\prod_{i \in I} A_i$ as the set of sequences (a_i), where $a_i \in A_i, i \in I$. If all A_i are equal to A, then $\prod_{i \in I} A_i$ is called a *Cartesian power* of A and is denoted by A^I. If $I = \{1, 2, \ldots, n, \ldots\}$, then we shall sometimes write $A_1 \times A_2 \times \ldots \times A_n \ldots$ instead of $\prod_{i \in I} A_i$.

A *binary relation* on A is a subset of $A^2 = \{(a_1, a_2) \mid a_1, a_2 \in A\}$. Sometimes instead of the notation $(a, b) \in \sigma$ we use $a\sigma b$. A binary relation σ is called

- *transitive* if $(a, b) \in \sigma, (b, c) \in \sigma \to (a, c) \in \sigma$ for every $a, b, c \in A$;
- *reflexive* if $(a, a) \in \sigma$ for every $a \in A$;
- *symmetric* if $(a, b) \in \sigma \to (b, a) \in \sigma$ for every $a, b \in A$;
- *anti-symmetric* if $(a, b) \in \sigma, (b, a) \in \sigma \to a = b$ for every $a, b \in A$;
- *equivalence* if it is transitive, reflexive and symmetric;
- *partial order* if it is transitive, reflexive and anti-symmetric;
- *total order* if it is a partial order and $(a, b) \in \sigma$ or $(b, a) \in \sigma$ for every $a, b \in A$.

Every equivalence relation on A divides A into disjoint *equivalence classes*; two elements a, b are in the same class if $(a, b) \in \sigma$.

Exercise 1.1.5. Prove that equivalence classes are disjoint and cover the whole A, i.e., they form a *partition* of A, and, conversely, every partition of A into disjoint subsets defines an equivalence relation on A.

The equivalence relation $=$ where every equivalence class is a singleton is called the *trivial* equivalence relation.

For every map $f \colon X \to Y$, the binary relation σ on X defined by $(a, b) \in \sigma$ if $f(a) = f(b)$ is an equivalence relation called the *kernel* of f. The map f is *injective* if the kernel is the trivial equivalence relation. The map f is *surjective* if $f(X) = Y$. In this case we can identify Y with the set of equivalence classes of the kernel of f, denoted by X/σ : we identify an equivalence class containing a with $f(a)$.

If σ is some set of pairs from A^2, then the smallest equivalence relation $\bar{\sigma}$ containing σ is called the *equivalence generated by* σ.

Exercise 1.1.6. Show that $\bar{\sigma}$ consists of all pairs (a, b) such that either $a = b$ or there exists a sequence of elements $x_0 = a, x_1, \ldots, x_n = b$ such that for every $i = 0, \ldots, n - 1$ either $(x_i, x_{i+1}) \in \sigma$ or $(x_{i+1}, x_i) \in \sigma$.

Recall also Zorn's lemma about partial orders [189].

Lemma 1.1.7. *Suppose $\leq$ is a partial order on a set A such that every totally ordered subset $B \subseteq A$ has an* upper bound, *i.e., an element $a \in A$ such that $b \leq a$ for every $b \in B$. Then A has a maximal element i.e., an element m such that $m \leq a \to m = a$ for every $a \in A$.*

Note that we are not using the full capacity of Zorn's lemma in this book: only sets of relatively small cardinalities appear here.

If f is a bijective map from X to X (i.e., a permutation of X), then we can consider positive and negative powers f^n of f, n an arbitrary integer, i.e., compositions of f or f^{-1} with itself several times. There is a useful equivalence relation on X associated with f: $x \sim y$ if some power f^n of f takes x to y. The equivalence classes of this relation are called the *orbits* of f.

We will also use, in a very limited manner, some basic notions from topology: distance functions (metrics), compact topological spaces, continuous functions. Sometimes we shall mention manifolds. But if the reader does not know what manifolds are, it is always enough to think of a torus (the surface of a donut) or a sphere (the surface of a ball). A couple of times, we will use polynomials in one variable and power series.

1.2 Words

1.2.1 The Origin of Words

In this book we consider *words* (i.e., finite strings of symbols), *infinite words* (i.e., infinite to the right strings of symbols), and *bi-infinite words* (i.e., infinite to both directions strings of symbols).

There are at least three different sources of words in mathematics: algebra, combinatorics and topology. Each of the following three examples showing various sources of words will be used later in the book.

Example 1.2.1. Take any semigroup S, and let X be a generating set of S (for definitions see Sections 1.4.1, 1.4.1.3). Then every element a of S is a product of elements of X: $a = x_1 x_2 \cdots x_n$. Therefore every element of S is represented by a word in the alphabet X. Important questions: What is the minimal length of a word representing a (i.e., the *length* of a)? Given two words, do they represent the same element of the semigroup (the *word problem*)? There are many more questions like these.

Example 1.2.2. Take any partition of the first m positive integers:

$$\{1,\ldots,m\} = P_1 \cup P_2 \cup \cdots \cup P_n.$$

Take an n-element alphabet $\{p_1, p_2, \ldots, p_n\}$. Label each number from 1 to m that belongs to P_i by the letter p_i. Now read these labels as you scan the numbers from 1 to m, and you get a word. For example, the word $p_1 p_2 p_1 p_2 p_1$ corresponds to the partition $\{1, 2, 3, 4, 5\} = \{1, 3, 5\} \cup \{2, 4\}$.

Example 1.2.3. Take any manifold M and divide it into n pieces M_1, M_2, ..., M_n. Again, associate a letter m_i with every piece M_i. Suppose that you are traveling on this manifold. Every hour write down a letter that corresponds to the region that you are visiting. After m hours you will get a word of length m. For example, you can drive along the US Interstate 80 from Chicago (IL) to Lincoln (NE) with a speed of 65 miles per hour and every hour write down the two letter abbreviation of the name of the state you are in. Then you will probably get the following word:

ILILIAIAIAIAIANENE

(the author has performed this experiment himself).

Example 1.2.2 is a particular case of Example 1.2.3: Take the interval $[1, m]$ of real numbers, divide it according to the partition P, and travel from 1 to m with a constant speed of 1 unit interval in 1 hour.

In fact all three ways of getting words are closely related. This will become clear later in the book.

1.2.2 The Free Semigroup

Any set X can be viewed as an alphabet. In this case the elements of X are called *symbols* or *letters*, and words with letters from X are called *words over the alphabet* X. Two words u and v are *equal*, denoted $u \equiv v$, if they are identical.

Given two words u and v, we can create a new word by first writing u and then v. The new word is called the result of *concatenating* u and v and is denoted by uv. For example, if $u \equiv abba$ and $v \equiv xyz$, then $uv \equiv abbaxyz$.

This operation is clearly associative (i.e., $(uv)w = u(vw)$ for every three words u, v, w), so the set of all words in X, usually denoted by X^*, is a *semigroup*. The most distinguished member of X^* is of course the *empty word*, i.e., the word containing no symbols. We denote the empty word by 1.

It is clear that $1u \equiv u1 \equiv u$ for any $u \in X^*$. Thus X^* is actually a *monoid* (a semigroup with an identity element). Sometimes we need to avoid the empty word. The set of all nonempty words over X is denoted by X^+. This is also a semigroup of course.

The monoid X^* is called the *free monoid over* X, and the semigroup X^+ is called the *free semigroup over* X. If n is a positive integer and u is a word, then the n-th *power* of u, u^n, is $\underbrace{uu \ldots u}_{n}$.

Exercise 1.2.4. Prove that if X consists of one element, then X^+ is isomorphic (see the definition of isomorphism in Section 1.4.1 below) to the additive semigroup of positive integers and X^* is isomorphic to the monoid of non-negative integers.

A word can be viewed geometrically as a *linear diagram* (a *labeled path*). Let u be a word of length n. Take a horizontal straight line $\mathbf{R}$. Take an interval on $\mathbf{R}$ and divide it into a union of n subintervals. Orient each of these subintervals from left to right, and label them by letters from u. The result is denoted by $\varepsilon(u)$. Later on we shall introduce 2-dimensional diagrams, which can be viewed as 2-dimensional words.

Let u be a word over an alphabet X. Then the *content* of u, $\text{cont}(u)$, of u is the set of all letters occurring in u. The *length* of u, denoted by $|u|$, is the number of letters in u (counting the repetitions). For example, if $u \equiv abbaa$, then the content of u is $\{a, b\}$ and $|u| = 5$.

If $u \equiv v_1 v_2 v_3$ for some words v_1, v_2, v_3, then v_1, v_2, v_3 are called *subwords* of u, v_1 is called a *prefix* of u, and v_3 is called a *suffix* of u. If v is a subword of u, then we shall write $v \leq u$. A word v can occur several times in u. Sometimes we need to specify a specific occurrence of v. An *occurrence* of a subword v in u is completely determined by the word v itself, a prefix of u, and a suffix of u. So if $u \equiv v_1 v v_2$, then the corresponding occurrence of v will be denoted by the triple (v_1, v, v_2). For example, the word $abbaabaa$ has two occurrences of the word aa: (abb, aa, baa) and $(abbaab, aa, 1)$.

We say that a nonempty word u *overlaps* with a nonempty word v if either one of these words is a subword of another one or a nonempty suffix of u (resp. v) is a prefix of v (resp. u). For example, the word ab overlaps with bc, aa and $abab$.

Any subset of X^* is called a *language*.

1.2.3 Orders on Words

If we order the letters in X, then we can compare arbitrary words over X. In fact there are several *order relations* on words that extend the order on X. For example, we can compare words lexicographically (as in any dictionary). The corresponding order is denoted by $<_{\text{lxg}}$. One can also compare words first by their lengths and then, if the lengths are equal, lexicographically. This order is called ShortLex, and is denoted by $<_{sl}$. For example, if $X = \{a, b\}$ and $a < b$, then $abba >_{lxg} aabbb$ and $abba <_{sl} aabbb$. The main difference between these orders is that the ShortLex order satisfies the *descending chain condition*, i.e., there are no infinite strictly descending sequence of elements, while the lexicographic order does not satisfy this condition: $bb >_{lxg} bab >_{lxg} baab >_{lxg} baaab >_{lxg} \cdots$.

Exercise 1.2.5. Prove that ShortLex satisfies the descending chain condition.

There are several other useful orders on words from X^*. In Section 4.4.2 we will use the following partial order on words called *Lex* and denoted by $\leq_\ell$. Again, fix any total order $\leq$ on X and modify the lexicographic order on X^+ by saying that we no longer compare different words when one of the words is a prefix of another. Formally $u \leq_\ell v$ if and only if either $u \equiv v$ or $u \equiv pxq$, $v \equiv pyr$ where p, q, r are (possibly empty) words, $x, y \in X$ and $x < y$.

For example, if $a < b < c \in X$, then $aaabccc <_\ell aaca$. Indeed, in this case $p = aa$, $x = a$, $q = bccc$, $y = c$, $r = a$.

Notice that every two words of the same length are Lex comparable.

Exercise 1.2.6. Show that if $u >_\ell v$, then $puq_1 >_\ell pvq_2$ for any words p, q_1, q_2.

In Section 5.1.4.3 we will use yet another order that was introduced by Dershowitz in [84]. Let, once again, $\leq$ be any total order on a finite alphabet X. The *recursive path ordering* $\geq_{rpo}$ on X^* is defined recursively as follows. Given two words $u \equiv su'$ and v from X^* where $s \in X$, we write $u \geq_{rpo} v$ if and only if either $u \equiv v$ or v is the empty word 1 or one of the following holds:

- $u' \geq_{rpo} v$,
- $v \equiv tv', t \in X$, and either

  - $s > t$ and $u >_{rpo} v'$, or
  - $s = t$ and $u' >_{rpo} v'$.

For example, if the alphabet consists of two letters a, b, and $a < b$, then $aba \geq_{rpo} ba$ because $ba \equiv ba$, and $baba >_{rpo} aaba$ because $b > a$ and $baba >_\rho aba$ (the latter inequality holds because $b > a$ and $baba >_{rpo} ba$, which, in turn, is true because $aba >_{rpo} a$, which is true because $ba >_{rpo} 1$).

Exercise 1.2.7. Show that $\geq_{rpo}$ is a total order satisfying the descending chain condition.

Exercise 1.2.8. Show that each of the four orders on words introduced in this section is *compatible* with the concatenation of words, that is, if $\leq$ is one of these orders and $u \leq v$, then for every two words p, q (in the same alphabet), we have $puq \leq pvq$.

1.2.4 Periodic Words

This entire book is about properties of words. Here we present just two basic and well-known facts about words.

Theorem 1.2.9. *Two words u and v commute (that is $uv \equiv vu$) if and only if they are both powers of a third word w.*

Proof. The "if" part of this statement is obvious. We give two proofs of the "only if" part. The first proof is well known, the second is due to Victor Guba. It is not as easy as the first one but it can be generalized to "nonlinear" words, diagrams (see Theorem 5.6.22).

(1) Induction on the length of uv. If $|uv| = 0$ (that is if both words are empty), the statement is obvious. In fact the statement is obvious if one of the words u, v is empty. Assume that the statement is true for any pair of words u, v with $|uv| < n$. Take two nonempty words u, v with $|uv| = n$ and assume that $uv \equiv vu$. Then either u is a prefix of v or v is a prefix of u. Without loss of generality assume the former is true, so $v \equiv uv'$ for some word v'. Then $uuv' \equiv uv'u$. Hence $uv' \equiv v'u$ (why?). Since $|uv'| < |uv|$, we can use the induction assumption. So we can find a word w such that $u \equiv w^k$, $v' \equiv w^\ell$ for some nonnegative integers k, ℓ. Then $v \equiv uv' \equiv w^{k+\ell}$, so both u and v are powers of w as desired.

(2) Let n be the length of uv. Consider the linear diagram $\Delta = \varepsilon(uv) = \varepsilon(vu)$. Let us denote the left end of Δ by $\iota(\Delta)$ and the right end of Δ by $\tau(\Delta)$.

Let us identify vertices $\iota(\Delta)$ and $\tau(\Delta)$. We get a circle subdivided into n labeled arcs. We can assume that all the arcs have the same length. The equality $uv \equiv vu$ gives us a rotation π of this circle preserving the partition (π maps the first arc of the first occurrence of u in Δ to the first arc of the second occurrence of u in Δ, etc.). It is easy to see that the group of all rotations of the circle preserving our partition is *cyclic*, i.e., generated by one element. Let ϕ be the generator of this group such that $\pi = \phi^k$ for some nonnegative $k < |uv|$, and let this k be the smallest possible. In other words, ϕ is the rotation that moves $\iota(\Delta)$ as little as possible in the clockwise direction. Let w be the word written on Δ between $\iota(\Delta)$ and $\phi(\iota(\Delta))$. Then it is easy to see that $u \equiv w^k$ and $v \equiv w^\ell$ for some nonnegative number ℓ.   □

Exercise 1.2.10. Unlike proof (1), proof (2) needs some polish (say, the fact that the group is cyclic needs a proof). Polish it!

Exercise 1.2.11. Prove using the idea from the second proof of Theorem 1.2.9 that some powers u^m and v^n of words u and v coincide if and only if u and v commute.

A word w' is called a *cyclic shift* of w if $w \equiv uv$, $w' \equiv vu$ for some words u, v. The cyclic shift is called *nontrivial* if u and v are not empty.

Exercise 1.2.12. Prove that a word w is a power of some shorter word if and only if w is equal to one of its nontrivial cyclic shifts.

Exercise 1.2.13. Suppose that u, v are words, $|v| > |u|$, u is not a power of a shorter word, and $u^m \equiv p_1 v q_1 \equiv p_2 v q_2$ for some p_1, p_2, q_1, q_2. Show that $|p_2| - |p_1|$ is a multiple of $|u|$. **Hint:** We can assume that v starts with u, so p_1 is empty. Then, looking at the prefix of v of length $|u|$ in the second occurrence of v in u^m, conclude that u is equal to its nontrivial cyclic shift and apply Exercise 1.2.12.

A word v is called *periodic with period u* if it is a subword of a power u^n for some $n \geq 1$ and $|v| \geq |u|$. For example, the word aba is periodic with periods ab and ba.

Clearly if a word w is periodic with period u, and v is a cyclic shift of u, then w is periodic with period v.

Exercise 1.2.14. Prove that if w is a periodic word with period u and $|u|$ divides $|w|$, then w is a cyclic shift of a power of u.

Exercise 1.2.15 (See [191, Section 3 in Chapter 11]). Prove that if u, v, w are three words and $uv \equiv vw$, then u, w are cyclic shifts of each other, and v is a subword of a power of u. Moreover, for some words u_1, w_1 and some $m \geq 0$, we have $u \equiv w_1 u_1, w \equiv u_1 w_1, v \equiv (w_1 u_1)^m w_1$. **Hint:** We have that for some $m \geq 0$, $v \equiv u^m w_1$ where w_1 does not start with u. Then $uu^m w_1 \equiv u^m w_1 w$. Hence $u w_1 \equiv w_1 w$. Since w_1 does not start with u, the word u must start with w_1. Thus $u \equiv w_1 u_1$ and $w_1 u_1 w_1 \equiv w_1 w$, hence $w \equiv u_1 w_1$.

The following theorem, due to Fine and Wilf [102], shows that periodic words containing many periods "know" their periods.

Theorem 1.2.16. *Let w be a periodic word with periods u and v. Suppose that $|w| \geq |u| + |v| - \gcd(|u|, |v|)$. Then w is periodic with period t of length dividing $\gcd(|u|, |v|)$.*

Proof. By taking cyclic shifts of u and v if necessary, we can ensure that both u and v are prefixes of w. We shall prove that u and v commute, hence by Theorem 1.2.9 the words u and v are powers of some word t; clearly the length of t divides $\gcd(|u|, |v|)$.

Enumerate all letters in the alphabet and with every word $u \equiv x_{i_0} x_{i_1} \dots x_{i_{n-1}}$ (where x_i is the ith letter in the alphabet) associate the formal series

$$f_u(x) = \sum_{j=0}^{\infty} i_{j \bmod n} x^j.$$

For example, if $u \equiv aba$ and a is the first letter, b is the second letter in the alphabet, then

$$f_u(x) = 1 + 2x + x^2 + x^3 + 2x^4 + x^5 + \dots.$$

It is clear that this series satisfies the equality

$$(1 - x^n)f_u(x) = P_u(x),$$

where $P_u(x)$ is a polynomial obtained by taking the first $n = |u|$ terms of $f_u(x)$. This polynomial has degree $|u| - 1$.

We have

$$f_u(x) = (1 - x^{|u|})^{-1} P_u(x)$$

and

$$f_v(x) = (1 - x^{|v|})^{-1} P_v(x)$$

where the degree of $P_u(x)$ does not exceed $|u| - 1$ and the degree of $P_v(x)$ does not exceed $|v| - 1$. Notice that the first $|w|$ coefficients of the series $f_w(x)$ coincide with the first $|w|$ coefficients of $f_u(x)$ and with the first $|w|$ coefficients of $f_v(x)$ (because both u and v are prefixes of w). Therefore the first $|w|$ coefficients of $f_u(x) - f_v(x)$ are zeroes. The polynomial $1 - x^{\gcd(|u|,|v|)}$ is a common factor of the polynomials $1 - x^{|u|}$ and $1 - x^{|v|}$ (prove it!). Therefore

$$f_u(x) - f_v(x) = (1 - x^{\gcd(|u|,|v|)})(1 - x^{|u|})^{-1}(1 - x^{|v|})^{-1}R(x),$$

where $R(x)$ is a polynomial of degree at most $|u| + |v| - \gcd(|u|,|v|) - 1$. Thus

$$R(x) = (1 - x^{\gcd(|u|,|v|)})^{-1}(1 - x^{|u|})(1 - x^{|v|})(f_u(x) - f_v(x)).$$

But the series on the right-hand side is divisible by $x^{|u|+|v|-\gcd(|u|,|v|)}$ and the polynomial on the left-hand side has degree at most $|u| + |v| - \gcd(|u|,|v|) - 1$. Therefore $R(x) = 0$, so $f_u(x) = f_v(x)$. Let u^∞ be the infinite power of u, $uuu\ldots$ and let v^∞ be the infinite power of v, $vvv\ldots$. We have proved that these two infinite words coincide letter by letter. Since $u^{|v|}$ and $v^{|u|}$ are finite prefixes of these words having the same length, $u^{|v|} \equiv v^{|u|}$. It remains to use Exercise 1.2.11. □

Example 1.2.17. Let $w \equiv aaabaaa$. Then w is periodic with period $u \equiv aaab$ of length 4 and with period $v \equiv aaabaa$ of length 6. But w is not periodic with period of length $2 = \gcd(6,4)$. This example shows that the inequality in Theorem 1.2.16 is optimal. Indeed, $|w| = 7$ while $|u| + |v| - \gcd(|u|,|v|) = 4 + 6 - 2 = 8$.

Exercise 1.2.18. Prove that if w is a periodic word with periods u and v, and $|w| \geq |u| + |v| - \gcd(|u|,|v|)$, then some cyclic shifts of u, v are powers of the same word t. **Hint:** Apply Theorem 1.2.16 to w and find a word t of length dividing $\gcd(|u|,|v|)$ such that w is periodic with period t. Use the fact that both u and v (being subwords of w) are periodic with period t, and apply Exercise 1.2.14.

Exercise 1.2.19. Prove that if u, v, w are three words and $w^n \equiv u^p v^q$ for some $n \geq 4, p \geq 2$ and $q \geq 2$, then u, v, w commute pairwise. **Hint:** Suppose without loss of generality that u^p has length $\geq \frac{n}{2}|w|$. Then u^p is a periodic word with periods u and w. The length of u^p is $p|u|$. Consider two cases $|u| \geq |w|$ and $|u| < |w|$, in each case apply Theorem 1.2.16 and deduce that there exists a word t of length dividing $\gcd(|u|, |w|)$ such that $u \equiv t^k$ for some k. Since u^p starts with w, and $|t|$ divides $|w|$, deduce that $w \equiv t^l$ for some l. Then v^p is also a power of t and we can apply Exercise 1.2.11.

Remark 1.2.20. It was proved by Lyndon and Schützenberger [211] that the conclusion of Exercise 1.2.19 holds even if $n \geq 2$ $(p, q \geq 2)$. Guba [129] and Harju and Nowotka [146] gave easier proofs. Many similar problems have been considered later. For example, it is proved in [147] that if $w, u_1, \ldots, u_n$ are words in a free semigroups, $w \neq u_i$ for all i, and $w^k \equiv u_1^{k_1} \ldots u_n^{k_n}$ where $2 \leq n \leq k$ and $k, k_1, \ldots, k_n \geq 3$, then all words $w, u_1, \ldots, u_n$ are powers of the same word t (hence these words pairwise commute). The proof also uses Theorem 1.2.16.

There are many other applications of Theorem 1.2.16. For example:

Theorem 1.2.21 (Guba [129]). *If the cube of a word u is a product of two periodic words whose periods are shorter than u, then u is a power of a shorter word.*

1.3 Graphs

1.3.1 Basic Definitions

A *(directed) graph* is a quadruple of sets and maps: the set of *vertices* V, the set of *edges* E, and two maps from E to V: the map $_-\colon e \mapsto e_-$ and the map $_+\colon e \mapsto e_+$. We shall also use $\iota(e)$ for e_- and $\tau(e)$ for e_+. The vertex e_- is called the *tail* of e, the vertex e_+ is called the *head* of e. We shall sometimes denote an edge e by $e_- \to e_+$ or $\iota(e) \to \tau(e)$ (although there might be several edges with the same tail and head; we call such edges *parallel*). The *out-degree* of a vertex v is the number of edges with tail v. The *in-degree* of a vertex v is the number of edges with head v. An undirected graph is a graph where for every edge $e \to f$ there is the opposite edge $f \to e$. In that case we identify the edges $e \to f$ and $f \to e$, and denote the new edge by the 2-element set $\{e, f\}$ (there can be several parallel edges connecting the same pair of vertices).

As usual, we represent graphs as pictures where each vertex is presented as a small circle and each edge e is shown as an arrow starting at e_- and pointing at e_+.

An *induced subgraph* of a graph (V, E) is a graph (V', E') where $V' \subseteq V$ and E' consists of all edges from E connecting vertices from V'

Two edges e and e' are said to be *consecutive* if $e_+ = e'_-$. A *path* p in a graph Γ is either a vertex or a nonempty sequence of consecutive edges. Its *tail* p_- is the tail of its first edge, and its *head* p_+ is the head of its last edge. A path consisting of one vertex is called an *empty path*, the vertex being its tail and its head. The *length* of a path is 0 if the path is empty and n if the path is the word $e_1 e_2 \ldots e_n$ of consecutive edges. We shall also use *infinite* and *bi-infinite* paths being respectively infinite or bi-infinite words of consecutive edges.

We say that a path $e_1 e_2 \ldots e_n$ *connects* two vertices v and v' if $v = \iota(e_1)$ and $\tau(e_n) = v'$. In an undirected graph, the relation $\sim$ between vertices defined by $x \sim y$ if x and y are connected by a path is an equivalence relation, and classes of $\sim$ are called *connected components* of the graph.

A (directed) graph Γ is called *strongly connected* if every two vertices of Γ are connected by a path. A path $e_1 e_2 \ldots e_n$ is called a *cycle* or a *closed path* if $\iota(e_1) = \tau(e_n)$. A *simple cycle* is a cycle $e_1 e_2 \ldots e_n$ such that vertices $\iota(e_1), \iota(e_2), \iota(e_n)$ are all different. A *simple path* is defined in a similar way.

Example 1.3.1. A *planar graph* is a graph whose vertices are (some) points on the plane, edges are oriented arcs connecting vertices; the arcs do not have common points except for the vertices.

1.3.2 Automata

1.3.2.1 Definitions and Basic Properties

In this book we shall often use *labeled* graphs, i.e., graphs (Q, E) where edges have labels, letters from some alphabet A (different edges may have the same label). Labeled graphs are also called *automata*. We normally denote such an automaton by (Q, A) suppressing the edge set in the notation. If (Q, E) is a finite graph, we call (Q, A) a *finite automaton*.

If $\mathcal{A}$ is an automaton, then the graph obtained from $\mathcal{A}$ by forgetting edge labels is called the *underlying graph* of $\mathcal{A}$. If Γ is a graph, then any automaton for which Γ serves as the underlying graph is said to be a *coloring* of Γ. Figure 1.1 shows a graph and two automata having that graph as the underlying graph.

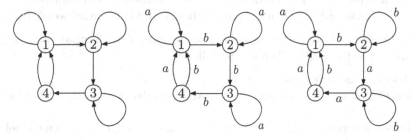

Figure 1.1 Two automata with the same underlying graph

An automaton is said to be *deterministic* if, for every vertex q, the edges with tail q have different labels. A deterministic automaton with the vertex set Q and the label alphabet A is called *complete* if each vertex $q \in Q$ has out-degree $|A|$.

An automaton is called *rooted* if we fix two sets of distinguished vertices Q_-, Q_+ called *input* and *output* vertices respectively. A word w is said to be recognized by a rooted automaton if it *labels a path* p with $p_- \in Q_-$ and $p_+ \in Q_+$ (that is the labels of the edges of that path spell w). A set of words (language) in an alphabet A is called a *rational language* (also called *regular*) if this is exactly the set of words recognized by some finite automaton. Thus if we denote the label of a path p by $\mathrm{Lab}(p)$, then the language recognized by the automaton (Q, A) is the set of words $\{\mathrm{Lab}(p) \mid p_- \in Q_-, p_+ \in Q_+\}$. Rational languages are some of the simplest possible languages.

Exercise 1.3.2. Show that every finite language is rational.

Remark 1.3.3. The set of rational languages is closed under many natural operations (see Eilenberg [93], Sakarovitch [275] or Lawson [194]). For example, if $L_1, L_2 \subseteq A^*$ are rational languages, then $L_1 \cup L_2, A^* \smallsetminus L_1, L_1 \smallsetminus L_2$, etc. are rational.

We shall use the following

Lemma 1.3.4. *The intersection of two rational languages $L_1, L_2 \subseteq A^*$ is rational.*

Proof. Let $\mathcal{A}_i = (Q_i, A)$ be an automaton recognizing $L_i, i = 1, 2$. Consider the *equalizer automaton* $\mathcal{A} = (Q_1 \times Q_2, A)$ where an edge $(q, r) \to (q', r')$ exists and is labeled by a if there exists an edge $q \to q'$ labeled by a in $\mathcal{A}_1$, and an edge $r \to r'$ labeled by a in $\mathcal{A}_2$. The set of input (output) vertices of $\mathcal{A}$ consists of pairs (q_1, q_2) where q_i is an input (output) vertex of $\mathcal{A}_i$. Then the language recognized by $\mathcal{A}$ is $L_1 \cap L_2$, hence $L_1 \cap L_2$ is a rational language. □

Exercise 1.3.5. Show that for every rational language $L \subseteq A^*$ each of the following languages is rational as well.

- the language $C(L)$ of words from A^* that contain subwords from L,
- the language $E(L)$ of words from A^* that do not contain subwords from L

Hint: Let $\mathcal{A} = (Q, A)$ be an automaton recognizing L with input vertices Q_- and output vertices Q_+. Then do the following

- Add two more vertices u, v to Q,
- for each $x \in A$ add an edge $u \to u$ and an edge $v \to v$, and label these edges by x,
- for each vertex $q \in Q_-$ and each $x \in A$, add an edge $u \to q$ and label it by x,
- for each vertex $q \in Q_+$ and each $x \in A$, add an edge $q \to v$ and label it by x,

- add u to the set of input vertices Q_- and v to the set of output vertices Q_+.[1]

Show that the resulting automaton recognizes $C(L)$.

One of the very basic facts about automata and rational languages is the following.

Theorem 1.3.6 (See [93, 275]). *Every rational language is recognized by a finite complete automaton with only one input vertex.*

1.3.3 Mealy Automata

If $\Gamma = (V, E, \iota, \tau)$ is a graph, and A is an alphabet, we can label edges of Γ by pairs from $A \times A$. We always assume that the resulting automaton is *strongly deterministic*, that is for every vertex q and every $a \in A$, there exists exactly one edge e whose tail is q and the label has the form (a, b) for some $b \in A$. Such an automaton is called a *Mealy automaton*.

Figure 1.2 shows a Mealy automaton with two vertices q_0, q_1 and two-letter alphabet $\{0, 1\}$.

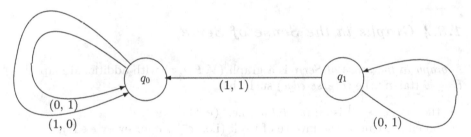

Figure 1.2 A Mealy automaton with two vertices and a 2-letter alphabet

If $M = (\Gamma, A)$ is a Mealy automaton, then for every word w in the alphabet A and every vertex q of Γ we can construct a new word $M_q(w)$ of the same length as w: start with vertex q, and read the first letter, say, a_1 of w. Find the edge of M with tail $q_1 = q$ labeled by a pair (a_1, b_1) (there exists exactly one such edge since M is strongly deterministic). Let q_2 be the head of that edge. Then the first letter of $M_q(w)$ is b_1, and the next vertex we need to consider is q_2. Let a_2 be the second letter of w. Find the edge with tail q_2 and label of the form (a_2, b_2). Let q_3 be the head of that edge. Then the second

---

[1] This long instruction is reminiscent of some recipes from a cook book. These usually end with something like "cook for 16 minutes at 350 degrees, flipping after every 4 minutes". That is not what you want to do with an automaton.

letter of $M_q(w)$ is b_2 and the next vertex is q_3. Continuing in that manner, we obtain $M_q(w)$ by scanning w from left to right and traveling along Γ at the same time. Thus M_q is a map $A^* \to A^*$.

Remark 1.3.7. One can easily extend M_q to a transformation of the set of all (bi-)infinite words in the alphabet A in the same way.

Thus with every Mealy automaton $M = (\Gamma, A)$ where $\Gamma = (V, E)$, we associate $|V|$ maps of $A^* \to A^*$. The semigroup (under composition of maps) generated by these maps is called the *semigroup of automatic maps defined by M*. An example of such a semigroup can be found in Section 3.7.2.4.

Exercise 1.3.8. Find the images of words 1010011 and 01001 under the maps M_{q_0} and M_{q_1} where M is a Mealy automaton of Figure 1.2.

If each M_q is invertible, then one can consider the group of permutations of A^* generated by all $M_q, q \in V$. It is called the *group of automatic permutations defined by M*. For an example of such a group see Section 5.7.4.

Exercise 1.3.9. Find a Mealy automaton such that the corresponding group of automatic transformations is a given cyclic (finite or infinite) group. **Hint:** For a finite cyclic group you can use a Mealy automaton with only one vertex, but for the infinite cyclic group you will need an automaton with two vertices and a 2-letter alphabet.

1.3.4 Graphs in the Sense of Serre

A *graph in the sense of Serre* is a graph (V, E, ι, τ) with additional map $^{-1}$: $E \to E$ (taking the *inverse edge*) such that

- the map $e \mapsto e^{-1}$ is an *involution*, i.e., $(e^{-1})^{-1} = e$,
- no edge e can be the inverse of itself (i.e., $e^{-1} \neq e$ for every $e \in E$),
- $\iota(e) = \tau(e^{-1})$ for every $e \in E$.

In that case we always represent E as a disjoint union of two subsets of equal sizes E^+ and E^- such that $(E^+)^{-1} = E^-$. Edges from E^+ (resp. E^-) are called *positive* (resp. *negative*). A path consisting of positive edges is called *positive*.

Remark 1.3.10. Note that one of the maps ι or τ in the definition of graph in the sense of Serre is redundant because given the map $\iota : E \to V$, we can define $\tau(e)$ as $\iota(e^{-1})$.

A graph in the sense of Serre is called a *tree* if it is connected and every cycle in it has two consecutive mutually inverse edges.

Exercise 1.3.11. (1) Show that every finite tree has vertices of degree 1 (these are called *leaves*). **Hint:** Show that all paths in the tree without mutually inverse consecutive edges have bounded lengths, consider such a path of maximal length. Its terminal vertex is a leaf.

(2) Show that the number of vertices m and the number of positive edges n of any tree satisfy $m = n + 1$. **Hint:** Use induction on m and the first part of the exercise.

1.3.5 Inverse Automata and Foldings

Let $\Gamma = (V, E, \iota, \tau, ^{-1})$ be a graph in the sense of Serre, and let A be an alphabet divided into two parts A^+, A^- (positive and negative letters) of equal sizes. Suppose we are also given an involution $^{-1}: A^+ \leftrightarrow A^-$. Then we say that labeling of Γ by A is *correct* if positive edges get positive labels and for every edge e with label a, the label of e^{-1} is a^{-1}. The correctness allows us to mention only positive edges and only positive labels when the labeling of a graph Γ is defined. If the labeling is correct and deterministic, that is no two edges with the same tail have the same label, the resulting automaton is called an *inverse automaton*. Suppose that a graph Γ in the sense of Serre with correct labeling is not deterministic, that is, there are two edges e, f with $\iota(e) = \iota(f)$ having the same label a. Then we can identify the edges e and f (and also edges e^{-1} and f^{-1}). The set of vertices V becomes smaller because the heads of the edges e and f are identified, and the functions ι, τ are redefined accordingly. This operation is called (*edge*) *folding*. After a number of foldings, the automaton Γ becomes an inverse automaton. We shall show (Exercise 1.7.13) that the inverse automaton does not depend on the order in which we fold edges in the original automaton.

1.4 Universal Algebra

1.4.1 Basic Definitions

A *(universal) algebra* is a set A with several (possibly infinite number of) functions $f_i : A^{n_i} \rightarrow A$, $i = 1, \ldots, r$ called (*basic operations*) The number n_i is called the *arity* of the operation f_i. An operation of arity 0 (a *nullary operation*) simply distinguishes an element in the algebra. The vector $(n_1, \ldots, n_r)$ is called the *type* or *signature* of the algebra. Usually when we consider algebras of a certain type, we fix the names of the operations. Usually operations are denoted by special symbols like $+, *, \cdot, \circ$, etc.

A *subalgebra* B of an algebra A is any subset of A closed under all basic operations of A, that is, if we apply a basic operation to elements of B, the result is also in B. If A is a semigroup (group, ring, etc.), then a subalgebra B of A is called subsemigroup (subgroup, subring, etc.). If $X \subseteq A$, then the subalgebra *generated* by X is the smallest subalgebra of A containing X. We denote this subalgebra by $\langle X \rangle$.

A map ϕ from an algebra A of type τ to an algebra B of the same type is called a *homomorphism* if it preserves all basic operations of A, that is, for every operation $f(x_1, \ldots, x_n)$ we have

$$\phi(f(x_1, \ldots, x_n)) = f(\phi(x_1), \ldots, \phi(x_n)).$$

A bijective homomorphism is called an *isomorphism*. In that case A and B are called *isomorphic*. Usually we do not distinguish isomorphic algebras. The *kernel* of a homomorphism $\phi : A \to B$, i.e., the equivalence relation $\mathrm{Ker}(\phi) = \{(a_1, a_2) \in A \times A \mid \phi(a_1) = \phi(a_2)\}$ *respects the operations* of A. That is, if $f(x_1, \ldots, x_k)$ is an operation from τ, and $(a_1, a_1'), \ldots, (a_k, a_k') \in \mathrm{Ker}(\phi)$, then $(f(a_1, \ldots, a_k), f(a_1', \ldots, a_k')) \in \mathrm{Ker}(\phi)$. Every equivalence relation σ on A that respects the operations from τ is called a *congruence* on A. Every congruence partitions A into a union of disjoint *congruence classes*. The set A/σ of all these congruence classes can be equipped with the structure of an algebra of type τ in the following natural way: $f(C_1, \ldots, C_m) = C$ if $C_1, \ldots, C_m$ are σ-classes and C is the σ-class containing the set $\{f(c_1, \ldots, c_m) \mid c_i \in C_i\}$. The algebra A/σ is called the *quotient (or factor-algebra) of A over the congruence σ* and is a homomorphic image of A under the *natural homomorphism* $A \to A/\sigma$ that sends each element $a \in A$ to the unique σ-class to which a belongs. By construction, σ is the kernel of this homomorphism. Thus, every congruence of A is the kernel of a homomorphism.

If $\{A_i \mid i \in I\}$ is a sequence of algebras of the same type, then the *Cartesian product* of this sequence, denoted by $\prod_{i \in I} A_i$, is the algebra of the same type defined on the Cartesian product of sets $A_i, i \in I$, where all basic operations act coordinate-wise.

Taking subalgebras, homomorphic images and Cartesian products is what algebraists do with algebras most often.

If the type is given and the names of the basic operations are fixed, one can define *terms* by composing the operations. For example, if the type is $(2, 1)$, and the basic operations are $\circ$, $^{-1}$, then the operation $((x \circ y)^{-1} \circ z)^{-1}$ is a term. The letters x, y, z are the *variables* of this term.

Every term $t(x_1, x_2, \ldots, x_n)$ of type τ defines a function $t_A : A^n \to A$ on any algebra A of type τ (the composition of functions corresponding to the basic operations). That function is again an operation on the algebra, which is called a *derived operation*. This allows one to consider the *value* of $t(x_1, \ldots, x_n)$ in A corresponding to any *substitution* $x_1 \to a_1, \ldots, x_n \to a_n$ of elements of A for variables of t.

1.4.1.1 Terms and Trees

Similar to words that are represented by linear diagrams, terms of arbitrary type can be represented by finite directed graphs, namely planar trees where leaf vertices have labels. For example, suppose that the type consists of one binary operation $\cdot$. If the term is just a variable, then the corresponding tree is a vertex labeled by that variable. Let p be a term that is not a variable. Then $p = p_1 \cdot p_2$ for some shorter terms p_1 and p_2. We can assume that the trees T_1 and T_2 corresponding to p_1 and p_2 are already constructed. To draw the tree T for p, start with a root o, label it by the symbol of the operation, $\cdot$. Draw two edges pointing down and left and down and right respectively. The heads of these edges are called *children* of the root (which is called the *parent* of these children). Then attach trees T_1 and T_2 identifying their roots with the children. For example, Figure 1.3 shows the tree representing the term $(x \cdot y) \cdot x$. Note that in the trees corresponding to terms in this type every vertex has out-degree 2 or 0. Such trees are called *full binary trees*.

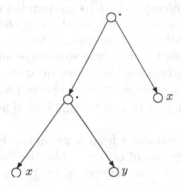

Figure 1.3 The tree representing $(x \cdot y) \cdot x$

Note also that since our type has only one operation, we do not need to label non-leaf vertices at all, and since all edges in the tree are pointing down we do not need to use directed edges to draw these trees.

Full binary trees are called *equal* if they correspond to equal terms. Another, more fancy way to say that is: binary trees are called equal if there exists a continuous deformation of the plane that takes one tree to another.

1.4.1.2 Identities and Varieties

An *identity* (also called a *law*) of algebras of type τ is a formal equality $t = t'$ between terms of type τ. We say that the identity $t = t'$ holds in an algebra A of type τ if the values of t and t' are the same for every substitution of elements of A for variables involved in t and t'. A *variety given by a set of identities* Σ is the class of all algebras of type τ satisfying identities of Σ.

1.4.1.3 Examples of Universal Algebras

A *semigroup* is an algebra of type (2) satisfying the *associativity identity* $(xy)z = x(yz)$.

A *monoid* is an algebra of the type $(2,0)$ (the nullary operation 1 is the identity element) satisfying the identities $(xy)z = x(yz), 1 \cdot x = x \cdot 1 = x$.

A *group* can be considered as an algebra of the type $(2,1,0)$ (the three operations are: the product, the operation of taking the inverse, and the identity element). In addition to being a monoid, it satisfies the identities $xx^{-1} = x^{-1}x = 1$. There are several important derived operations on groups. Here are two of these operations that will be used later. If a, b are elements of a group, then we can *conjugate* a by b to produce the element $b^{-1}ab$ usually denoted by a^b. Elements a and a^b are called *conjugate*. Also one can form a *commutator* $[a, b] = a^{-1}b^{-1}ab$. The intuition behind these operations is that two conjugate elements are in some sense similar (recall the notion of similar matrices from Linear Algebra) and the commutator $[a, b]$ measures the degree of non-commutativity of a and b because $[a, b]$ is just a quotient of ab and ba.

Exercise 1.4.1. (1) Show that $[a, b] = 1$ if and only if $ab = ba$ and if and only if $a^b = a$.

(2) Show that for every element t from a group G, the map $a \mapsto a^t$, where $a \in G$, is an *automorphism* of G (i.e., a bijective homomorphism from G to G). In particular, for every integer n, $(a^n)^t = (a^t)^n$.[2]

(3) Show that if a and b are conjugate in a group G and a has *exponent* n (that is $x^n = 1$ and $x^m \neq 1$ for every $0 < m < n$), then b has exponent n.

(4) Show that for every three elements a, b, c in a group G we have the following Hall identities (the last one is also called the *Hall–Witt* identity)

- $[a, b] = [b, a]^{-1}$ (the *anti-commutativity law for commutators*),
- $a^b = a[a, b]$,
- $[ab, c] = [a, c]^b[b, c] = [a, c][[a, c], b][b, c]$ (the *distributive law for group commutators*),
- $[[a, b^{-1}], c]^b[[b, c^{-1}], a]^c[[c, a^{-1}], b]^a = 1$ (the *group analog of the Jacobi identity*).

The following exercise gives some examples of groups.

Exercise 1.4.2 (Requires some knowledge of Linear Algebra). Let n be a natural number.

(1) (The general linear group) Show that the set of all $n \times n$-matrices with real entries and non-zero determinant is a group with operation – the usual product of matrices. This group is denoted by $\mathrm{GL}(n, \mathbb{R})$. Show that the map $A \mapsto \det(A)$ is a homomorphism from $\mathrm{GL}(n, \mathbb{R})$ to the multiplicative group of non-zero real numbers.

---

[2] Note that one needs to distinguish here $a^n = \underbrace{a \cdot \ldots \cdot a}_{n}$ from $a^t = t^{-1}at$.

(2) (The special linear group) Show that the set $SL(n, \mathbb{R})$ of all $n \times n$-matrices with real entries and determinant 1 is a subgroup of $GL(n, \mathbb{R})$.

(3) (The special orthogonal group) Consider the set $SO(n, \mathbb{R})$ of all matrices $B \in SL(n, \mathbb{R})$ such that $BB^T = 1$ (where B^T is the transpose of B). Show that it is a subgroup of $SL(n, \mathbb{R})$.

An *inverse semigroup* is a semigroup with an additional unary operation $^{-1}$ satisfying the identities $(x^{-1})^{-1} = x$, $xx^{-1}x = x$, $xx^{-1}yy^{-1} = yy^{-1}xx^{-1}$. Clearly every group is an inverse semigroup.

Exercise 1.4.3. Prove that every inverse semigroup satisfies the identity $(xy)^{-1} = y^{-1}x^{-1}$. **Hint for a syntactic proof:** (1) Prove that in every inverse semigroup, for every element a, the elements aa^{-1}, $a^{-1}a$ are *idempotents*, i.e., $(aa^{-1})^2 = aa^{-1}$, $(a^{-1}a)^2 = a^{-1}a$. (2) Prove that in an inverse semigroup, every idempotent coincides with its inverse, i.e., $e = e^2 \to e = e^{-1}$. (3) Prove that in every inverse semigroup, any two idempotents commute: $e^2 = e \& f^2 = f \to ef = fe$. (4) Show that in every inverse semigroup, for every element a there exists a unique element b ($= a^{-1}$) such that $aba = a, bab = b$. (5) Show that in every inverse semigroup, for every x, y, we have $(xy)(y^{-1}x^{-1})(xy) = xy$ and $(y^{-1}x^{-1})(xy)(y^{-1}x^{-1}) = y^{-1}x^{-1}$. Finally, deduce from (5) and (4) that $(xy)^{-1} = y^{-1}x^{-1}$. **Hint for a semantic proof:** Use the Wagner–Preston Theorem (Exercise 1.4.23).

Exercise 1.4.4. Give an example of a 2-element semigroup with an unary operation satisfying the first and third identities of inverse semigroups, but not the second.

Exercise 1.4.5. Let X be a totally ordered set with order relation $\leq$. Define the operations
$$x \cdot y = \begin{cases} x & \text{if } x \leq y \\ y & \text{if } y \leq x, \end{cases}$$
and $x^{-1} = x$. Show that X with these two operations is an inverse semigroup.

Exercise 1.4.6. Let X be a set. Consider the set $B(X)$ consisting of all partial maps $x \mapsto y$ of X whose domains (and ranges) are *singletons* (subsets with one element) and the symbol 0. Define the multiplication by $(x \mapsto y)(y \mapsto z) = x \mapsto z$, all other products are equal to 0. Define also $(x \mapsto y)^{-1} = y \mapsto x$. Show that $B(X)$ with these two operations is an inverse semigroup. The inverse semigroup $B(X)$ is called the *Brandt semigroup*, and it will appear later in this book several times.

Exercise 1.4.7. Find all inverse semigroups with at most 3 elements.

A *ring* R is an algebra with two binary operations, addition + and multiplication $\cdot$ (which we shall omit in the terms), satisfying two axioms: (a) R is a commutative group with respect to + and (b) $x(y + z) = xy + xz$, $(y + z)x = yx + zx$ (the distributivity of the product with respect to the sum). A ring is called *associative* (resp. *commutative*) if the product is associative (commutative). If a ring is associative, then it is a semigroup with respect to multiplication, this is called the *multiplicative semigroup* of the ring. Some examples of associative rings include the ring of integers $\mathbb{Z}$, the ring $\mathbb{Z}/n\mathbb{Z}$ of integers modulo n (where n is a positive integer), the ring of square matrices of size n with real entries.

Exercise 1.4.8. Let X be a commutative group with operation + and identity element 0. Define a multiplication operation on X by $x \cdot y$. Show that X becomes an associative ring.

Exercise 1.4.9. Let R be an associative ring, $n \geq 1$. Let $M_n(R)$ be the set of all square $n \times n$-matrices with entries from R with operations of addition and multiplication of matrices. Show that $M_n(R)$ is an associative ring. **Hint:** Use the standard definitions of matrix addition and multiplication. For example,

$$\begin{pmatrix} 1 & 2 & 3 \\ 4 & 5 & 6 \\ 7 & 8 & 9 \end{pmatrix} + \begin{pmatrix} 1 & 2 & 3 \\ 1 & 2 & 3 \\ 1 & 2 & 3 \end{pmatrix} = \begin{pmatrix} 2 & 4 & 6 \\ 5 & 7 & 9 \\ 8 & 10 & 12 \end{pmatrix},$$

$$\begin{pmatrix} 1 & 2 & 3 \\ 4 & 5 & 6 \\ 7 & 8 & 9 \end{pmatrix} \begin{pmatrix} 1 & 2 & 3 \\ 1 & 2 & 3 \\ 1 & 2 & 3 \end{pmatrix} = \begin{pmatrix} 6 & 12 & 18 \\ 15 & 30 & 45 \\ 24 & 48 & 72 \end{pmatrix},$$

A ring with the identity element 1 is a *field* if it is associative, commutative, and every non-zero element a has an inverse a^{-1} (i.e., $aa^{-1} = 1$). The ring of rational (respectively real, complex) numbers $\mathbb{Q}$ (respectively $\mathbb{R}$, $\mathbb{C}$) is a field.

Exercise 1.4.10. Let p be a prime number. Show that the ring $\mathbb{F}_p$ of integers modulo p is a field.

The *characteristic* of a field F is the smallest non-zero number p such that

$$\underbrace{1 + \ldots + 1}_{p} = 0$$

(note that in that case

$$px = \underbrace{x + \ldots + x}_{p} = 0$$

for every element x of the field). If such p does not exist, we say that the characteristic is 0.

Exercise 1.4.11. Show that the characteristic of a field is either 0 or a prime number. The characteristic of $\mathbb{Q}$ is 0, and the characteristic of the field $\mathbb{F}_p$ is p.

Exercise 1.4.12 (Newton binomial). Show that if a, b are commuting elements of a ring R, then for every $n > 0$,

$$(a + b)^n = \sum_{i=0}^{n} \binom{n}{i} a^i b^{n-i}.$$

Hint: Use induction and the identity from Exercise (1.1.4).

A *vector space* V over a field K is an algebra with one binary operation $+$ and one unary operation $\alpha\cdot$ for every $\alpha \in K$, so that V is a commutative group with respect to $+$, and for each $\alpha, \beta \in K$, $x, y \in V$, we have $(\alpha + \beta)x = \alpha x + \beta x$, $\alpha(x + y) = \alpha x + \alpha y$, $\alpha(\beta(x)) = (\alpha\beta)x$, $1x = x$.

We shall need the standard notions from Linear Algebra. Let us recall these here. Let V be a vector space over a field K and U be a *subspace* of V (i.e., a subalgebra of V). A set of elements S of V is called *linearly independent* if whenever the sum $k_1 s_1 + \ldots + k_n s_n$ is 0, where $k_i \in K$, s_i are pairwise different elements of S, all coefficients k_i must be equal to 0. If U is a subspace generated by S, we say that U is *spanned* by S. Any smallest (under inclusion) subset of V that spans V is called a *basis* of V. The number of elements in a basis (which does not depend on the basis) is called the *dimension* of V.

The following two exercises are straight from a Linear Algebra book.

Exercise 1.4.13. (1) Show that every non-zero element of a finite dimensional vector space belongs to a basis.

(2) Show that a set S of vectors of a finite-dimensional vector space V is a basis if these vectors are linearly independent and $|S|$ is equal to the dimension of V.

3. Let L, R be two vector spaces over a field K, B be a basis of L. Show that every map $\phi : B \to R$ extends to a homomorphism $\bar{\phi} : L \to R$ by $\bar{\phi}(\sum \alpha_i l_i) = \sum \alpha_i \phi(l_i)$.

Recall that for every two subspaces V_1, V_2 of a vector space V, their sum $\{v_1 + v_2 \mid v_1 \in V_1, v_2 \in V_2\}$ and their intersection are again subspaces of V.

Exercise 1.4.14 (Modular law for subspaces). Let V be a vector space over a field K. Let V_1, V_2, U be subspaces of V such that $V_1 \subseteq U$. Then

$$(V_1 + V_2) \cap U = (V_1 \cap U) + (V_2 \cap U).$$

Exercise 1.4.15. Show that all vector spaces over the same field of the same dimension are isomorphic.

Exercise 1.4.16. Show that if a field K has q elements, then a vector space over K of dimension n has q^n elements.

A *linear algebra* R over an associative and commutative ring K is a ring with one unary operation $\alpha\cdot$ for each $\alpha \in K$, satisfying the following axioms:

(a) $\alpha(x + y) = \alpha x + \alpha y$,
(b) $(\alpha + \beta)x = \alpha x + \beta x$,
(c) $(\alpha\beta)x = \alpha(\beta x)$, and
(d) $x(\alpha y) = (\alpha x)y = \alpha(xy)$

for every $\alpha, \beta \in K$ and $x, y \in R$.

If K has an identity element, then we assume, in addition, that $1x = x$.

An *associative algebra* R *over an associative and commutative ring* K is a linear algebra over K where the product is associative.

As in the case of groups, there are several important derived operations in rings and algebras. For example, for every two elements of a ring, one can define their *commutator* $(a, b) = ab - ba$. If we consider an associative ring as universal algebra with operations of addition and commutator, we obtain a *Lie ring*. In fact by the famous Poincaré–Birkhoff–Witt theorem [162] one can define Lie rings as subalgebras of associative rings with operations of addition, subtraction, and commutator.

Exercise 1.4.17. Show that in every associative ring the commutator satisfies the following properties (compare with Exercise 1.4.1):

• $(a + b, c) = (a, c) + (b, c)$;
• $(a, b) = -(b, a)$;
• $((a, b), c) + ((b, c), a) + ((c, a), b) = 0$ (the *Jacobi identity*).

Exercise 1.4.18. Let K be a field of characteristic $p > 0$, A be an associative algebra over K, a, b be commuting elements of A. Use Exercise 1.1.3 to show that

$$(a + b)^p = a^p + b^p.$$

1.4.1.4 Congruences in Groups, Rings and Vector Spaces

For several classical types of algebras each congruence is completely determined by one congruence class. Such are groups, rings and vector spaces.

Let σ be a congruence on a group (resp. ring) A. Let N be the class containing 1 (resp. 0). Show that N is a subgroup (resp. subring) such that for every $x \in N$, $a \in A$, $a^{-1}xa \in N$ (resp. $ax, xa \in N$). Such a subgroup is called a *normal subgroup* (resp. *ideal*) of A.

Exercise 1.4.19. Show that, conversely, for every normal subgroup (resp. ideal) N of A there exists a unique congruence σ on A such that the class of σ containing 1 (resp. 0) is N.

In this case we write A/N instead of A/σ, and say that A is an *extension of* N *by* A/N. If σ is the kernel of some homomorphism ϕ, then we also call N the *kernel* of ϕ. If two elements a, b of a group (resp. ring) A are in the same congruence class of σ (i.e., their images in A/σ are equal), then we say that $a = b$ *modulo* N and write $a \equiv b \pmod{N}$.

Exercise 1.4.20. Prove that this is the case if and only if $aN = bN$ (resp. $a + N = b + N$). Moreover, the congruence classes in this case are all of the form aN (resp. $a + N$), $a \in A$

The sets aN (resp. $a + N$) are called *cosets* of N. The number of cosets is called the *index* of N. If the index is finite, we say that N is of *finite index* in A.

Exercise 1.4.21. Show that for every vector space V, every congruence is uniquely determined by the congruence class N containing 0.

The quotient space is denoted by V/N. We say that a set of elements S is *linearly independent modulo* N if their images in V/N are linearly independent.

Exercise 1.4.22. Give examples of congruences on semigroups that are not uniquely determined by any of its congruence classes. Is every congruence on every inverse semigroup determined by any of its congruence classes?

1.4.1.5 Algebras of Transformations

The important classes of algebras considered above were defined syntactically. But in fact most of these classes can be also defined semantically as algebras of certain transformations. The next exercise gives some examples.

Exercise 1.4.23. (1) Prove Cayley's theorem (see [142]): every group is isomorphic to a group of some permutations of a set (i.e., a subgroup of the group of all permutations of the set). **Hint:** Every element g of a group G gives rise to a permutation of G: $x \mapsto gx$.

(2) Prove that every semigroup (monoid) is isomorphic to a semigroup (monoid) of some not necessarily injective transformations of a set (see [68, Volume 1]) **Hint:** Embed S into a monoid (denoted S^1) by adding a formal identity element. Then every $g \in S$ induces a transformation $x \mapsto gx$ of S^1.

(3) Prove the Wagner-Preston's theorem (see [68, Volume 1]) that every inverse semigroup is isomorphic to a semigroup of bijections between some subsets of a set: the multiplication is composition of (partial) maps, the operation $^{-1}$ sends each bijection to its inverse. **Hint:** Every element g of an inverse semigroup S induces a map from $g^{-1}gS$ to $gg^{-1}S$: $\phi_g: x \mapsto gx$ (use the fact that by definition of inverse semigroups $gS = gg^{-1}S$). The map ϕ_g is a bijection and ϕ_{gh} is a composition of ϕ_g and ϕ_h. Show that for different elements g, h the maps ϕ_g, ϕ_h are different.

(4) Prove that every associative ring is isomorphic to a ring of some *endomorphisms*, i.e., homomorphisms into itself, of some commutative group with natural operation of addition $(f + g)(x) = f(x) + g(x)$ and composition of maps as multiplication, see [27]. **Hint:** Embed the ring R into a ring R^1 with an identity element. Every element $g \in R$ induces an endomorphism of the additive group of R^1: $x \mapsto gx$.

(5) Prove that every associative algebra over a field is isomorphic to the algebra of some endomorphisms (called *linear operators*) of a vector space over the same field, see [27]. **Hint:** The same proof as in Part (4).

1.4.2 Free Algebras in Varieties

If we fix a type τ, the names of operations, $f_1, \ldots, f_r$, and a set of variables X, then the set F_X of all terms of this type containing these variables can be turned into an algebra of type τ: the operation f_i of arity n_i applied to a vector of terms $(g_1, \ldots, g_{n_i})$ gives the term $f_i(g_1, \ldots, g_{n_i})$. This is the *absolutely free algebra of type τ over the set X of free generators*. Being *free over X* in a given class $\mathcal{K}$ of algebras means that:

- $F_X = F_X(\mathcal{K})$ belongs to $\mathcal{K}$,
- F_X is generated by X and
- every map ϕ from X into any algebra $A \in \mathcal{K}$ is uniquely extendable to a homomorphism $\bar{\phi}$ from F_X into A.

The following theorem is a fundamental (but not difficult) result about varieties of algebras.

Theorem 1.4.24 (See Mal'cev [218]). *Every variety contains a free algebra over any set X. The free algebra is determined up to isomorphism by the cardinality of X.*

Free algebras in a variety are called *relatively free*. If F is a relatively free algebra over X, then X is called a set of *free generators* of F and the cardinality of X is called the *rank* of F.

Relatively free algebras can be described without reference to varieties (see [218]).

Theorem 1.4.25. *An algebra F generated by a set X is relatively free with a set of free generators X if and only if every map $X \to F$ extends to a homomorphism $F \to F$.*

If F is a relatively free algebra in a variety $\mathcal{V}$ with the set of free generators X and A is an algebra in the same variety, then any map $\phi : X \to A$ is called a *substitution*. If $\phi(X)$ generates A, then every element a of A has a pre-image $f \in F$ under $\bar{\phi}$, and we say that f *represents* a in A.

Exercise 1.4.26. Suppose that F is a relatively free algebra in a variety $\mathcal{V}$ with free generating set $\{x_1, \ldots, x_m\}$, and the equality

$$R_1(x_1, \ldots, x_m) = R_2(x_1, \ldots x_m)$$

is true in F where R_1, R_2 are terms of the signature of $\mathcal{V}$. Prove that $R_1 = R_2$ is an identity of $\mathcal{V}$.

1.4.3 The Garrett Birkhoff Theorem

Any variety is clearly closed under taking subalgebras, homomorphic images and Cartesian products. The converse also holds:

Theorem 1.4.27 (Garrett Birkhoff [218]). *Any class of algebras closed under taking subalgebras, homomorphic images and Cartesian products is a variety.*

Thus varieties may be defined in a "syntactic" way (by identities) and in a "semantic" way (as classes closed under these three most popular algebraic constructions).

A similar situation may be found in other parts of mathematics. For example, a manifold can be defined ("syntactically") by equations and ("semantically") as a locally Euclidean topological space.

The fact that a manifold is locally Euclidean means that if we are on this manifold, and cannot go very far (or if we can not memorize big volumes of data), then we won't be able to distinguish the manifold from the ordinary Euclidean space.

The fact that a variety of universal algebras is closed under the three constructions means that we can "live" inside a variety, use these constructions and never need any algebras outside the variety.

If T is a class of universal algebras of the same type, then the *variety generated by T*, $\operatorname{var} T$ is the intersection of all varieties containing T. Equivalently, $\operatorname{var} T$ is the variety given by all identities that hold in all algebras of T.

Let T be a class of universal algebras. Then $H(T)$ (resp. $S(T)$, $P(T)$) denotes the class of homomorphic images (resp. subalgebras, Cartesian products) of algebras from T. It is not difficult to deduce from Theorem 1.4.27 the following formula (due to Garrett Birkhoff) (see [218]):

$$\operatorname{var} T = HSP(T). \tag{1.4.1}$$

Exercise 1.4.28. Prove (1.4.1). **Hint:** Prove that for every class U, we have

$$SH(U) \subseteq HS(U), PH(U) \subseteq HP(U), PS(U) \subseteq SP(U),$$
$$HH(U) \subseteq H(U), SS(U) \subseteq S(U), PP(U) \subseteq P(U).$$

Note that all classes of algebras from Section 1.4.1.3 are varieties except for the class of all fields: it is not closed under Cartesian products.

1.4.4 Relatively Free Algebras May Not Know Their Ranks. The Jónsson–Algebras

Consider the signature consisting of one binary operation · and two unary operations α, β. The variety $\mathcal{J}$ is given by three identities

$$\alpha(a \cdot b) = a, \beta(a \cdot b) = b, \alpha(a) \cdot \beta(a) = a.$$

An example of algebra from this variety can be constructed as follows. Consider any countable set X. Then $X \times X$ is also countable, so there exists a bijection $f: X \times X \to X$. Now let us define $x \cdot y = f(x, y)$, and define $\alpha(x)$ and $\beta(x)$ as the first and the second coordinates of the pair $f^{-1}(x)$ respectively. Clearly the algebra we obtain this way belongs to $\mathcal{J}$. Conversely, for every algebra A in $\mathcal{J}$ we can define a bijection between $A \times A$ and A by $f(a, b) = a \cdot b$. Hence every algebra in $\mathcal{J}$ is either trivial or infinite.

The variety $\mathcal{J}$ is called the Jónsson–Tarski variety, it also satisfies the following remarkable property [163].

Theorem 1.4.29. *All finitely generated relatively free algebras in $\mathcal{J}$ with more than one element are isomorphic.*

Nevertheless this "pathology" does not happen in any "natural" variety because of another theorem from [163].

Theorem 1.4.30. *Let $\mathcal{V}$ be a variety of universal algebras that contains a finite nontrivial algebra. Then every relatively free algebra with $k \geq 0$ free generators cannot be generated by fewer than k elements.*

Proof. Indeed, let B be a nontrivial finite algebra in $\mathcal{V}$, and let A be a relatively free algebra in $\mathcal{V}$ with a finite set X of free generators. Then, by definition of relatively free algebras, every map $X \to B$ extends to a homomorphism from A to B. Since all these homomorphisms are different (their restrictions to X are different), there are exactly $|B|^{|X|}$ homomorphisms from A onto B by Exercise 1.1.1. If A can be generated by a set Y with fewer than $|X|$ elements, then the number of homomorphisms from A to B is at most $|B|^{|Y|}$ (every homomorphism from A to B is uniquely determined by its restriction to Y), a contradiction.   □

Theorem 1.4.30 applies to all varieties considered later in this book: semigroups, inverse semigroups, rings and groups.

1.4.5 Locally Finite Varieties

Theorem 1.4.31. *Let C be a class of algebras with finitely many operations. Then the following conditions are equivalent.*

(1) *Every finitely generated algebra in* var C *is finite, in other words, every algebra in* var C *is* locally finite.
(2) *For every natural number m there exists a number $n = n(m)$ such that the* order *(i.e., the number of elements) of every m-generated subalgebra of any algebra from C does not exceed n.*

The second condition means that the orders of all m-generated subalgebras of algebras from C are bounded from above.

Exercise 1.4.32. (a) Prove that the class of all finite semigroups does not satisfy condition (2) of Theorem 1.4.31, but any class consisting of finite number of finite algebras satisfies this condition.
(b) Give an example of a class C consisting of one infinite algebra that satisfies condition (2).

Proof of Theorem 1.4.31. (1) $\rightarrow$ (2). Suppose that every finitely generated algebra in var C is finite. Let us take any number m and any m-generated subalgebra $S = \langle X \rangle$ of an algebra from C. Since every variety is closed under taking subalgebras, $S \in$ var C. By Theorem 1.4.24, var C contains relatively free algebras with any number of generators. Let us take the relatively free algebra F_m with m generators. We may suppose that F_m is generated by the same set X. Since every finitely generated algebra in var C is finite, F_m is finite. Let n be the number of elements in F_m. We know that every relatively free algebra is free "outside", that is every map from the set of generators X to any other algebra A in the variety var C is extendable to a homomorphism $F_m \rightarrow A$. Therefore there exists a homomorphism $\phi : F_m \rightarrow S$ that is identical on X. The image of this homomorphism is a subalgebra of S. It contains the set of generators X, hence it contains all elements of S. Therefore S is an image of F_m, so the number of elements in S does not exceed n (the number of elements in F_m).

(2) $\rightarrow$ (1). Suppose that for every m we have found a number n such that the order of any m-generated subalgebra of any algebra in C does not exceed n. We have to prove that every finitely generated algebra in var C is finite. Let S be an m-generated algebra in var C. By formula (1.4.1) S is a homomorphic image of a subalgebra T of a Cartesian product $\prod_i A_i$ of algebras from C.

Suppose S is infinite. Then T is also infinite. Notice that we may assume that T is also m-generated. Indeed, we can take a pre-image of each generator or S in T, and generate a subalgebra by these preimages; S will be a homomorphic image of this subalgebra.

Recall that the Cartesian product $\prod_i A_i$ consists of vectors whose i-th coordinate is from A_i. The *projection* π_i of the Cartesian product onto A_i is a homomorphism. $\pi_i(T)$ is generated by m elements (images of generators of T). Therefore the order of $\pi_i(T)$ does not exceed some number $n = n(m)$ (all algebras A_i are from C).

There exists only finitely many algebras of order $\le n$ (we have only finitely many operations, each of them is defined by the "multiplication table" and there are only finitely many "multiplication tables").

Thus there exist only finitely many images $\pi_i(T)$.

For every finite algebra A there exist only finitely many homomorphisms from T to A. Indeed, each homomorphism is determined by the images of generators of T, we have finitely many generators and A is also finite.

Therefore the number of different kernels of homomorphisms π_i in T is finite. Recall that the kernel of a homomorphism is the partition of T, which glues together elements that go to the same place under this homomorphism. Each of these partitions has only finitely many classes. Since T is infinite, there exist two different elements t_1 and t_2 in T, which belong to the same class of each of these kernels. Thus $\pi_i(t_1) = \pi_i(t_2)$ for every i. Therefore these two vectors have the same coordinates. This means that they are equal, a contradiction.   □

An important case, when Theorem 1.4.31 trivially applies, is when C consists of one finite algebra. In this case we say that the variety $\operatorname{var} C$ is *finitely generated*. The case when C consists of finitely many finite algebras $A_1, \ldots, A_n$ is not more general because then $\operatorname{var} C = \operatorname{var}(A_1 \times \ldots \times A_n)$.

1.4.6 The Burnside Problem for Varieties of Algebras

For every class of algebras one can ask the question first asked in 1902 by Burnside [60] for groups where the orders of all subgroups generated by one element are bounded: *Is every algebra in that class locally finite?* That question is sometimes formulated in a more philosophical (and less precise) way: *What makes an algebra in this class finite?* For some classes of algebras, this question is easy, for others it is very complicated, but this question is always in the center of attention for any class of algebras, basically because generally finite algebras are much easier than infinite. Among various classes of algebras where the Burnside question was studied, varieties are of particular interest. The original (bounded) Burnside problem solved by Novikov and Adian [3] is about the variety of groups $\mathcal{B}_n$ given by one identity $x^n = 1$.

1.4.7 Finitely Based Finite Algebras

One of the most important properties of a variety is being *finitely based*, that is, being given by a finite number of identities. We say that an algebra is *finitely based* if the variety it generates is finitely based.

For each of the types of algebras above, it is known that "most" varieties are not finitely based. Still there are important classes of varieties consisting of finitely based varieties. Here are some of the results. We will talk more about some of them later.

Theorem 1.4.33 (Oates and Powell [248, 252], Kruse [188], L'vov [208]). *Every finite group (finite associative ring) is finitely based.*

Note that the proofs of Theorem 1.4.33 in cases of groups and associative rings (and also in the case of finite Lie rings [20] not discussed in this book) are very similar. Thus we shall present (in Section 4.5.2) only the proof for rings, which is easier. All proofs are based on the following universal algebra idea used first by Oates and Powell.

We call an algebra D a *divisor* of an algebra A if D is a homomorphic image of a subalgebra B of A. The divisor D is said to be *proper* if either B is a proper subalgebra of A (that is, $B \neq A$) or the homomorphism of B onto D is not an isomorphism. A finite algebra A is called *critical* if A does not belong to the variety generated by all of its proper divisors.

Lemma 1.4.34. *Every locally finite variety of algebras is generated by its critical algebras.*

Proof. Let $\mathcal{V}$ be a locally finite variety and let $\mathcal{V}'$ be the variety generated by all critical algebras in $\mathcal{V}$. Assume that $\mathcal{V}' \subsetneq \mathcal{V}$. Then there exists an identity $t_1 = t_2$ that holds in $\mathcal{V}'$ but fails in some algebra $A \in \mathcal{V}$. If n is the number of variables in the terms t_1 and t_2, then the identity $t_1 = t_2$ fails already in some n-generated subalgebra B of the algebra A. Since $\mathcal{V}$ is locally finite, B is finite, and we conclude that there exists a finite algebra in $\mathcal{V}$ but not in $\mathcal{V}'$, i.e., in $\mathcal{V} \smallsetminus \mathcal{V}'$.

Now let C be an algebra with the smallest number of elements from $\mathcal{V} \smallsetminus \mathcal{V}'$. Then every proper divisor of C must lie in $\mathcal{V}'$ hence C does not belong to the variety generated by all of its proper divisors. We see that C is critical. But then C must belong to $\mathcal{V}'$ by the definition of $\mathcal{V}'$, a contradiction. □

A variety $\mathcal{V}$ is said to be a *Cross* variety if it has the following three properties:

(a) $\mathcal{V}$ is locally finite;
(b) $\mathcal{V}$ is finitely based;
(c) $\mathcal{V}$ has only finitely many critical algebras.

Proposition 1.4.35. *Every Cross variety has only finitely many subvarieties and each of them is a Cross variety.*

Proof. Let $\mathcal{V}$ be a Cross variety. The fact that $\mathcal{V}$ has only finitely many subvarieties immediately follows from the property (c) in the definition of a Cross variety and from Lemma 1.4.34. In order to show that every subvariety $\mathcal{W} \subseteq \mathcal{V}$ is itself a Cross variety, it suffices to verify that $\mathcal{W}$ is finitely based as the properties (a) and (c) are clearly inherited by subvarieties. If $\mathcal{W} = \mathcal{V}$,

there is nothing to prove; otherwise by Lemma 1.4.34 and property (c) there
is a maximal finite decreasing chain of subvarieties

$$\mathcal{V} = \mathcal{V}_0 \gneqq \mathcal{V}_1 \gneqq \ldots \gneqq \mathcal{V}_n = \mathcal{W} \tag{1.4.2}$$

between $\mathcal{V}$ and $\mathcal{W}$. We show that $\mathcal{V}_i$ is finitely based by induction. By Property (b), the claim holds for $i = 0$. Now suppose that $i > 0$ and $\mathcal{V}_{i-1}$ has a finite
basis of identities Σ. Take any identity σ_i that holds in $\mathcal{V}_i$ but fails in $\mathcal{V}_{i-1}$.
Then the system of identities $\Sigma \cup \{\sigma_i\}$ defines a variety that contains $\mathcal{V}_i$ and
is strictly contained in $\mathcal{V}_{i-1}$ (why?). From the maximality of the chain (1.4.2)
there is no variety strictly between $\mathcal{V}_{i-1}$ and $\mathcal{V}_i$ hence $\Sigma \cup \{\sigma_i\}$ must define
$\mathcal{V}_i$. Thus $\mathcal{V}_i$ is also finitely based.   □

Now we can state the results by Oates–Powell, Kruse and L'vov more
precisely:

Theorem 1.4.36. *Every finite group (finite ring) generates a Cross variety.*

Note that semigroups, monoids and inverse semigroups are missing from
Theorems 1.4.33, 1.4.36. That is because for this types of algebras the theorem
fails dramatically (see Sections 3.6).

Note also that by the celebrated theorem of McKenzie [225] (solving a
well-known Tarski problem) the class of finite universal algebras with one
binary operation that generate finitely based varieties is not *recursive*, i.e.,
there is no algorithm,[3] which recognizes if a finite universal algebra generates
a finitely based variety.

1.4.8 Inherently Non-finitely Based Finite Algebras: The Link Between Finite and Infinite

A locally finite variety of algebras $\mathcal{V}$ is called *inherently non-finitely based* if
every locally finite variety containing $\mathcal{V}$ is not finitely based. In other words,
$\mathcal{V}$ is inherently non-finitely based if for every finite set Σ of identities of $\mathcal{V}$
the Burnside question for var Σ has negative answer: var Σ contains finitely
generated infinite algebras. If A is a finite algebra, then A is called *inherently
non-finitely based* if var A is inherently non-finitely based. If a finite algebra
B is such that $A \in \text{var } B$ (say if A is a subalgebra or a homomorphic image
of B) and A is inherently non-finitely based, then so is B. Thus finding
one inherently non-finitely based finite algebra gives many finite algebras

---

[3] We do not give a definition of algorithm because everybody knows what it is, but
nobody can define it precisely.

without finite bases of identities (and generating different varieties) and also gives many finitely based varieties where the Burnside question has a negative answer.

A connection between inherently non-finitely based algebras and the Burnside question is more apparent in this theorem.

Theorem 1.4.37. *A locally finite variety V with finite number of operations is inherently non-finitely based if and only if there exists a sequence of finitely generated infinite algebras $A_n, n \geq 1$ such that every n-generated subalgebra in A_n belongs to V (and hence is finite).*

Proof. Indeed, suppose that V is inherently non-finitely based. Consider the relatively free algebra F_n in V. It is finite by our assumption. Let $f_1, \ldots, f_m$ be all elements in F_n represented by some terms in the generators $x_1, \ldots, x_n$ of F_n. The algebra F_n is completely determined by its "multiplication tables". Each entry of the table corresponding to an operation γ looks like

$$\gamma(f_{i_1}, \ldots) = f_j.$$

We can rewrite each of these equalities in terms of $x_1, \ldots, x_n$, obtaining a *relation* $w = w'$. Let us denote by Σ the finite set of these relations. By Exercise 1.4.26 each relation from Σ is an identity in V. Let W be the variety given by identities Σ. Since V is inherently non-finitely based and Σ is finite, W contains a finitely generated infinite algebra A_n. Since every n-generated subalgebra $A' = \langle a_1, \ldots, a_n \rangle$ of A_n satisfies all the identities from Σ, it satisfies all the relations from the "multiplication tables" of F_n (with x_i replaced by a_i). Hence A' is a homomorphic image of F_n (the map $x_i \mapsto a_i, i = 1, \ldots, n$, extends to a homomorphism), and hence is finite and belongs to V.

Conversely, suppose that a sequence of algebras A_n, $n \geq 1$, as in the formulation of the theorem has been found. Consider any variety W containing V and given by identities in at most n variables Σ. Since every n-generated subalgebra A' of A_n belongs to V, A' satisfies Σ. That means that A_n satisfies Σ as well (every substitution of elements of A_n for variables of Σ maps the variables inside one of these A'). Hence $A_n \in W$, and W is not locally finite. $\square$

Thus having a finite inherently non-finitely based algebra of certain type implies that we can construct a sequence of infinite finitely generated algebras of that type, which give stronger and stronger negative answers to the Burnside question: these algebras, while all infinite, become more and more finite.

It turned out [228, 243, 267] that many finite algebras are inherently non-finitely based, and in fact the set of all finite inherently non-finitely based finite algebras (even with one binary operation) is not recursive [225]. Here is a relatively easy example.

Definition 1.4.38 (Shallon [290]). Let $G = (V, E)$ be a (directed) graph. The *graph algebra* $A(G)$ is the algebra with the set of elements $V \cup \{\infty\}$ and one binary operation

$$x \cdot y = \begin{cases} x & \text{if } (x, y) \in E, \\ \infty & \text{otherwise.} \end{cases}$$

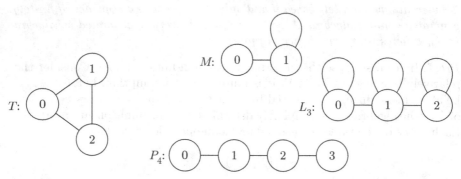

Figure 1.4 Four graphs

A complete description of inherently non-finitely based finite graph algebras is given by the following theorem.

Theorem 1.4.39 (Baker, McNulty, Werner [18]). *For every finite graph G the following conditions are equivalent.*

(1) $A(G)$ *is not finitely based.*
(2) $A(G)$ *is inherently non-finitely based.*
(3) G *contains one of the graphs shown on Figure 1.4 as an induced subgraph.*

We are not going to give a full proof of Theorem 1.4.39, but we shall prove the implication (3) → (2) by showing that the graph algebras corresponding to graphs in Figure 1.4 are inherently non-finitely based. Note that algebra M first appeared in the paper [242], where the fact that it is not finitely based was proved. This algebra has 3 elements, which is the smallest number of elements in any non-finitely based finite algebra by a result of Lyndon [209].

Implication (3) → (2) is a corollary of the following general theorem. The theorem connects the finite basis question with some properties of subshifts.

Let B be a finite algebra. The infinite Cartesian power $B^{\mathbb{Z}}$ consists of bi-infinite words over B with coordinate-wise product. Let T be the map that shifts each bi-infinite word one letter to the right (in Section 1.6.2 we shall call the pair $(B^{\mathbb{Z}}, T)$ a *full shift*). By the HSP theorem, $B^{\mathbb{Z}}$ belongs to var B.

Theorem 1.4.40 (Baker, McNulty, Werner [19]). *Let B be a finite algebra with one binary operation and zero denoted by ∞. Suppose that there exists a bi-infinite word α of non-zero elements of B such that*

(a) in $B^{\mathbb{Z}}$, the product of any two shifts of α, $T^n(\alpha) \cdot T^m(\alpha)$, is either a shift $T^i(\alpha)$ of α or a bi-infinite word containing ∞.

(b) there are only finitely many $m, n \in \mathbb{Z}$ such that $\alpha \cdot T^m(\alpha) = T^n(\alpha)$ or $T^m(\alpha) \cdot \alpha = T^n(\alpha)$.

(c) there exists t such that $\alpha \cdot T^t(\alpha) = T(\alpha)$ or $T^t(\alpha) \cdot \alpha = T(\alpha)$.

Then B is inherently non-finitely based.

Proof. Consider the subalgebra U of $B^{\mathbb{Z}}$ generated by all shifts $T^i(\alpha)$, $i \in \mathbb{Z}$. There is a congruence $\sim$ on U with one class consisting of all bi-infinite words containing ∞ and all other classes singletons (in the case of semigroups we shall call such congruences the Rees congruences corresponding to an ideal). The quotient $V = U/\sim$ clearly belongs to var B. Its elements, by (a), are all shifts $T^i(\alpha)$, $i \in \mathbb{Z}$, and symbol $\widetilde{\infty}$ that denotes the $\sim$-class consisting of bi-infinite words containing the letter ∞. The map T induces an automorphism of V (the shift), which we shall also denote T. Note that T has exactly two orbits, $\{\widetilde{\infty}\}$ and the rest of V. For every $k \geq 1$, the map T^k has exactly $k + 1$ orbits.

Note that (b), (c) imply that α is not periodic, i.e., $T^k(\alpha) \neq \alpha$ for any $k > 0$. Indeed, suppose that $T^k(\alpha) = \alpha$ for some $k > 0$. By (c), we have that for some t one of the two equalities from (c) hold. Suppose that the first equality holds (the other case is very similar): $\alpha \cdot T^t(\alpha) = T(\alpha)$. Then $T^m(\alpha) \cdot T^{m+t}(\alpha) = T^{m+1}(\alpha)$ for every m. Taking $m = k\ell$ for any ℓ, we get (since T is an automorphism) $T^{k\ell}(\alpha) \cdot T^{k\ell+t}(\alpha) = T^{k\ell+1}(\alpha)$ or $\alpha \cdot T^{k\ell+t}(\alpha) = T(\alpha)$ for every ℓ, which contradicts (b).

We will use the following easy observation several times. Let q be the maximal number that appears in (b).

Remark 1.4.41. Note that if k is bigger than q, then $T^s(\alpha)T^t(\alpha) = \widetilde{\infty}$ in V whenever $|t - s| \geq k$.

Let Z be the semigroup of all integers with operation $m * n = \max(m, n)$. It has the automorphism $\tau: i \mapsto i - 1$. We construct two algebras W_1 and W_2.

The algebra W_1 consists of all pairs (a, b) where $a \in V \smallsetminus \{\widetilde{\infty}\}$, $b \in Z$ and the symbol ∞_1 (which will play the role of zero). The product is $(a, b)(a', b')$ is $(aa', b * b')$ if $aa' \neq \widetilde{\infty}$ and ∞_1 otherwise. (This construction is called the 0-*direct product* of V and Z.)

Exercise 1.4.42. Prove that the algebra W_1 belongs to var B. **Hint:** Indeed, let $u = v$ be any identity of B. Prove that if the identity $u = v$ is *normal* (i.e., u and v contain the same letters) then $u = v$ holds on Z, and so it holds on W_1, which is a homomorphic image of $B^{\mathbb{Z}} \times Z$. If $u = v$ is not normal, and, say, letter x appears in u but not in v, set $x = \infty_1$ and deduce that identities $v = \infty_1$ and $u = \infty_1$ both hold in B. Then both $u = \infty_1$ and $v = \infty_1$ hold identically in W_1, hence, again, W_1 satisfies $u = v$.

Now take a "very large" integer k. It will be clear how large k should be, but we certainly assume that k is bigger than $2q$.

The algebra W_2 consists of orbits of the automorphism (T^k, τ) on the set $(V \smallsetminus \{\infty\}) \times Z$, and symbol ∞_2 (which plays the role of zero), with multiplication of orbits defined by

$$O_1 \cdot O_2 = \begin{cases} O_3, & \text{if } O_1 \text{ contains a pair } (a,b), O_2 \text{ contains a pair } (c,d) \\ & \text{such that } (ac, b*d) \in O_3, \\ \infty_2, & \text{otherwise.} \end{cases}$$

Note that the partition into orbits of (T^k, τ) is *not* a congruence on V so the following exercise is not quite trivial.

Exercise 1.4.43. Check that the operation on W_2 is well defined that is the product O_1O_2 is uniquely determined by O_1, O_2. **Hint:** This is one of the places in the proof where Remark 1.4.41 is used.

We claim that

(1) The algebra W_2 is not locally finite.
(2) If ℓ is much less than k, then every ℓ-generated subalgebra in W_2 is a quotient of a subalgebra of W_1 and so it belongs to var B by Exercise 1.4.42.

By Theorem 1.4.37, this would imply that B is inherently non-finitely based.

To show (1), let $\beta = T^k(\alpha)$, $\gamma = T^t(\alpha)$. Since $\alpha\gamma = T(\alpha)$ (the case $\gamma\alpha = T(\alpha)$ is similar) in V by (c), we have

$$\begin{aligned}
(\ldots(\beta T^k(\gamma))T^{k+1}(\gamma)\ldots)T^{2k-1}(\gamma) &= (\ldots(T^k(\alpha)T^k(\gamma))T^{k+1}(\gamma)\ldots)T^{2k-1}(\gamma) \\
&= (\ldots(T^{k+1}(\alpha)T^{k+1}(\gamma))\ldots)T^{2k-1}(\gamma) \\
&= \ldots \qquad\qquad\qquad\qquad\qquad\qquad (1.4.3) \\
&= T^{2k}(\alpha) \\
&= T^k(\beta).
\end{aligned}$$

Remark 1.4.44. If in (1.4.3) we replace β by the pair (β, i) (for some $i \geq 0$), and we replace each $T^j(\gamma)$ by $(T^j(\gamma), 0)$, then the result will be the pair $(T^k(\beta), i)$, since $0 * i = i = i * 0$ in Z.

Consider the subalgebra Q of W_2 generated by the elements (orbits)

$$\overline{(\beta, 0)}, \overline{(T^k(\gamma), 0)}, \overline{(T^{k+1}(\gamma), 0)}\ldots, \overline{(T^{2k-1}(\gamma), 0)}.$$

By Remark 1.4.44, Q contains $\overline{(T^k(\beta), 0)} = \overline{(\beta, 1)}$. Applying Remark 1.4.44 for $i = 1$, we conclude that Q also contains $\overline{(T^k(\beta), 1)} = \overline{(\beta, 2)}$. Continuing in that manner, we obtain that Q contains all the elements $\overline{(\beta, s)} = \overline{(T^{ks}(\beta), 0)}$, $s \geq 0$. All these elements are different (prove it using the fact that α is not a periodic word!), hence Q is infinite and finitely generated, and W_2 is not locally finite. This proves Claim (1).

To prove Claim (2), pick any natural number ℓ that is much less than k, and take ℓ elements $\overline{(T^{i_j}(\alpha), 0)}$, $j = 1, \ldots, \ell$ (note that every element of W_2 is of that form or is equal to ∞_2).

Let D be the union of orbits of elements $T^{i_j}(\alpha)$, $j = 1, \ldots, \ell$, in V under the automorphism T^k, i.e., the set of all elements of the form $T^{i_j + ks}(\alpha)$, $j = 1, \ldots, \ell$, $s \in \mathbb{Z}$. Let S be the subalgebra of V generated by D. Let L be the maximal size of an ℓ-generated subalgebra in V.

Let ϕ be the map $S \setminus \{\infty\} \to \mathbb{Z}$ that takes $T^t(\alpha)$ to t. Then ϕ is injective since α is not periodic. Also $\phi(D)$ is a union of ℓ arithmetic progressions with step k (and k is much bigger than ℓ).

By (b), if $\delta_1, \delta_2 \in S$ and $\delta_3 = \delta_1 \delta_2 \neq \infty$, then $|\phi(\delta_i) - \phi(\delta_j)| \leq q$, $i, j = 1, 2, 3$. Therefore for every product $\rho \neq \infty$ of at most L elements of D, involving $\delta \in D$, $\phi(\rho)$ is at distance at most qL from $\phi(\delta)$. Thus the set of all these numbers $\phi(\omega)$ is contained in the interval $I_\delta = [\phi(\delta) - qL, \phi(\delta) + qL]$. Moreover all elements $\delta' \in D$ that can be involved in the products representing elements from $\phi^{-1}(I_\delta) \cap S$ are themselves in $\phi^{-1}(I_\delta)$. Since we can assume that $k > 2qL + 1$ (the number of integers in the interval I_δ), each orbit of T^k from D intersects $\phi^{-1}(I_\delta)$ at most once. Thus $\phi^{-1}(I_\delta) \cap S$ is contained in a subalgebra of V generated by at most ℓ elements. By the definition of L, then $S_\delta = (\phi^{-1}(I_\delta) \cap S) \cup \{\infty\}$ is a subalgebra of V generated by at most ℓ elements.

If δ and δ' belong to the same orbit of T^k, then $S_{\delta'} \setminus \{\infty\}$ is obtained from $S_\delta \setminus \{\infty\}$ by applying T^{ks} for some s (why?). If $\epsilon \in S_\delta$, $\epsilon' \in S_{\delta'}$ then either $\epsilon \epsilon' = \infty$ or $\epsilon \epsilon' \in S_\delta \cap S_{\delta'}$ (because both δ and δ' are involved in products representing $\epsilon \epsilon'$). Therefore S is the union of all S_δ.

Consider the map ψ from $\mathbb{Z}$ to the cyclic group C_k of integers modulo k, which takes any number z to $z \mod k$. Then $\psi(\phi(S))$ is contained in the union of at most ℓ intervals of natural numbers from $[0, k-1]$ of length at most $2wq + 1$. Since we can assume that $k > (2wq + 1)\ell$, there exists a natural number $i \leq k - 1$ that does not belong to any of these intervals. Consider then the interval of natural numbers $I = [i, i + k - 1]$. We have that

- $S_0 = (\phi^{-1}(I) \cap S) \cup \{\infty\}$ is a union of several subalgebras S_δ,
- S_0 is a subalgebra of S,
- S is a union of $T^{ks}(S_0)$, $s \in \mathbb{Z}$ (setting $T(\infty) = \infty$) and
- the product of an element from $T^{ks}(S_0)$ and an element from $T^{ks'}(S_0)$, $s \neq s'$, is ∞.

Exercise 1.4.45. Prove these properties.

Consider $(S_0 \setminus \{\infty\}) \times Z \subseteq W_1$. The image in W_2 of $(S_0 \setminus \{\infty\}) \times Z$ under the map $\chi: (\gamma, i) \mapsto \overline{(\gamma, i)}$ coincides with the image of $(S \setminus \{\infty\}) \times Z$ under χ (why?).

Exercise 1.4.46. Prove that the set $S_0' = ((S_0 \setminus \{\infty\}) \times Z) \cup \{\infty_1\}$ is a subalgebra of W_1. Show that if we set $\chi(\infty_1) = \infty_2$, the map $\chi: S_0' \to W_2$ becomes a homomorphism and the image of χ is finite. **Hint:** The only thing to be checked is that if the product of two sequences $\mu = T^i(\alpha)$, $\nu = T^j(\alpha)$ in

$S_0 \smallsetminus \{\infty\}$ is ∞, then the same is true for any shifts of μ, ν by powers of T^k (so the product of the corresponding orbits from W_2 is ∞_2). But that follows from the fact that $|i - j|$ is much smaller than k.

The set $((S_0 \smallsetminus \{\infty\}) \times \{0\}) \cup \{\infty_1\}$ is a subalgebra of $S_0' \subseteq W_1$ that is isomorphic to S_0 (why?). The image of this subalgebra contains the elements $\overline{(T^{i_j}(\alpha), 0)}, j = 1, \ldots, \ell$, so the subalgebra generated by these elements is a quotient of a subalgebra of W_1 and belongs to var B, as desired.   □

Now in order to prove that graph algebras corresponding to graphs on Figure 1.4 are inherently non-finitely based, it is enough to find a bi-infinite word α for each of these algebras satisfying the three conditions of Theorem 1.4.40 (see [19]).

Exercise 1.4.47. For each of the algebras corresponding to the graphs on Figure 1.4 check that the following bi-infinite word satisfies the conditions of Theorem 1.4.40.

- For graph T, $\alpha = \ldots 020210202 \ldots$.
- For graph M, $\alpha = \ldots 111101011011101111 \ldots$.
- For graph P_4, $\alpha = \ldots 01012323 \ldots$.
- For graph L_3, $\alpha = \ldots 0001222 \ldots$.

1.5 Growth of Algebras

Let W be an infinite algebra generated by a finite set X. The growth function of W provides a way of measuring the infiniteness of W. It can be defined as follows. Every element of W can be represented as the value of some term in X. Say, in the case of groups or semigroups, the term is a word in the alphabet X, and in the case of associative algebras over a field K terms are non-commutative polynomials in X (i.e., formal *linear combinations* $\sum k_i u_i$ of words u_i in the alphabet X with coefficients k_i from K). With each term we assign its *degree*, which in the case of semigroups or groups is simply the length of the word, in the case of associative algebras is the length of the longest word of the linear combination. For other types of algebras the degree can be more complicated. For every natural n consider the set B_n of elements of W of degree at most n. In the case of semigroups or groups it is a finite set, in the case of algebras over a field, B_n is a finite dimensional subspace. In these and many other cases, we can talk about the *size* of B_n: in the case of groups and semigroups the size is the number of elements, in the case of associative algebras over fields it is the dimension of the subspace. Thus we have a function f_X from the set of natural numbers to itself sending each n to the size of B_n. It is the *growth function* of W with respect to the generating set X, one of the main numerical invariants of an algebra. For "finite" algebras (we put "finite" in quotation marks because it includes here the case of finite dimensional associative algebras, for example), the growth function with respect to any generating set is bounded by a constant. For infinite algebras it is unbounded. Thus growth functions distinguish finite from infinite, and allow one to talk about different grades of infiniteness. Originally the first growth functions were investigated by Schwartz in the case of groups [310]. He noticed that if M is a compact Riemannian manifold and $\bar{M}$ is its universal cover, then the volume growth function of $\bar{M}$ is *equivalent* to the growth function of the fundamental group of M with respect to some (hence any) finite generating set. Here the volume growth for $\bar{M}$ is defined as the function that sends every natural number to the volume of the ball of radius n around some fixed point of $\bar{M}$. Two functions $f, g \colon \mathbb{N} \to \mathbb{N}$ are called *equivalent* if for some constant $C > 1$ and every sufficiently large $n \in \mathbb{N}$ we have

$$\frac{1}{C} f\left(\frac{n}{C}\right) \le g(n) \le C f(Cn).$$

(For example, every two exponential functions $a^n, a > 1$, are equivalent, any polynomial function $a_k n^k + \ldots + a_0$ with real coefficients, $a_k > 0$, is equivalent to n^k, etc.)

Exercise 1.5.1. Show that for any finitely generated group (semigroup, linear algebra) the growth functions with respect to any two finite generating sets are equivalent.

It follows that for a compact manifold M the growth functions of its fundamental group with respect to any two finite generating sets are equivalent.

Together with the growth function f_X it is natural to consider the *spherical growth function* s_X and the *growth series*. The function s_X is defined by $s_X(n) = f_X(n) - f_X(n-1)$ for every $n > 0$ and $s_X(0) = 1$ (thus in the case of semigroups or groups, it is the number of elements of length n). The growth series is defined by

$$P_X(z) = \sum_{n=0}^{\infty} s_X(n) z^n.$$

Similarly, one can define the growth function, the spherical growth function, and growth series of any language. We shall see that analytic and algebraic properties of the function represented by that series reflect syntactic and semantic properties of the algebra or the language (see Sections 1.8.9, 5.9.2.3).

1.6 Symbolic Dynamics

1.6.1 Basic Definitions[4]

A *topological dynamical system* in general is a compact topological space X with a semigroup S of continuous transformations $X \to X$.

Since we will study only concrete dynamical systems in this book, we do not define "compact topological space" (and assume that the notion of "continuous map" is known). The most popular semigroups are the infinite cyclic semigroup $\mathbb{N}$ or the group of integers $\mathbb{Z}$ (the so-called discrete dynamical systems), the group of reals $\mathbb{R}$ and the semigroup of positive reals (continuous dynamical systems). The dynamical system is denoted by (X, S). If $S = \mathbb{Z}$ or $S = \mathbb{N}$ is generated by one element T, then we write (X, T) instead of $(X, \langle T \rangle)$.

For example, let M be a compact manifold without boundary (like a sphere or a torus) and let F be a continuous tangent vector field on it. This vector field determines a flow on M. For every point x in M and for every real number r we can find the point $\alpha_r(x)$ where x will be in r seconds if we start at the point x. The transformations α_r form a group isomorphic to the additive group of real numbers (if some natural analytic conditions on M and F hold). This is a typical continuous dynamical system. The transformations α_n corresponding to the integers form a discrete dynamical system.

If we take this discrete dynamical system, divide M into finite number of parts as we did in Section 1.2.1, then with a discrete trajectory of a point x in M we can associate a bi-infinite word of labels of regions, which are visited by the point x in $\ldots, -3, -2, -1, 0, 1, 2, \ldots$ seconds. The set of all these bi-infinite words approximates the original dynamical system. This approximation is in general better if the areas of the regions are smaller (this is similar to approximating solutions of differential equations). An important observation made first probably by Hadamard, says that under some (not very restrictive) conditions this set of words may be considered as a dynamical system itself.

One of the basic general results about (discrete) dynamical systems is the following recurrence theorem (here we assume that the semigroup S is the group $\mathbb{Z}$).

Definition 1.6.1. Let (D, T) be a dynamical system. Assume that T is a bijection. A point $x \in D$ is called *uniformly recurrent* (also called *almost periodic*) if for every neighborhood U of x in D there exists a number $N = N(U)$ such that for every integer k one of the points

$$T^k(x), T^{k+1}(x), \ldots, T^{k+N-1}(x)$$

belongs to U.

---

[4] Warning: Some knowledge of elementary topology is required to read this subsection.

Theorem 1.6.2 (George Birkhoff recurrence theorem [115, 238]). *Every discrete dynamical system* (D,T) *contains a uniformly recurrent point.*

Proof. Let (D,T) be a dynamical system; that is, D is compact, and $T:D \to D$ is a continuous bijection. Consider all subsystems (D',T) where D' is a closed subset of D stable under T, T^{-1}. Since the intersection of any decreasing sequence $D_1 \supset D_2 \supset \ldots$ of subsystems is again a dynamical subsystem of (D,T), and D is compact, there exists a minimal (under inclusion) subsystem (D',T).[5] We claim that every point x in D' is uniformly recurrent. Let U be a neighborhood of x in D'. Then $V = \bigcup_{i \in \mathbb{Z}} T^i(U)$ is an open subset of D', and $T(V) = T$. Therefore $D' \smallsetminus V$ is a closed subset and a subsystem of D'. By the minimality of D', $D' \smallsetminus V$ is empty. Hence $D' = V = \bigcup_{i \in \mathbb{Z}} T^i(U)$. By compactness, there exists L such that $D' \subseteq \bigcup_{i=-L}^{L} T^i(U)$. Let $N = 2L + 1$. Then for every $k \in \mathbb{Z}$ there exists $j \in \{-L, \ldots, L\}$ such that $T^{k+L}(x) \in T^j(U)$, hence $T^{k+L-j}(x) \in U$ as required since $k \le k + L - j \le k + N$. $\square$

1.6.2 Subshifts

1.6.2.1 The Definitions

Let A be a finite alphabet. Consider the set $A^{\mathbb{Z}}$ of all bi-infinite words in the alphabet A. If $\alpha \in A^{\mathbb{Z}}$, and $m \le n$ are integers, then $\alpha(m,n)$ is the subword of α starting at the position number m and ending at the position number n. One can define a distance function on $A^{\mathbb{Z}}$ by the following rule. Let $\alpha, \beta \in A^{\mathbb{Z}}$. Let n be the largest number such that the word $\alpha(-n,n)$ coincides with the word $\beta(-n,n)$ (if the letters $\alpha(0,0)$ and $\beta(0,0)$ are different, set $n = -1$). Then the distance $\text{dist}(\alpha, \beta)$ between α and β is equal to $\frac{1}{2^n}$ (in particular, if $\alpha = \beta$, then $\text{dist}(\alpha, \beta) = \frac{1}{2^\infty} = 0$ as required by the definition of metric).

Exercise 1.6.3. Prove that for every $\alpha \in A^{\mathbb{Z}}$ the *open ball of radius* $\epsilon > 0$ around α, that is, the set $B(\alpha, \epsilon) = \{\omega \mid \text{dist}(\omega, \alpha) < \epsilon\}$ is equal to the *closed ball* $\bar{B}(\alpha, \epsilon') = \{\omega \mid \text{dist}(\omega, \alpha) \le \epsilon'\}$ for some $\epsilon' < \epsilon$. Hence every open ball of $A^{\mathbb{Z}}$ is closed. Show that if $\epsilon \le 1$, then $\bar{B}(\alpha, \epsilon)$ consists of all bi-infinite words ω such that $\omega(-n,n) = \alpha(-n,n)$ where $n = -\lfloor \log_2 \rfloor$ where $\lfloor x \rfloor$ denotes the biggest integer not exceeding x.

This metric makes $A^{\mathbb{Z}}$ a compact topological space (this is a standard elementary topology fact). Let T be a shift on $A^{\mathbb{Z}}$ to the right, that is $T(\alpha)(i,i) = \alpha(i-1,i-1)$ for all $i \in \mathbb{Z}$. It is easy to prove that T and T^{-1} are continuous maps of $A^{\mathbb{Z}}$ onto itself (prove it!).

Therefore $(A^{\mathbb{Z}}, T)$ is a dynamical system.

---

[5] Note that we are using Zorn's lemma here. The partially ordered set to which Zorn's lemma applied is the set of all subsystems of (D,T), so that $(D',T) \le (D'',T)$ if and only if $D' \supseteq D''$.

Definition 1.6.4. The system $(A^{\mathbb{Z}}, T)$ and all its subsystems (that is, closed subsets of $A^{\mathbb{Z}}$ which are stable under T, T^{-1}) are called *subshifts*.

One can define subshifts without using any topology as follows:

Definition 1.6.5. A subset X of $A^{\mathbb{Z}}$ is a subshift if

- $T(X) = X$ and
- if $\omega \in A^{\mathbb{Z}}$ and all finite subwords of ω are subwords of some of the bi-infinite words from X, then $\omega \in X$.

Exercise 1.6.6 (Requires knowledge of some topology). Prove that Definitions 1.6.4 and 1.6.5 are equivalent.

Thus a subshift (X, T) is completely determined by the language $L(X)$ of (finite) words in the alphabet A of the subshift X that do not appear as subwords in the bi-infinite words from X, i.e., by the set of forbidden subwords. It is clear that $L(X)$ is an *ideal* of the free monoid A^*, that is a subsemigroup of A^* such that $ab \in L(X)$ provided either a or b is in $L(X)$ (prove that!).

If $L(X)$ is *finitely generated as an ideal*, i.e., consists of a finite set of words L_0 and all words from A^* containing words from L_0, then (X, T) is called a subshift of *finite type*. Thus subshifts of finite type are determined by a finite set of forbidden subwords L_0. Note that in this case we can assume that all words from L_0 are of the same length. Indeed, if m is the maximal length of words in L_0, consider the set L'_0 of all words of length m each of which contains a subword from L_0. Then the subshifts determined by L'_0 and L_0 coincide (prove it!). If all words of the forbidden set L_0 are of length $m + 1$, then the subshift is called an *m-step subshift*. Note that every m-step subshift is also n-step for every $n \geq m$ (prove it!).

The subshift $(A^{\mathbb{Z}}, T)$ is called the *full shift*. Another concrete example is the *golden mean* subshift $\mathcal{F}$, the set of all bi-infinite words over $A = \{a, b\}$ such that every occurrence of b is followed by an occurrence of a (i.e., $\{bb\}$ is the set of forbidden subwords). Both the full shift and the golden mean subshift are subshifts of finite type. The full shift is 0-step (one can even argue that it is -1-step), in fact every 0-step subshift is equal to the full shift over some alphabet (prove it!). The golden mean subshift is a 1-step subshift (by definition).

1.6.2.2 The Complexity Function of a Subshift

Let $(X, T) \subseteq (A^{\mathbb{Z}}, T)$ be a subshift. Let L be the language of all subwords of the bi-infinite words in X and $f(n)$ be the spherical growth function of that language (see Section 1.5). Thus for every n, $f(n)$ is the number of words of length n that are subwords of the bi-infinite words from X. The function $f: \mathbb{N} \to \mathbb{N}$ is called the *complexity function* of (X, T).

Exercise 1.6.7. Show that the complexity function of the full shift $(A^{\mathbb{Z}}, T)$ is $f(n) = |A|^n$.

Exercise 1.6.8. Show that the complexity function $f(n)$ of the golden mean subshift $\mathcal{F}$ satisfies $f(1) = 2, f(2) = 3, f(n+1) = f(n) + f(n-1)$, that is $f(n)$ is the $n + 2$nd Fibonacci number. **Hint:** Induction on n. Use the fact that every word of length $n + 1$ that appears as $\omega(i, n+i)$ for some $\omega \in \mathcal{F}$ either ends with a (and then $\omega(i, i+n-1)$ can be arbitrary subword of length n in a bi-infinite word from $\mathcal{F}$) or ends with ab (and $\omega(i, i+n-2)$ can be arbitrary subword of length $n - 1$ in a bi-infinite word from $\mathcal{F}$).

Exercise 1.6.9. Let u be a (finite) word in the alphabet $\{a, b\}$. Let $X_u \subset A^{\mathbb{Z}}$ be the *u-periodic subshift*, that is the set of all bi-infinite words ω in $A^{\mathbb{Z}}$ such that all finite subwords of ω are periodic with period u.

(1) The subshift X_u is *finite* (i.e., consists of finite number of bi-infinite words).
(2) Show that the complexity function of any finite subshift is bounded from above by a constant.
(3) Prove that every finite subshift is a subshift of finite type.

Exercise 1.6.10. Show that for every infinite subshift (X, T_X), the complexity function satisfies $f(n) \geq n + 1$. **Hint:** If the alphabet contains only one letter, the subshift is finite (consists of one element), so we can assume that the alphabet contains more than one letter. Then $f(1) \geq 2$. Clearly, $f(n + 1) \geq f(n)$ for all n. If $f(n+1) = f(n)$ for some n, then every subword of length n of a bi-infinite word from the subshift uniquely determines the letter next to the right of the subword. Show that then any bi-infinite word in the subshift is periodic with period of length at most $n + 1$, and the subshift is finite (since there are only finitely many words of length $\leq n + 1$ over a finite alphabet).

1.6.2.3 Sturmian Subshifts

The following example is a simple 1-dimensional version of the real life example from Section 1.2.1. Consider the unit circle, that is the interval $[0, 1]$ with 0 and 1 identified. Let α be an irrational number from the interval $(0, 1)$. Subdivide the circle into two parts $[0, 1 - \alpha), [1 - \alpha, 1)$. Label the first part by N (Nebraska) and the second part by I (Iowa). Start with any number t between 0 and 1. For every $n \in \mathbb{Z}$ record the label of the part of the circle containing $t + n\alpha$, we get a bi-infinite word ω_t. The set of all such bi-infinite words is a subshift (prove it!), which is called the *Sturmian subshift* corresponding to the number α. The Sturmian subshifts (introduced by Morse and Hedlund [239]) have many remarkable properties and are very well studied (see, for example, [207]).

Here is perhaps the main property of Sturmian subshifts: these subshifts have the smallest complexity functions among infinite subshifts.

Theorem 1.6.11 (See [207], compare with Exercise 1.6.10). *The complexity function of every Sturmian subshift is* $f(n) = n+1$. *Conversely, every subshift with complexity function* $f(n) = n + 1$ *is a Sturmian subshift.*

1.6.2.4 The Entropy and the Conjugacy Problem for Subshifts

A homomorphism between subshifts (X, T_X) and (Y, T_Y) is a continuous map $\phi: X \to Y$, which commutes with T:

$$\phi(T_X(x)) = T_Y(\phi(x)).$$

As with the definition of subshifts themselves, homomorphisms of subshifts can be defined purely syntactically.

Let (X, T_X) be a subshift with alphabet A and let B be an alphabet.

Fix a nonnegative integer n. Let $\mathcal{A}$ be the (finite) set of all words over A of length $2n + 1$. Fix a map $\psi: \mathcal{A} \to B$. Then consider the map $\bar{\psi}$ from X to the full shift $B^{\mathbb{Z}}$ taking every bi-infinite word ω to the bi-infinite word ω' where for every $i \in \mathbb{Z}$,

$$\omega'(i, i) = \psi(\omega(i - n, i + n)).$$

Thus in order to produce ω', we scan ω and for every i record the letter from B corresponding to the window of length $2n + 1$ around the i-th letter of ω.

It is not difficult to check that the map $\bar{\psi}$ is a homomorphism. It turns out (see Hedlund [149] and Lind–Marcus [203]) that the converse statement is true too.

Theorem 1.6.12. *Every homomorphism* χ *from a subshift* (X, T_X) *to a subshift* (Y, T_Y) *is of the form* $\bar{\psi}$ *for some* ψ *(and* n*).*

Proof. For every $b \in B$ let X_b be the set of bi-infinite words ω in X such that $\chi(\omega)$ has b at the 0-th coordinate. Then all X_b are disjoint and their union is the whole X. Moreover each X_b is a closed subset of X. Indeed, if $\omega \in X_b$, then

$$X_b = \{\alpha \mid \operatorname{dist}(\chi(\alpha), \chi(\omega)) \le 1\}\}.$$

Thus X_b is the preimage of a closed ball of radius $1/2$ in Y (the preimage of a closed set under a continuous map is closed). In a compact metric space, any two closed subsets U, V are distance-separated, that is for some $\epsilon > 0$, $\operatorname{dist}(u, v) > \epsilon$ for every $u \in U, v \in V$. Thus there exists k such that $\operatorname{dist}(\alpha, \beta) > \frac{1}{2^k}$ for every $\alpha \in X_b, \beta \in X_{b'}$ where $b \ne b'$. Let $\mathcal{B}$ be the set of all words in A of length $2k + 1$. If a word $u \in \mathcal{B}$ appears as $\omega(-k, k)$ in some $\omega \in X$, then all bi-infinite words $\omega' \in X$ with $\omega'(-k, k) = u$ belong to the same X_b with ω. Thus we define $\psi(u) = b$. If u does not appear as $\omega(-k, k)$ in any $\omega \in X$, we can define $\psi(u) \in B$ arbitrarily. It remains to solve the following exercise.

Exercise 1.6.13. Show that the map $\bar{\psi}$ coincides with χ.

□

Exercise 1.6.14. Let $\mathcal{F}$ be the golden mean subshift over $A = \{a, b\}$. Let $n = 1$, and let $\mathcal{B}$ be the set of all words of length 3: $aaa, aab, aba, abb, baa, bab, bba, bbb$. Let ψ assign letters a, b, a, b, a, b, a, b to these words (that is ψ sends each of the 8 words to its last letter). Show that ψ is a homomorphism from $\mathcal{F}$ to $\mathcal{F}$, which coincides with T^{-1}.

Exercise 1.6.15 (Requires some set theory). Show that every subshift has only countably many homomorphisms.

Two subshifts $(X, T_X), (Y, T_Y)$ are called *conjugate* if there exists a bijective homomorphism $\phi: X \to Y$ (called *conjugacy*). For example, T_X is a conjugacy from (X, T_X) to itself, i.e., an *automorphism* of (X, T_X). By Exercise 1.6.15 every subshift has only countably many conjugate subshifts. Note that a conjugacy is always a *homeomorphism*, that is its inverse is also continuous (hence a homomorphism). The easiest way to prove that it is a homeomorphism is by using the following general topology fact: if a continuous map from a compact space is bijective, then the inverse map is continuous too.

A major open problem of symbolic dynamics is the *conjugacy problem* for subshifts of finite type:

Problem 1.6.16. *Is there an algorithm that given two finite collections of words U, V decides whether the subshifts of finite type corresponding to U and V (that is, the subshifts consisting of all bi-infinite words that contain subwords from U and V, respectively) are conjugate?*

One of the main characteristics of a subshift (X, T) is its (topological) entropy $h(X)$ defined as

$$h(X) = \lim_{n \to \infty} \frac{1}{n} \ln f(n) \qquad (1.6.1)$$

where $f(n)$ is the complexity function of the subshift.

Exercise 1.6.17. Prove that the limit in (1.6.1) exists for every subshift S. **Hint:** First prove that the complexity function $f(n)$ is *submultiplicative*, i.e., it satisfies the inequality $f(m + n) \leq f(m)f(n)$ for every $m, n \geq 0$ and hence $g(n) = \ln f(n)$ is a *subadditive* function, i.e., $g(m + n) \leq g(m) + g(n)$ for every m, n. Then show that for every subadditive function $g(n)$ the limit $\lim_{n \to \infty} \frac{g(n)}{n}$ exists and is equal to the infimum of all the numbers $\frac{g(n)}{n}$. For this, fix an integer $m > 0$ and for every $n > m$ set $n = mq + r$ where $0 < r \leq m$. Then $\frac{g(n)}{n} \leq \frac{g(qm)}{qm} + \frac{g(r)}{n} \leq \frac{g(m)}{m} + \frac{g(r)}{n}$. Deduce that $\limsup_{n \to \infty} \frac{g(n)}{n} \leq \frac{g(m)}{m}$, which is true for every m.

Exercise 1.6.18. Prove that

- the entropy of the full shift on a k-letter alphabet, is $\ln k$,
- the entropy of the golden mean subshift is $\ln \phi$ where ϕ is the golden mean $\frac{1 + \sqrt{5}}{2}$,
- the entropy of any Sturmian subshift and every finite subshift is 0.

Exercise 1.6.19. Prove that the entropy of a subshift is a *conjugacy invariant* that is conjugate subshifts have the same entropy.

Exercise 1.6.20. Show that there are uncountably many pairwise non-conjugate subshifts of entropy 0. **Hint:** Consider the Sturmian subshifts and use Exercise 1.6.15.

Exercise 1.6.21. Show that there are infinitely many pair-wise non-conjugate subshifts of finite type that have entropy 0. **Hint:** Consider finite subshifts.

1.6.3 Edge Subshifts

Let $\Gamma = (V, E)$ be a finite graph. Then the set of all bi-infinite paths on Γ is a subshift with the alphabet E. This subshift is called the *edge subshift* of Γ.

Exercise 1.6.22. Prove that every edge subshift of a finite graph is a 1-step subshift.

Thus every edge subshift of a finite graph is a subshift of finite type. It turns out that up to conjugacy there are no other subshifts of finite type.

Theorem 1.6.23. *Every subshift of finite type is conjugate to an edge subshift of a finite graph.*

Proof. Let (X, T) be a subshift of finite type of M-step for some M. Consider the set $V = B_M$ all words of length M and the set $E = B_{M+1}$ of words of length $M + 1$ that are subwords in bi-infinite words from X. Let Γ be the graph with vertex set V and edge set E where for every $e \in E$, e_- is the prefix of e of length M, and e_+ is the suffix of e of length M (this is called the *de Bruijn graph*). The identity map from B_{M+1} to E defines a homomorphism ψ from X to the edge subshift Ω of the graph Γ.

Exercise 1.6.24. Show that ψ is a bijection.

$\square$

1.7 Rewriting Systems

1.7.1 Basic Definitions

A *rewriting system* is any graph in the sense of Serre. If we consider a graph as a rewriting system, then vertices are called *objects*, edges are called *moves* and positive (negative) edges are called *positive (negative)* moves.

If there exists a positive move of a rewriting system Γ with tail a and head b, then we say that a can be *positively rewritten* into b in one step and denote this situation by $a \to_\Gamma b$. A path in Γ is called a *derivation*. A path consisting of positive (negative) moves is called a *positive* (negative)derivation. If there exists a positive derivation from a to b, we say that a *can be rewritten to* b, and denote this by $a \xrightarrow{*}_\Gamma b$.

A rewriting system Γ is called *terminating* if every sequence $a_1 \rightarrow_\Gamma a_2 \rightarrow_\Gamma \ldots \rightarrow_\Gamma a_n \rightarrow_\Gamma \ldots$ stabilizes after a finite number of moves.

Example 1.7.1. The rewriting system whose objects are natural numbers and a positive move takes any perfect square to its positive square root is terminating while the rewriting system with the same objects and positive moves $n \rightarrow n + 1$ is not terminating.

Notice that if Γ is terminating, then $\xrightarrow{*}_\Gamma$ is always a partial order on the set of objects of Γ: if $a \xrightarrow{*}_\Gamma b$, then b is "better" than a. This order satisfies the descending chain condition, hence every set $Y \subseteq X$ of objects of Γ contains a "best" object i.e., an object that cannot be rewritten into some other object from Y by a positive move.

An object of a rewriting system Γ that cannot be changed by any positive move will be called *terminal*.

Example 1.7.2. In the terminating rewriting system from Example 1.7.1, the terminal elements are numbers that are not perfect squares.

1.7.2 Confluence

One of the main purposes of the theory of rewriting systems is to study the equivalence relations they generate. If Γ is a rewriting system, then the *equivalence relation generated* by Γ, that is the smallest equivalence relation containing $\rightarrow_\Gamma$ (see Section 1.1), will be denoted by $\xleftrightarrow{*}_\Gamma$.

A rewriting system Γ is called *confluent* if for every three objects a, b, c such that $a \xrightarrow{*}_\Gamma b$ and $a \xrightarrow{*}_\Gamma c$ there exists an object d such that $b \xrightarrow{*}_\Gamma d$, $c \xrightarrow{*}_\Gamma d$. This notion is illustrated by the following picture (Figure 1.5) where arrows pointing down symbolize positive moves.

Finding the object d and the derivations to d from b and c is called *completion of a diamond*.

A confluent and terminating rewriting system is called *Church–Rosser* (this concept was first introduced in [66]).

We call a rewriting system Γ *deterministic* if for every object d there is at most one positive rule applicable to d.

Exercise 1.7.3. Every deterministic rewriting system is confluent.

Exercise 1.7.4. Let Γ be a rewriting system where objects are natural numbers and moves are of two types. The moves of the first type take any natural number $n \geq 2$ to $n - 1$, the moves of the second type take every even natural number n to $n/2$. Check that this rewriting system is Church–Rosser. What are the terminal objects of this rewriting system?

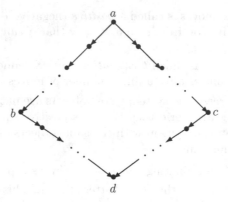

Figure 1.5 Confluence

Exercise 1.7.5. Let Γ be the rewriting system where objects are natural numbers and moves are of two types. The moves of the first type take every number to its sum of decimal digits (for example, $4567 \to 22 \to 4$). The moves of the second type take every odd number $n \geq 3$ to $\frac{n-1}{2}$. Prove that this system is not Church–Rosser.

Exercise 1.7.6 (The 3n+1-game). Let Γ be the rewriting system where objects are natural numbers and moves are of two types. The move of the first type is applicable only to even numbers, it divides the number by 2. The move of the second type is applicable only to odd numbers > 1, it multiplies the number by 3 and adds 1. For example, $10 \to 5 \to 16 \to 8 \to 4 \to 2 \to 1$. Prove that Γ is confluent. (The question whether Γ is Church–Rosser, i.e., terminating, is a major open problem.)

Exercise 1.7.7 (Requires some knowledge of Linear Algebra). Let Γ be the rewriting system where objects are square matrices over a field K, and moves are row operations:

- if row number i starts with s zeroes followed by $a \neq 0$ and row number j ($j > i$) starts with s zeroes followed by $b \neq 0$, then subtract row number i multiplied by b/a from row number j,
- switch rows number i and number j where $i > j$ and row number i has more leading zeroes than row number j,
- if the first non-zero entry of row number i is a, then multiply this row by $1/a$,
- if $i < j$, row number j starts with s zeroes followed by 1, and entry number $s + 1$ of row number i is $a \neq 0$, then subtract row number j multiplied by a from row number i.

Show that Γ is Church–Rosser. Show that the terminal objects in this rewriting system are matrices in the reduced row echelon form. Deduce the well-

known important theorem from Linear Algebra that every matrix has only
one reduced row echelon form.

Lemma 1.7.8. *For every two objects a, b of a confluent rewriting system Γ,
if $a \overset{*}{\longleftrightarrow}_\Gamma b$, then there exists an object c such that $a \overset{*}{\to}_\Gamma c$ and $b \overset{*}{\to}_\Gamma c$.*

Proof. Since $a \overset{*}{\longleftrightarrow}_\Gamma b$ there exists a sequence of positive or negative moves
connecting a and b. In other words we have a picture like the one on Figure 1.6

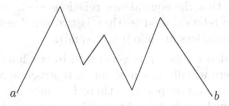

Figure 1.6 A typical derivation

This picture means that there are several negative moves at the beginning
(moving the object uphill), followed by some positive moves (moving the
object downhill), followed by some negative moves, etc. Thus there may be
several "peaks" between a and b. Using the confluence we can remove the
"peaks" one-by-one (this is called the *peak reduction*) and make one deep
canyon instead of several peaks (as on Figure 1.7).

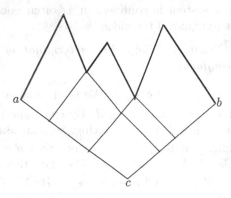

Figure 1.7 Turning peaks into canyons

$\square$

Lemma 1.7.8 implies that if Γ is confluent, then every equivalence class of $\overset{*}{\longleftrightarrow}_\Gamma$ contains at most one terminal object, and if Γ is also terminating, then every equivalence class contains exactly one terminal object. In this case terminal objects can be considered *canonical representatives* (or *normal forms*) of the equivalence classes of $\overset{*}{\longleftrightarrow}_\Gamma$.

Exercise 1.7.9. Let Γ_1 be the rewriting system where objects are natural numbers and moves take a number to its sum of decimal digits. Let Γ_2 be the rewriting system where objects are natural numbers and moves take a number $n > 9$ to $n - 9$. Prove that the equivalence relations $\overset{*}{\longleftrightarrow}_{\Gamma_1}$ and $\overset{*}{\longleftrightarrow}_{\Gamma_2}$ coincide with the congruence relation mod 9 (that is two numbers are equivalent if and only if their remainders modulo 9 are equal).

Thus if an equivalence relation is generated by a Church–Rosser rewriting system, then there usually is a nice set of representatives of equivalence classes. What is even more important, there is a nice algorithm to check if two objects are equivalent: just apply positive moves to both objects until you get the same object or until you get two different terminal objects. In the first case the objects are equivalent, in the second case they are not equivalent. This procedure may be long and boring but it always works (provided the rewriting system is Church–Rosser).

It is not always easy to check if a rewriting system Γ is confluent. It is easier to check that it is *locally confluent* that is for every three objects a, b, c such that $a \to_\Gamma b$ and $a \to_\Gamma c$ there exists an object d such that $b \overset{*}{\to}_\Gamma d$ and $c \overset{*}{\to}_\Gamma d$. In many cases local confluence implies confluence. The following classical result is sometimes called the Diamond Lemma. It shows that in order to check that a system is confluent it is often enough to complete diamonds with short (one move) top sides.

Theorem 1.7.10 (Newman, [250]). *Every terminating locally confluent rewriting system is confluent.*

Proof. Suppose that $a \overset{*}{\to}_\Gamma b$ and $a \overset{*}{\to}_\Gamma c$. We need to show that there exists an object d such that $b \overset{*}{\to}_\Gamma d$ and $c \overset{*}{\to}_\Gamma d$. By contradiction, suppose that it is not always true. Since Γ is terminating, we can assume that a is a "best" counterexample, that is every $a' \neq a$ such that $a \overset{*}{\to}_\Gamma a'$ is no longer a counterexample. Clearly $b \neq a$ and $c \neq a$. Therefore there exist $b' \neq a$ and $c' \neq a$ such that $a \to_\Gamma b' \overset{*}{\to}_\Gamma b$ and $a \to_\Gamma c' \overset{*}{\to}_\Gamma c$. By local confluence there exists an object a' such that $b' \overset{*}{\to}_\Gamma a'$ and $c' \overset{*}{\to}_\Gamma a'$. Since b' and c' are "better" than a, we can find objects b'' and c'' such that $b \overset{*}{\to}_\Gamma b''$, $a' \overset{*}{\to}_\Gamma b''$, $a' \overset{*}{\to}_\Gamma c''$ and $c \overset{*}{\to}_\Gamma c''$. Finally since a' is "better" than a, we can find an object d such that $b'' \overset{*}{\to}_\Gamma d$ and $c'' \overset{*}{\to}_\Gamma d$. This d is what we need because $b \overset{*}{\to}_\Gamma b'' \overset{*}{\to}_\Gamma d$ and $c \overset{*}{\to}_\Gamma c'' \overset{*}{\to}_\Gamma d$ (see Figure 1.8).

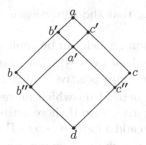

Figure 1.8 Local confluence implies confluence

□

Exercise 1.7.11. Let Γ be the rewriting system where objects are natural numbers and moves are of two types. The moves of the first type take every perfect square to its positive square root. The moves of the second type take every even number n to $\frac{n}{2}$. Show that this system is locally confluent and terminating (so it is confluent by the Diamond Lemma). Describe all terminal objects of this rewriting system.

Exercise 1.7.12. Let Γ be the rewriting system where objects are all sequences of letters a and b and a positive move replaces two neighbor letters b, a by a, b. For example,

$$(b,b,b,a,a,b,a) \to_\Gamma (b,b,b,a,a,a,b) \to_\Gamma (b,b,a,b,a,a,b) \to_\Gamma$$
$$(b,a,b,b,a,a,b) \to_\Gamma (b,a,b,a,b,a,b) \to \ldots$$

Prove that this rewriting system is Church–Rosser. Describe all terminal objects of this rewriting system.

Exercise 1.7.13. Consider the following rewriting system. The objects are all finite automata Γ that are graphs in the sense of Serre with correct labeling (see Section 1.3.5). Positive moves are the edge foldings. Prove that this rewriting system is Church–Rosser, so for every automaton Γ there exists a unique inverse automaton obtained from Γ by a series of foldings.

1.7.3 What If a Rewriting System Is Not Confluent? The Art of Knuth–Bendix

In "real life" we often have an equivalence generated by some rewriting system Γ and we want to find some easy canonical representatives in the equivalence classes. If Γ is Church–Rosser, then we know how to find these representatives. But what if it is not Church–Rosser? The idea is to find a Church–Rosser rewriting system Γ' that has the same set of objects X and is *equivalent* to Γ, meaning that the equivalence relations $\overset{*}{\longleftrightarrow}_\Gamma$ and $\overset{*}{\longleftrightarrow}_{\Gamma'}$ coincide.

There are certain operations, which can be applied to any rewriting system Γ without changing the equivalence $\overset{*}{\longleftrightarrow}_\Gamma$. For example, we can change the partition of the set of moves into positive and negative moves. Clearly the equivalence $\overset{*}{\longleftrightarrow}_\Gamma$ does not depend on which moves are called positive and which moves are called negative. Also if there exists a derivation connecting objects a and b, then we can add a move $a \to b$ to Γ, and this does not change $\overset{*}{\longleftrightarrow}_\Gamma$ either. Conversely if Γ contains a move $m : a \to b$ and there exists a derivation from a to b that does not involve the move m, then we can remove m from Γ.

Using these three transformations one needs to transform the rewriting system Γ into an equivalent Church–Rosser rewriting system. Clearly it is always possible to do. Indeed, we can simply choose one representative from each class of $\overset{*}{\longleftrightarrow}_\Gamma$ and consider a rewriting system with the same set of objects as Γ and moves that rewrite every object a into the representative of the class containing a. This rewriting system is clearly Church–Rosser.

Of course this method amounts to cheating. If we can find a good system of representatives of the equivalence classes, we do not need to change the rewriting system anyway. But at least it gives us some hope that the goal is achievable.

The simplest real procedure of converting a bad rewriting system into a Church–Rosser one is the following. It is one of the simplest versions of the so-called Knuth–Bendix procedure (originated in [182]). First equip the set X of objects of Γ with a total order $\leq$ satisfying the descending chain condition. Remove all moves of the form $a \to a$ (this does not change the equivalence relation). Then choose the set of positive moves in such a way that if $a \to b$ is a positive move, then $b < a$ (if $m : a \to b$ is a positive move and $a < b$, then put m into E^- and put m^{-1} into E^+). This will make Γ terminating.

Thus assume that Γ is already terminating. Suppose that Γ is not confluent. Then there exists a pair of derivations $a \overset{*}{\to}_\Gamma b$ and $a \overset{*}{\to}_\Gamma c$, which is impossible to complete to a diamond. Since Γ is terminating, then we can assume that b and c are terminal objects. Let $b < c$. Then add a positive move $c \to b$ to Γ. Clearly Γ will remain terminating and the equivalence $\overset{*}{\longleftrightarrow}_\Gamma$ will not change.

If we are smart enough or just lucky enough, then after a (possibly infinite) number of such transformations, Γ will turn into a nice Church–Rosser rewriting system. If we are not smart or not lucky, then the rewriting system will become more and more complicated. The hardest part here is choosing the order $\leq$. Different choices of orders can lead to drastically different results.

1.7.4 String Rewriting

A *string rewriting system* is a rewriting system where objects are words over some alphabet X, and moves are determined by a set of *rules* $u_i \rightarrow v_i$, $i = 1, 2, \ldots$, where $u_i, v_i \in X^*$. A move corresponding to the rule $u \rightarrow v$ replaces subword u by v. The tail of this move has the form puq and the head has the form pvq where $p, q \in X^*$. We will denote this move by the triple $(p, u \rightarrow v, q)$.

The string rewriting system Γ with alphabet X and the set of rules $\mathcal{R}$ will be denoted by $\mathrm{sr}\langle X \mid \mathcal{R} \rangle$. It is also called a *Thue system*.

A move $(p, u \rightarrow v, q)$ in Γ can be illustrated by the Figure 1.9.

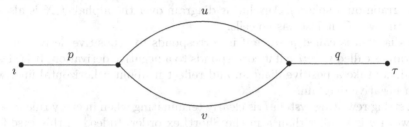

Figure 1.9 An elementary diagram

This picture is called the *elementary diagram* corresponding to the move $(p, u \rightarrow v, q)$. It is a plane graph with the *initial vertex* $\imath$, the *terminal vertex* τ, every edge is oriented from left to right and is labeled by a letter from X in such a way that on the *top path* of this diagram we can read the word puq and on the *bottom path* we can read the word pvq. The *cell* is bounded by two paths: the top path is labeled by u and the bottom path is labeled by v.

With every derivation $w_0 \rightarrow_\Gamma w_1 \rightarrow_\Gamma \cdots \rightarrow_\Gamma w_n$ one can associate a (w_0, w_n)-*diagram* Δ *over* Γ that is obtained by taking the elementary diagrams $\Delta_1, \ldots, \Delta_n$ corresponding to the moves and gluing the top of Δ_2 to the bottom of Δ_1, the top of Δ_3 to the bottom of Δ_2, etc. This diagram is also a plane labeled graph with the *initial vertex* $\imath$, the *terminal vertex* τ, every edge is oriented from left to right and is labeled by a letter from X in such a way that on the *top path* of Δ we can read the word w_0 and on the *bottom path* we can read the word w_n. Each *cell* corresponds to a rule $u_i \rightarrow v_i$ or $v_i \rightarrow u_i$. It is clear that Δ has exactly n cells.

Example 1.7.14. Let Γ be the string rewriting system with the alphabet $\{a, b\}$ and one rule $ba \rightarrow ab$ (like in Exercise 1.7.12). Then the following *diagram* (Figure 1.10) corresponds to the derivation (consisting of negative moves):

$$aabb \equiv a \cdot ab \cdot b \rightarrow a \cdot ba \cdot b \equiv ab \cdot ab \rightarrow ba \cdot ab \rightarrow ba \cdot ba \equiv baba \qquad (1.7.1)$$

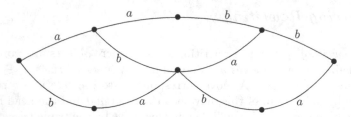

Figure 1.10 The diagram corresponding to a derivation

If the top and the bottom paths of a diagram have the same labels, then the diagram is called *spherical* (because by identifying these two paths, we get a graph on a sphere). Any linear diagram over the alphabet X is also a diagram over Γ: it just has no cells.

A diagram is called *positive* if it corresponds to a positive derivation. A diagram is called *negative* if it corresponds to a negative derivation. It is clear that if we take a positive diagram and reflect it about a horizontal line, we get a negative diagram.

A string rewriting system Γ is clearly terminating when in every rule $u \to v$ the word v is smaller than u in the ShortLex order. Indeed in this case for every move $(p, u \to v, q)$ the head is ShortLex smaller than the tail, and the ShortLex order satisfies the descending chain condition.

Example 1.7.15. The string rewriting system from Exercise 1.7.12 is terminating because $ab <_{sl} ba$ if we assume that $a < b$ (and we can assume that because we can order letters in any way we like).

If a string rewriting system with finitely many rules is terminating, then checking whether it is confluent is a "finite" problem. This means that we need to complete only finitely many diamonds.

Theorem 1.7.16. *Let $\Gamma = sr\langle X \mid \mathcal{R} \rangle$ be a terminating string rewriting system. Then Γ is confluent if and only if the following two conditions hold:*

(1) *Suppose that $xy \to s, yz \to t$ are two positive rules in Γ, where y is not empty and let $w_1 \equiv sz, w_2 \equiv xt$. Then there exists a word w such that $w_1 \overset{*}{\to}_\Gamma w, w_2 \overset{*}{\to}_\Gamma w$.*

(2) *Suppose that $xyz \to s, y \to t$ are two positive rules from Γ and let $w_1 \equiv s, w_2 \equiv xtz$. Then there exists a word w such that $w_1 \overset{*}{\to}_\Gamma w, w_2 \overset{*}{\to}_\Gamma w$.*

In other words any 2-cell diagram of one of the two forms on Figure 1.11.

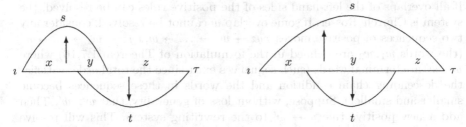

Figure 1.11 Pictures of overlaps

can be extended to a spherical (w, w)-diagram with the same initial and terminal vertices, which is cut by the path labeled by xyz into a negative (top) diagram and a positive (bottom) diagram as in Figure 1.12

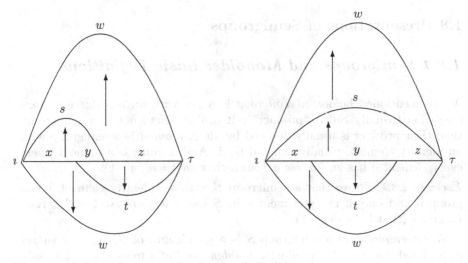

Figure 1.12 Extending an overlap into a spherical diagram

(arrows in these pictures indicate the direction of positive derivation).

Exercise 1.7.17. Prove this theorem.

The word xyz from parts 1 and 2 of Theorem 1.7.16 is called an *overlap*.

Constructing this diagram (or, equivalently, two positive derivations with the same terminal word) is called a *resolution of an overlap*.

In particular, the Knuth–Bendix procedure for a string rewriting system $\mathrm{sr}\langle X \mid \mathcal{R} \rangle$ can be organized as follows. First choose a total order $>$ on X^* that respects the concatenation and satisfies the descending chain condition. Make sure that the left-hand sides of the positive rules from R are bigger in that order than the right-hand sides (if not, replace the positive rule by its inverse).

If all overlaps of the left-hand sides of the positive rules can be resolved, the system is Church–Rosser. If some overlap w cannot be resolved, consider any two sequences of positive moves $xyz \to w_1 \to \ldots \to w, xyz \to w_2 \to \ldots \to w'$ (the words w_1, w_2 are defined in the formulation of Theorem 1.7.16) where $w \neq w'$ are terminal words (such sequences exist since the total order satisfies the descending chain condition and the words in these sequences become smaller and smaller). Suppose, without loss of generality, that $w > w'$. Then add a new positive rule $w \to w'$ to the rewriting system. This will resolve the overlap xyz, but perhaps new overlaps are introduced (possibly involving the new rule), so repeat the procedure. If there are no more unresolvable overlaps, the procedure stops and we get a Church–Rosser rewriting system. The origin of this useful procedure is hard to trace. It was probably first used by Cohn in [70, 71] (see also Bergman [44] and the introduction to Book and Otto [49]).

1.8 Presentations of Semigroups

1.8.1 Semigroups and Monoids: Basic Definitions

As we mentioned before, a *semigroup* is a set with an associative binary operation (usually called "product"). If a, b are elements in a semigroup, then their product is usually denoted by ab. A *monoid* is a semigroup with an identity element, usually denoted by 1. And a *group* is a monoid where every element a has an *inverse* a^{-1} such that $aa^{-1} = a^{-1}a = 1$.

Exercise 1.8.1. Prove that a semigroup S with an identity element 1 is a group if and only if every element a in S has a *left inverse* b and a *right inverse* c (that is $ba = ac = 1$).

A *subsemigroup* of a semigroup S is a subalgebra of S, i.e., any subset of S closed under taking products. A *submonoid* of a monoid S is any subsemigroup of S containing the identity element of S. One can also consider submonoids of semigroups. Then we do not require the submonoids to contain the identity element of the semigroup S. The semigroup S may not contain an identity element at all, and even if it has one, the submonoid may contain its own identity element. A subgroup of a semigroup S is a subsemigroup that is also a group.

Exercise 1.8.2. (1) Prove that the free monoid X^* is generated as a monoid by X and is generated as a semigroup by $X \cup \{1\}$.
(2) Prove that the group S_n of all permutations of the set $\{1, 2, \ldots, n\}$, $n \geq 2$ is generated by the transposition

and the cycle

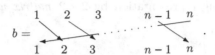

(3) Prove that the semigroup $\mathcal{F}_n$ of all functions from $\{1, 2, \ldots, n\}$ $(n \geq 2)$ to itself is generated by the transposition a, the cycle b from Part (2) and the map

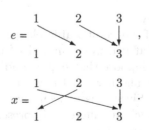

Exercise 1.8.3. Let $a_1, \ldots, a_k$ be natural numbers, let

$$d = \gcd\{a_1, \ldots, a_k\}$$

and let S be the subsemigroup of the additive semigroup of natural numbers generated by these numbers. Then clearly every number in S is divisible by d. Prove that S contains all but finitely many natural numbers divisible by d.

1.8.2 Free and Non-free Semigroups

If S is any semigroup, then any map $\phi : X \to S$ extends uniquely to a *homomorphism* $\bar{\phi} : X^+ \to S$ (it takes every word $w \equiv x_1 \ldots x_n$, $x_i \in X$, to the product $\phi(x_1) \ldots \phi(x_n)$ in S). Recall that this property is the (universal algebraic) reason why X^+ is called *free*.

Similarly if S is a monoid, then every map $\phi : X \to S$ extends uniquely to a homomorphism $\bar{\phi} : X^* \to S$ (the empty word maps to 1).

Thus if S is a semigroup (monoid) generated by X and ϕ is the identity map $X \to X$, then for every $s \in S$ there exists at least one word w such that $\bar{\phi}(w) = s$. In this case we say that s is *represented* by the word w.

Exercise 1.8.4. Let B be the submonoid of $\mathcal{F}_3$ generated by two functions

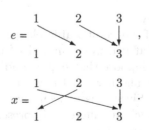

Prove that B consists of six elements and find the shortest words representing each of these elements. This is the so-called 6-element *Brandt monoid*, it is usually denoted by B_2^1. If we remove the identity element 1 from B_2^1, we get the 5-element *Brandt semigroup* B_2. (see Exercise 1.4.6) Prove that semigroup B_2 has the following representation by 2×2 *matrix units*: $\begin{pmatrix} 1 & 0 \\ 0 & 0 \end{pmatrix}$, $\begin{pmatrix} 0 & 1 \\ 0 & 0 \end{pmatrix}$, $\begin{pmatrix} 0 & 0 \\ 1 & 0 \end{pmatrix}$, $\begin{pmatrix} 0 & 0 \\ 0 & 1 \end{pmatrix}$, $\begin{pmatrix} 0 & 0 \\ 0 & 0 \end{pmatrix}$.

Exercise 1.8.5. By Exercise 1.8.2 the semigroup $\mathcal{F}_n$ is generated by three functions a, b, c described there. Find a word in a, b, c that represents the following function.

(1)

(2)

1.8.3 A Characterization of Free Semigroups

A semigroup S is called *cancellative* if $ab = ac$ or $ba = ca$ always implies $b = c$. We say that an element x in a semigroup S is *indecomposable* if x is not equal to a product of any two other elements. Every free semigroup is cancellative and all its elements are products of indecomposable elements.

The following theorem gives a (more semantic) characterization of free semigroups

Theorem 1.8.6 (Levi, [200]). *Suppose that S is a cancellative semigroup without an identity element, in which every element is a product of indecomposable elements, and for every four elements a, u, v, c from S the equality $au = vc$, implies either $u = c$ or $u = bc$ or $c = bu$ for some b. Then S is a free semigroup and the set of indecomposable elements of S is its free generating set.*

Proof. Let X be the set of indecomposable elements of S. By assumption, X generates S. Suppose that two words w and w' in X are equal in S. We need to show that $w \equiv w'$. Suppose that $w \not\equiv w'$ and w' is a shortest possible such word that there exists another word w with $w = w'$ in S (in particular w is not shorter than w'). Since S is cancellative, we can assume that the first (last) letters of w and w' are different. Represent $w \equiv au, w' \equiv vc$ where

$a, c \in X$. Then $u \not\equiv c$ and u, v are not empty (why?). By the assumption of the theorem, there exists an element b such $u = bc$ in S because the case $c = bu$ is not possible for $c \in X$. Then $abc = vc$, so $ab = v$ in S (by cancellation). If b is represented by a word z in X, then the words az and v represent the same element in S but are not the same words because the first letter of v is not a. This contradicts the assumption that w' was a shortest possible because v is shorter than w'.   □

1.8.4 Congruences, Ideals and Quotient Semigroups

A *cyclic semigroup* is a semigroup generated by one element. Each cyclic semigroup is isomorphic to a quotient semigroup of the free semigroup $\{x\}^+$ (by the main property of free semigroups).

Exercise 1.8.7. Describe all congruences on the additive semigroup of positive integers. Show that any cyclic semigroup generated by x either is equal to $\{x\}^+$ or is finite.

If a cyclic semigroup generated by an element x is finite, then $x^m = x^{m+p}$ for some $m, p \geq 1$. The smallest such m (denoted m_x) is called the *index* and the smallest n such that $x^{m_x} = x^{m_x+n}$ is called the *period* or *exponent* of x.

Exercise 1.8.8. Show that the subsemigroup $\langle x \rangle$ generated by x is a group if and only if the index of x is 1.

Exercise 1.8.9. Show that if $x^m = x^{m+n}$ for some element x of a semigroup S, $m, n \geq 1$, then m is at least as large as the index of x and n is divisible by the period of x.

Exercise 1.8.10. Let a be an element of a finite group G. Prove that the exponent of a divides $|G|$. **Hint:** The exponent of a is the number of elements in the subgroup C generated by a. The group G is subdivided into a union of disjoint cosets aC, each of which is of size $|C|$, hence $|C|$ divides $|G|$.

Exercise 1.8.11. Prove that every subgroup of a cyclic group is cyclic. Prove that every subsemigroup of a cyclic semigroup is finitely generated. **Hint:** The first part is an application of the Euclidean algorithm. For the second part only the infinite cyclic semigroup $\mathbb{N}$ needs being considered (by Exercise 1.8.7). Let H be a subsemigroup of $\mathbb{N}$. Let d be the greatest common divisor or numbers from H. By Exercise 1.8.3 H contains all numbers that are divisible by d and greater than some number nd, $n \in \mathbb{N}$. Let S be the finite set of numbers from H that do not exceed nd. Show that the set S together with a finite set of multiples of d generate H.

Exercise 1.8.12. Describe all congruences on the semigroup $\mathcal{F}_2$.

Recall that a subset I of a semigroup S is called an *ideal* if for every $a \in I$, $b \in S$ we have $ab, ba \in I$.

Exercise 1.8.13. Let I be an ideal of S. Consider the partition σ_I of S with one class I and each of the other classes consisting of one element. Prove that this partition is a congruence.

The congruence described in Exercise 1.8.13 is called the *Rees congruence* associated with the ideal I, the corresponding quotient semigroup S/σ_I is called the *Rees quotient semigroup* of S over the ideal I and is denoted by S/I.

A semigroup is called *nilpotent of class* $k,k \geq 1$, if the product of every k elements of it is zero. Note that under this (not very standard but convenient) definition, a semigroup that is nilpotent of class k is also nilpotent of class $k+1$.

Exercise 1.8.14. Show that the class I plays the role of *zero* in S/I, that is $Ix = xI = I$ for every $x \in S/I$.

Exercise 1.8.15. Let k be a natural number. Let I_k be the set of all words in X^+ with length $\geq k$. Prove that I_k is an ideal of X^+. Prove also that the semigroup X^+/I_k is nilpotent of class k.

1.8.5 String Rewriting and Presentations

The equivalence generated by a string rewriting system $\Gamma = \mathrm{sr}\langle X \mid R \rangle$ is a congruence on X^+ (or X^*). Indeed, if there exists a derivation $w = w_1 \to_\Gamma \ldots \to_\Gamma w_n = w'$ from w to w', then for every two words p, q we have a derivation $pwq = pw_1q \to_\Gamma \ldots \to_\Gamma pw_nq = pw'q$. Geometrically this means that for every (w, w')-diagram Δ over Γ we can construct a $(pwq, pw'q)$-diagram by attaching linear diagrams $\varepsilon(p)$ and $\varepsilon(q)$ to $\imath(\Delta)$ and $\tau(\Delta)$ as on Figure 1.13

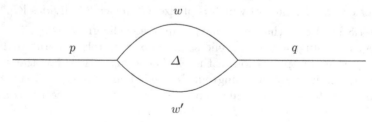

Figure 1.13 Attaching linear diagrams to a diagram

Conversely every congruence σ on X^+ is generated by a string rewriting system. For example, one can pick a representative x_C in each class C of σ and consider the rewriting system with (infinitely many) rules $z \to x_C$ where C is the σ-class containing z. This rewriting system clearly generates σ. Therefore every semigroup S is isomorphic to the quotient semigroup X^+/σ of X^+ over some congruence generated by a string rewriting system $\Gamma = \mathrm{sr}\langle X \mid \mathcal{R} \rangle$. In this case the rewriting system Γ is called a *presentation* of S *by the set of generators* X *and the set of defining relations* $\mathcal{R}$ and we shall write $S = \mathrm{sg}\langle X \mid \mathcal{R} \rangle$. Similarly one can define monoid presentations (replacing X^+ by X^*). If S is a monoid presented by $\mathrm{sr}\langle X \mid \mathcal{R} \rangle$, then we shall write $S = \mathrm{mn}\langle X \mid \mathcal{R} \rangle$.

The monoid X^*/σ comes with the natural homomorphism $X^* \to X^*/\sigma$. Notice that two words w and w' over X represent the same element in S (we shall say that these words are *equal in* S) if and only if there exists a derivation of w from w'. Geometrically this means that w and w' are equal in S if and only if there exists a (w, w')-diagram over Γ. In this case we shall write $w = w' \pmod{\mathcal{R}}$ or $w =_S w'$. If we consider $(u \to v) \in \mathcal{R}$ as a defining relation of S then we shall use "$=$" instead of $\to$ and write $u = v$. Notice that all equalities from $\mathcal{R}$ are true in S and all other relations (i.e., equalities between products of generators of S), follow from the defining relations. If $S = \mathrm{mn}\langle X \mid \mathcal{R} \rangle$ happens to be a group, then we shall write $S = \mathrm{gp}\langle X \mid \mathcal{R} \rangle$. In that case we will omit obvious relations $xx^{-1} = 1, x^{-1}x = 1, 1x = x, x1 = x$.

Exercise 1.8.16. Let $S = \mathrm{mn}\langle X \mid \mathcal{R} \rangle$ and let T be any monoid. Prove that a map $\phi : X \to T$ extends to a homomorphism $S \to T$ if and only if the equalities $\phi(\mathcal{R})$ hold in T. Here $\phi(\mathcal{R})$ is the set of equalities obtained from $\mathcal{R}$ by replacing each letter $x \in X$ with $\phi(x)$.

When we consider monoids or semigroups presented by generators and defining relations Γ, one of the main problems is the *word problem* that asks when two words in X represent the same element in S. In general there is no algorithm to solve this problem.

But if Γ is finite and Church–Rosser (i.e., the semigroup is given by a finite Church–Rosser presentation), then such an algorithm exists, as we have mentioned in Section 1.7. Every congruence class contains a unique terminal word usually called *canonical*. The canonical words have a nice description as exactly those words that do not contain the left parts of the rewriting rules as subwords. Thus canonical words are defined in terms of forbidden subwords (as subshifts in Section 1.6.2). In particular, by Exercise 1.3.5, we have the following lemma:

Lemma 1.8.17. *If a Church–Rosser string rewriting system is finite or, more generally, the set of left-hand sides of the positive rules forms a rational language, then the language of canonical words is rational.*

Remark 1.8.18. The converse statement (that every rational language is the language of canonical words of some finite Church–Rosser string rewriting system) was a long standing open problem. It was proved in 2012 by Diekert, Kufleitner, Reinhardt, and Walter [87]. Quite surprisingly, the Church–Rosser presentation they construct defines a finite monoid.

It is usually convenient to assume that the monoid (semigroup) presented by a Church–Rosser rewriting string system Γ consists of the canonical words with the natural multiplication: the product of two canonical words w and w' is the canonical word in the class containing the word ww'.

Exercise 1.8.19. Prove that the monoid (semigroup) A consisting of all words of the form $a^m b^n$, $m \geq 0, n \geq 0$ (resp. $m \geq 0, n \geq 0, m + n > 0$) with operation $a^m b^n \cdot a^p b^q = a^{m+p} b^{n+q}$ is isomorphic to the monoid (semigroup) $\mathrm{mn}\langle a, b \mid ba = ab \rangle$ (resp. $\mathrm{sg}\langle a, b \mid ba = ab \rangle$). Generalize this result to the monoid $A_n = \mathrm{mn}\langle a_1, \ldots, a_n \mid a_i a_j = a_j a_i, \ i > j \rangle$, $n > 1$ (semigroup $A'_n = \mathrm{sg}\langle a_1, \ldots, a_n \mid a_i a_j = a_j a_i, i > j \rangle$). The monoid A_n (semigroup A'_n) is called the *free commutative monoid* (*free commutative semigroup*) of rank n.

Exercise 1.8.20. Consider the monoid $B = \mathrm{mn}\langle a, b \mid aba = a, \ bab = b, \ bb = aa, \ aaa = aa, \ aab = aa, \ baa = aa \rangle$. Prove that it is isomorphic to the Brandt monoid B_2^1 from Exercise 1.8.4.

Exercise 1.8.21. Consider the semigroup $P = \mathrm{sg}\langle a, b \mid aa = a, \ ab = b, \ bb = ba \rangle$. Find the number of elements in P and construct the multiplication table of P. Prove that P is isomorphic to a subsemigroup of the Brandt monoid B_2^1.

Exercise 1.8.22. Let $n > 1$. Consider the monoid $D_{2n} = \mathrm{mn}\langle a, b \mid a^2 = 1, b^n = 1, ab = b^{n-1}a \rangle$, $n > 1$. Prove that the presentation is Church–Rosser and defines a group of order $2n$. Show that D is isomorphic to the group of all symmetries of the regular n-gon on the plane. This is the so-called *dihedral group* of order $2n$.

Exercise 1.8.23. Consider the monoid $Q = \mathrm{mn}\langle a, b \mid a^4 = 1, a^2 = b^2, aba = b \rangle$. Show that this presentation defines a group of order 8 that is not isomorphic to D_4. This group is called the group of *quaternions*. **Hint:** Use the order $a > b$ on the set of generators, and use the Knuth–Bendix procedure to find a Church–Rosser presentation and canonical words.

Exercise 1.8.24 (Page 182, [195]). Let $n \geq 2$. Consider the monoid

$$S_n = \mathrm{mn}\langle a_1, \ldots, a_{n-1} \mid a_i^2 = 1 \ (i \geq 1),$$
$$a_j a_i = a_i a_j \ (j \geq i + 2),$$
$$a_{i+k} a_{i+k-1} \cdots a_{i+1} a_i a_{i+k} = a_{i+k-1} a_{i+k} a_{i+k-1} \cdots a_{i+1} a_i \ (i \geq 0, \ k \geq 1) \rangle.$$

Show that S_n is isomorphic to the group of all permutations of the set $\{1, \ldots, n\}$, i.e., the *symmetric group* of order $n!$. **Hint:** Show that the presentation is Church–Rosser, find canonical forms, and the number of elements of S_n. Then consider a map sending a_i to the *transposition* that is the permutation switching i and $i + 1$ and leaving other numbers fixed. Show that this map is a homomorphism onto the symmetric group. Then show that the homomorphism must be injective by counting the number of canonical words in S_n (that is the most interesting part of the exercise).

Exercise 1.8.25. Consider the monoid

$$\mathrm{mn}\langle a_1, \ldots, a_n, \bar{a}_1, \ldots, \bar{a}_n \mid a_i \bar{a}_i = 1, \bar{a}_i a_i = 1, a_i a_j = a_j a_i \rangle.$$

Find a Church–Rosser presentation of that monoid and prove that it is a commutative group that is isomorphic to the Cartesian product of n copies of $\mathbb{Z}$. Under this isomorphism any word $a_1^{m_1} \ldots a_n^{m_n}$ corresponds to the vector $(m_1, \ldots, m_n)$. Show that this is a relatively free group in the variety of commutative groups.

Exercise 1.8.26. Consider the monoid

$$\mathrm{mn}\langle\, t, \bar{t}, x_i, \bar{x}_i, i \in \mathbb{Z} \mid x_i \bar{x}_i = 1, \bar{x}_i x_i = 1, t\bar{t} = 1, \bar{t}t = 1, x_i x_j = x_j x_i,$$
$$\bar{t} x_i t = x_{i+1}, i, j \in \mathbb{Z} \,\rangle.$$

Find a Church–Rosser presentation of this monoid; show that it is a group generated by t and x_0 and the subgroup generated by $x_i, i \in \mathbb{Z}$ is free commutative. Find an infinite Church–Rosser presentation of this group in generators $x_0, t, \bar{x}_0, \bar{t}$. This group is called the *wreath product* of $\mathbb{Z}$ and $\mathbb{Z}$ and is denoted by $\mathbb{Z} \wr \mathbb{Z}$.

1.8.6 The Free Group: The Syntactic Definition

We know that every variety of algebras contain relatively free algebras of arbitrary ranks. Here we give an explicit (syntactic) definition of free groups (i.e., relatively free groups in the variety of all groups). More semantic definitions are given in the next two sections.

Let X be an alphabet, X_- be a copy of X, and let $x \mapsto \bar{x}$ be a bijection $X \leftrightarrow X_-$. Consider the following presentation $\mathrm{mn}\langle\, X \cup X_- \mid x\bar{x} = 1, \bar{x}x = 1, x \in X \,\rangle$.

Exercise 1.8.27. Check that the string rewriting system corresponding to this presentation is Church–Rosser and that the resulting monoid F_X is a group. Prove that for every group G, every map $X \to G$ uniquely extends to a homomorphism $F_X \to G$.

The canonical words of this string rewriting system are precisely the words in $X \cup X_-$ without subwords of the form $x\bar{x}, \bar{x}x, x \in X$. These words are called *freely reduced*. Thus every word in the alphabet X is equal to a unique freely reduced word in F_X.

The group F_X is the *free group* with free generating set X_+ (prove it!).

Exercise 1.8.28. Show that the spherical growth function of the free group with free generating set $x_1, \ldots, x_k$ is $s(n) = 2k(2k-1)^{n-1}, n \geq 1$.

Theorem 1.8.29. *Two elements u, v in the free group F_X commute if and only if u and v are powers of another element w in F_X, in particular some powers of u and v coincide.*

Proof. Induction on the length of u. If u is empty, the statement is clear. Suppose that the length of u is the smallest possible for a counterexample. Suppose first that $u \equiv a^{-1}u'a$ for some letter a. Then the equality $uv = vu$

is equivalent to $a^{-1}u'av = va^{-1}u'a$ or $u'ava^{-1} = ava^{-1}u'$. Since u' is shorter than u, there exists a word w such that $u' = w^k, ava^{-1} = w^\ell$ in the free group. Then $a^{-1}u'a = (a^{-1}w'a)^k, v = (a^{-1}wa)^\ell$, as required. Hence $u \not\equiv a^{-1}u'a$ for any letter a (i.e., u is *cyclically reduced*).

Note that $uv \equiv u_1v_1$ where $u \equiv u_1p, v \equiv p^{-1}v_1$, that is p is the maximal suffix of u that cancels with a prefix of v. Then the equality $uv = vu$ becomes $u_1v_1 = p^{-1}v_1u_1p$ or $(pu_1)(v_1u_1) = (v_1u_1)(pu_1)$ where we can assume that all the words in parentheses are reduced. There is no cancellation between the two words in parentheses on the left and right sides of this equality. Hence this equality holds in the free semigroup. By Theorem 1.2.9 then pu_1 and v_1u_1 are powers of another word q in the free semigroup. Conjugating by u_1^{-1}, we conclude that $u = u_1p$ and $uv = u_1v_1$ are powers of $u_1qu_1^{-1}$ in the free group. Hence v is a power of $u_1^{-1}qu_1^{-1}$ too.   □

Theorem 1.8.29 and Exercise 1.8.11 immediately imply

Corollary 1.8.30. *Every commutative subgroup of the free group F is cyclic.*

Exercise 1.8.31. Prove this corollary. **Hint:** You will need to prove first that every subgroup of a cyclic group is cyclic, see Exercise 1.8.11.

1.8.7 The Free Group and Ping-Pong

One of the best semantic ways to show that a group with two generators is free is to let the generators play ping-pong. The game has the following setting (it was invented by Felix Klein).

Theorem 1.8.32. *Let X be a set (call it a ping-pong table), on which the group $G = gp\langle a, b\rangle$ acts, i.e., there exists a homomorphism from G to the group of bijections from X to X that takes every $g \in G$ to a bijection $\cdot g : X \to X$. Suppose that X is a disjoint union of two nonempty subsets $X_a \cup X_b$ (the two sides of the table). Suppose also that $X_a \cdot a^n \subseteq X_b$ and $X_b \cdot b^n \subseteq X_a$, $n \in \mathbb{Z}, n \neq 0$ (the player a sends the ball to b's side of the table and vice versa). Then the group G is free.*

Proof. Let F_2 be the free group with free generators x, y. Then the map $x \mapsto a, y \mapsto b$ extends to a homomorphism of F_2 onto G. We need to show that this homomorphism is injective. For this, we need to take a canonical form of a non-identity element of F_2 and show that the corresponding element in G is not an identity. Let w be such a canonical word $w = x^{\epsilon_1}y^{\delta_1} \ldots x^{\epsilon_n}y^{\delta_n}$ where $\epsilon_i, \delta_i \in \mathbb{Z}$ and ϵ_1, δ_n may be equal to 0, other ϵ_i, δ_j are not zero. Note that if $w(a, b) = 1$ in G, then $a^{-1}w(a, b)a = 1$ in G as well. Therefore we can assume that $\epsilon_1 \neq 0, \delta_n = 0$. Now take any element z in X_a, and apply the bijection corresponding to $w(a, b)$ to z. The result is an element in X_b. Indeed, $\cdot a^{\epsilon_1}$ sends z to $z_1 \in X_b$, then $\cdot b^{\delta_1}$ sends z_1 back to X_a, etc., until finally $\cdot a^{\epsilon_n}$ sends the element to X_b. Therefore the bijection $\cdot w(a, b)$ is not the identity function. Hence $w(a, b)$ is not the identity in G.   □

Exercise 1.8.33. Consider two matrices $a = \begin{bmatrix} 1 & 2 \\ 0 & 1 \end{bmatrix}$, $b = \begin{bmatrix} 1 & 0 \\ 2 & 1 \end{bmatrix}$. The matrices have determinant 1, and so they generate a subgroup of the multiplicative group $SL_2(\mathbb{R})$ of all 2×2-matrices with real entries and determinant 1. Show that this subgroup is free. **Hint:** Let $\mathbb{C}^*$ be the set of all complex numbers with infinity ∞ adjoined.

Define a homomorphism from $SL(2, \mathbb{R})$ to the group of all bijections of $\mathbb{C}^*$ to $\mathbb{C}^*$ that takes every matrix $\begin{bmatrix} p & q \\ r & s \end{bmatrix}$ to the *Möbius transformation*

$$z \mapsto \frac{pz + q}{rz + s}.$$

If $z = \infty$ or $rz + s = 0$ then the image of z is defined naturally, say, $1/0 = \infty$, $5 \cdot \infty / 3 \cdot \infty = 5/3$ (show that this is indeed a homomorphism). Let X_a be the set $\{z \mid |z| < 1\}$, $X_b = \{z \mid |z| > 1\}$. Show that matrices a and b play ping-pong on X.[6]

Exercise 1.8.34. Consider the monoid $S = \mathrm{mn}\langle a, b \mid b^6 = 1, a^2 = b^3 \rangle$. Show that S is a group and find a Church–Rosser presentation of that group. Show that elements from $\{1, a, a^2, a^3\}$ are different and all elements from $\{1, b, b^2, b^3, b^4, b^5\}$ are different in that group. Show that this group is isomorphic to the group $SL(2, \mathbb{Z})$ of all 2×2 matrices with integer entries and determinant 1 (compare with Exercise 1.4.2) and the map $a \mapsto \begin{bmatrix} 0 & -1 \\ 1 & 0 \end{bmatrix}$, $b \mapsto \begin{bmatrix} 1 & -1 \\ 1 & 0 \end{bmatrix}$ extends to an isomorphism. **Hint:** The last statement is a more difficult part of the exercise, you may have to modify the ping-pong Theorem 1.8.32 to prove that.

1.8.8 Free Groups and Rings of Formal Infinite Linear Combinations

A yet another way to construct a free group is the following. Let A be an alphabet. Let K be any field. Consider the set $K\langle\!\langle A \rangle\!\rangle$ of all formal infinite linear combinations

$$\sum_{u \in A^*} \alpha_u u$$

where $\alpha_u \in K$. One can define addition and multiplication on that set as follows:

---

[6] The set $\{z \mid |z| = 1\}$ that separates X_a from X_b is of course the ping-pong net.

$$\sum_{u\in A^*} \alpha_u u + \sum_{u\in A^*} \beta_u u = \sum_{u\in A^*} (\alpha_u + \beta_u)u,$$

$$\Big(\sum_{u\in A^*} \alpha_u u\Big)\Big(\sum_{u\in A^*} \beta_u u\Big) = \sum_{u\in A^*, v\in A^*} (\alpha_u \beta_v)uv.$$

Exercise 1.8.35. Prove that $K\langle\!\langle A\rangle\!\rangle$ is a ring (in fact an associative algebra over the field K) with the identity element $\sum_{u\in A^*} \alpha_u u$ where $\alpha_1 = 1$ and all other α_u are equal to 0.

Note that all elements $1+a$, $a \in A$ are invertible in $K\langle\!\langle A\rangle\!\rangle$. Indeed, $(1+a)^{-1} = 1 - a + a^2 - a^3 + \ldots$ (the sum of a geometric progression!).[7] Hence $1 + a$, $a \in A$, and their inverses generate a subgroup G of the multiplicative semigroup of the ring $\mathbb{Q}\langle\!\langle A\rangle\!\rangle$. The following result was proved by Magnus [215].

Theorem 1.8.36. *The group G is free with free generating set $\{1+a \mid a \in A\}$.*

Proof. Consider any nontrivial reduced word w in $A \cup A^{-1}$ and its image $\phi(w)$ in G under the substitution $a \mapsto 1+a, a \in A, a^{-1} \to (1+a)^{-1}$. We need to show that $\phi(w) \neq 1$. Since w is reduced and nontrivial, $w = a_1^{n_1} a_2^{n_2} \ldots a_m^{n_m}$ where $a_i \in A$, $a_i \neq a_{i+1}$, $n_i \in \mathbb{Z}, n_i \neq 0$. Suppose that the characteristic p of K is not 0. Then for every $i = 1, \ldots, m$ let q_i be the maximal power of p dividing n_i. If the characteristic of K is 0, then let $q_i = 1$, $i = 1, \ldots, m$. By Exercise 1.4.18 $(1 + a_i)^{q_i} = 1 + a_i^{q_i}$ in G for every i. Let u be the word $a_1^{q_1} a_2^{q_2} \ldots a_m^{q_m} \in A^+$. Then u is a nontrivial reduced word and the coefficient of u in the linear combination $\phi(w)$ is – up to the sign – the product of integers $\frac{n_i}{q_i}$, i.e., it is equal to

$$\frac{n_1}{q_1} \frac{n_2}{q_2} \ldots \frac{n_m}{q_m}$$

(check it!), which is not equal to 0 in K. Hence, indeed, $\phi(w) \neq 1$ in G □

As a corollary of Theorem 1.8.36 we deduce

Theorem 1.8.37. *There exists a total order $\leq$ on any free group F_A that is compatible with the product, i.e., $g \leq h \to tg \leq th, gt \leq ht$ for every $g, h, t \in F_A$.*

Proof. Let $K = \mathbb{Q}$, the field of rational numbers. Let $g \in F_A$. Consider the linear combination $\phi(g)$. We can represent $\phi(g)$ as $1 + \alpha_w w + \sum_{u\neq w} \alpha_u u$ where w is the ShortLex smallest word in A^+ whose coefficient in $\phi(g)$ is not 0. Let us call α_w the *chief* coefficient of $\phi(g)$. For example, the chief coefficients of $1 + 3a, a \in A$ is 3, and the chief coefficient of $(1+a)^{-1}$ is -1. Now we can write $g \leq h$ if either $g = h$ or the chief coefficient of $\phi(g^{-1}h)$ is positive.

Exercise 1.8.38. Show that $\leq$ is a total order that is compatible with the product in the free group F_A.

                                                                                            □

---

[7] Moreover, an infinite linear combination of elements from A^+ is invertible if and only if the coefficient of 1 is not 0.

1.8.9 The Growth Function, Growth Series and Church–Rosser Presentations

Suppose that a rewriting system $\mathcal{P} = \mathrm{sr}\langle\, X \mid u_i \to v_i,\ i \in I \,\rangle$ is Church–Rosser (but possibly infinite) and $|v_i| \le |u_i|$ for every $i \in I$. Then the canonical words for this rewriting system are shortest representatives in their equivalence classes. Hence in order to compute the growth function of the semigroup presentation $S = \mathrm{sg}\langle\, X \mid u_i = v_i, i \in I \,\rangle$, one needs to count the number of canonical words of length $\le n$, $n \in \mathbb{N}$. Since the rewriting system is Church–Rosser, the canonical words are precisely the words that do not contain subwords u_i, $i \in I$.

Let P_X be the growth series of the semigroup S. It turns out that P_X is often a *rational function* (a quotient of two polynomials).

For example, this is the case if the language of the left-hand sides of the rewriting system is rational. This follows from Lemma 1.8.17 and the following classical result by Chomsky and Schützenberger about rational languages (see [93, 275]).

Theorem 1.8.39. *The growth series of every rational language is a rational function.*

Proof. (Requires some knowledge of Linear Algebra.) Indeed, let L be a rational language. By Theorem 1.3.6, there exists a finite deterministic automaton (Q, A) with one input vertex q_0 and a set of output vertices Q_+ that recognizes L. Since (Q, A) is deterministic, the number $c(k)$ of words of length k in L is the same as the number of paths of length k that start at q_0 and end in Q_+. For every $q \in Q_+$ let $c_q(k)$ be the number of paths p of length k with $p_- = q_0, p_+ = q$. Let f_q be the corresponding series $\sum c_q(k) z^k$. Then the growth series of L is a finite sum $\sum_{q \in Q_+} f_q$. Thus it is enough to prove that each f_q is a rational function. Let M be the *adjacency matrix* of the underlying graph of (Q, A), that is the $|Q| \times |Q|$-matrix with (i, j)-entry equal the number of edges $i \to j$. A standard Linear Algebra exercise gives that $c_q(k)$ is the (q_0, q)-entry of M^k (prove it!). Let $\vec{e}$ be the row vector from $\mathbb{R}^{|Q|}$ with q_0-coordinate 1 and other coordinates 0, and $\vec{f}$ be the column vector from $\mathbb{R}^{|Q|}$ with q-coordinate 1 and the other coordinates 0. Then

$$c_k(q) = \vec{e} M^k \vec{f}.$$

Therefore

$$f_q(z) = \sum \vec{e} M^k \vec{f} z^k = \vec{e} \left(\sum (Mz)^k \right) \vec{f}.$$

The sum $\sum (Mz)^k$ is equal to $(1 - Mz)^{-1}$ (the sum of a geometric progression!). Hence

$$f_q(z) = \vec{e}(1 - Mz)^{-1} \vec{f},$$

i.e., the (q_0, q)-entry of $(1 - Mz)^{-1}$. That entry (also from Linear Algebra) is the quotient of the (q, q_0)-cofactor of the matrix $1 - Mz$ by the determinant of $1 - Mz$. Both the numerator and the denominator of this fraction are polynomials in z, hence $f_q(z)$ is a rational function. □

Exercise 1.8.40. Represent the growth series of the free group with respect to its free generating set as a rational function. **Hint:** Use Exercise 1.8.28.

Exercise 1.8.41. Compute the rational function represented by the growth series of the standard presentation of the free commutative semigroup

$$\mathrm{sg}\langle a_1, \ldots, a_n \mid a_i a_j = a_j a_i, i > j \rangle$$

with n generators.

1.8.10 The Cayley Graphs

Suppose that a semigroup S is generated by a set X. Consider the following directed graph $\Gamma_X(S)$. Its vertices are all elements of S, and edges are $s \to sx$, $s \in S, x \in X$. This graph is called the (right) *Cayley graph* of S relative to the generating set X. We can label each edge $s \to sx$ by the letter x and obtain the *labeled Cayley graph* of S with respect to X, denoted by $\Gamma_X^l(S)$.

If S is a group and X is *symmetric*, that is $X^{-1} = X$, then every edge $s \to sx$ of $\Gamma_X(S)$ has an inverse edge $sx \to s$, so $\Gamma_X(S)$ becomes a graph in the sense of Serre.

Words in the alphabet X label paths on $\Gamma_X^l(S)$. Given a vertex s and a word u by $s \cdot u$ we denote the terminal vertex of the path starting at s and labeled by u. If two words u, v over X are equal in X, then $s \cdot u = s \cdot v$ for every $s \in S$. If S is a monoid, then the converse is true (take $s = 1$). In particular, the words that are equal to 1 in a group S are precisely the words that label loops in the labeled Cayley graph $\Gamma_X^l(S)$.

Exercise 1.8.42. Show that the Cayley graph of the free group F_X with respect to X is a *tree* with every vertex of degree $2|X|$, and the Cayley graph of the free semigroup is a tree with distinguished vertex (root) $\emptyset$, each vertex having out-degree $|X|$ and all vertices except the root having in-degree 1, the in-degree of the root is 0. On Figure 1.14, the graph on the left is a part of the Cayley graph of the free group $\mathrm{gp}\langle x, y \rangle$. All edges there are pointed away from the origin O, horizontal edges are labeled by a, vertical edges are labeled by b. The graph on the right is a part of the Cayley graph of the free monoid $\mathrm{mn}\langle a, b \rangle$. All edges are pointed down (away from the root O). The edges pointing down and left are labeled by a, the edges pointing down and right are labeled by b. Note that the graph is the infinite full binary tree without leaves. Every full binary tree is a subtree of this tree containing the root.

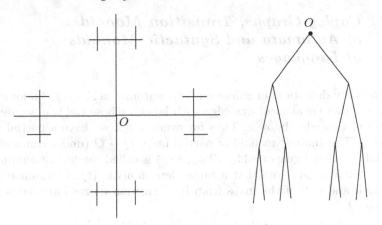

Figure 1.14 The Cayley graphs of the free group and free monoid with 2 generators

Exercise 1.8.43. Find the Cayley graph of the group of all permutations of $\{1, 2, 3\}$ with respect to the generating set consisting of permutations

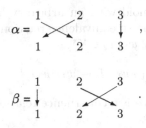

More generally, one can define the (right) Cayley graph of a *semigroup act* (Y, S), i.e., a set Y with unary operations $\cdot s$, $s \in S$ satisfying the identities

$$(y \cdot s) \cdot t = y \cdot st,$$

$y \in Y, s, t \in S$ (in this case we shall say that S is a semigroup *acting* on a set Y). If S is generated by a set A, then the *Cayley graph* of the act (Y, S) with respect to the generating set A is the directed graph with vertex set Y and edge set $y \to y \cdot a$, $y \in X$, $a \in A$.

For example, if S is a group, H is a subgroup of S, then S acts on the set of all *right cosets* Hg of H: $Hg \cdot t = Hgt$. The corresponding Cayley graph of the act (with respect to a generating set of S) is called a (right) *Schreier graph* of the subgroup H [286]. One can similarly introduce the *labeled Cayley graph* of an act and the *labeled Schreier graph* of a subgroup (corresponding to a generating set).

Exercise 1.8.44. Show that the labeled Cayley graph of every finite act of the free monoid A^* is a complete automaton with out-degree $|A|$ of every vertex. Show that, conversely, every finite complete automaton (Q, A) is the labeled Cayley graph of a finite act of A^*.

1.8.11 Cayley Graphs, Transition Monoids of Automata and Syntactic Monoids of Languages

Recall that, by definition, in a deterministic automaton $\mathcal{A} = (Q, A)$, for every letter $a \in A$ there is at most one edge with label a whose tail is q; we denote the head of this edge by $q \cdot a$. Thus for every $a \in A$ we have a partial map $\phi_a : Q \to Q$. The (finite!) monoid of partial maps $Q \to Q$ (under composition of partial functions) generated by all $\phi_a, a \in A$ is called the *transition monoid* of the automaton $\mathcal{A}$. Note that a finite deterministic (Q, A) automaton is complete if and only if the maps from its transition monoid are defined on the whole Q.

Exercise 1.8.45. Prove that the transition monoid of a complete automaton $\mathcal{A}$ is a group if and only if the automaton obtained by reversing the orientation of all edges of $\mathcal{A}$ (but keeping the labels) is also complete.

Exercise 1.8.46. Draw an automaton whose transition monoid is the cyclic group of integers modulo p.

One can associate a monoid with an arbitrary language as follows. Let $L \subseteq X^*$ be a language. Define an equivalence relation on the free monoid X^*: $u \sim v$ if and only if for every $w_1, w_2 \in X^*$,

$$w_1 u w_2 \in L \leftrightarrow w_1 v w_2 \in L.$$

Exercise 1.8.47. Prove that $\sim$ is a congruence relation on X^*, hence $X^*/\sim$ is a monoid.

The monoid $X^*/\sim$ is called the *syntactic monoid* of the language L.

Exercise 1.8.48. What is the syntactic monoid of the language of all words from X^* of even length?

Theorem 1.8.49. *The following conditions are equivalent.*

(1)   *A language $L \subseteq A^*$ is rational;*
(2)   *The syntactic monoid of L is finite;*

Proof. (1)$\to$ (2) Suppose that L is recognized by a finite deterministic automaton (Q, A) with input vertex ι and the set of output vertices Q_+.

Consider the homomorphism ϕ from A^* to the transition monoid of (Q, A) that takes each $a \in A$ to the partial function on Q defined by the letter a. Suppose that u and v are in different classes of the equivalence relation $\sim$, that is for some $w_1, w_2 \in A^*$ we have $w_1 u w_2 \in L$, but $w_1 v w_2 \notin L$. This means $\iota \cdot w_1 u w_2 \in Q_+$ but $\iota \cdot w_1 v w_2 \notin Q_+$. Therefore the partial maps $Q \to Q$ defined by the words u, v are different. Thus the number of $\sim$-classes cannot exceed the number of elements in the transition monoid of (Q, A), and the syntactic monoid of L is finite.

(2)$\to$ (1) Suppose that the syntactic monoid $M = A^*/\sim$ of a language $L \subseteq A^*$ is finite. Note that by the definition of $\sim$, L is a union of $\sim$-classes (indeed, if $u \sim v$ and $u \in L$, then $1u1 \in L$, hence $v = 1v1 \in L$). Let $M' = L/\sim$ be the set of all $\sim$-classes contained in L.

Let (Q, A) be the Cayley graph of M with respect to the generating set A considered as an automaton with input vertex 1 and output vertices M'. Let us show that the language recognized by (Q, A) is L. By the definition of M', we have $1 \cdot L \in M'$, therefore the language recognized by (Q, A) contains L. Suppose that for some $u \in A^*$, we have $1 \cdot u \in M'$. That means u belongs to one of the $\sim$-classes contained in L, hence $u \in L$, so the language recognized by (Q, A) is equal to L.   □

Exercise 1.8.50. Prove that every finite monoid is the transition monoid of a finite automaton. **Hint:** Let M be a finite monoid with generating set A. Consider the corresponding Cayley graph (Q, A) of M as a finite automaton. Then the transition monoid of that automaton is M.

Remark 1.8.51. Not every finite monoid is the syntactic monoid of some rational language. For example, let M be the monoid $\{1, \alpha, \beta, \gamma\}$ with multiplication $m_1 m_2 = m_2$ whenever $m_2 \neq 1$ (i.e., M is the 3-element right zero semigroup with adjoined identity, see Section 3.1). Then M is not the syntactic monoid of any language [194]. On the other hand, every finite group is the syntactic monoid of some rational language [226].

Not every language is rational, that is not every language has a finite syntactic monoid. This easily follows from the fact that the set of all languages is uncountable while the set of all finite automata is countable. But here is a concrete example.

Exercise 1.8.52. Let $A = \{a_1, \cdots, a_r\}$ and let L be the set of words w_n, in which each a_i occurs exactly n times. Then the free commutative monoid over A of rank n (see Exercise 1.8.19) is isomorphic to the syntactic monoid of L (in particular L is not a rational language).

For another example, consider the alphabet consisting of two symbols – the left and right parentheses "(" and ")". The *Dyck language* consists of balanced sequences of parentheses. More precisely we define *Dyck words* by induction: the empty word is a Dyck word, () is a Dyck word. If we insert a subword () in any place of a Dyck word, we get a Dyck word again. One can obtain all Dyck words by taking all terms of the signature with one binary operation and removing all variables and operation symbols (leaving the parentheses only):

$$((x \cdot y) \cdot (z \cdot t)) \cdot (((z \cdot y) \cdot x) \cdot t) \to (()())((()))\,.$$

Exercise 1.8.53. Show that the syntactic monoid of the Dyck language is infinite and has two generators p, q satisfying $pq = 1$, $qp \neq 1$.

We shall meet this semigroup (called the *bicyclic semigroup*) again in Section 3.8.1.

Chapter 2
Words that Can Be Avoided

We start with an old problem from number theory whose solution leads to the first example of an infinite cube-free word, thus showing that the word x^3 is avoidable. Then we shall describe all avoidable words.

Theorems proved in this Chapter include

- Theorem 2.1.2 of Thue, Hedlund, Morse and Arshon about cube-free words.
- Theorem 2.3.4 of Thue characterizing square-free substitutions.
- Theorem 2.5.14 of Bean, Ehrenfeucht, McNulty and Zimin characterizing avoidable words.

2.1 An Old Example

In 1851 Eugène Prouhet [271] (see also [5]) studied the following number theory problem, which on the first glance has nothing to do with combinatorial algebra.

Question. Are there arbitrary big numbers m such that some intervals $[1, M]$, M depends on m, of natural numbers can be divided into 2 disjoint parts P_1, P_2 such that the sum of all elements in P_1 is the same as in P_2, the sum of all squares of elements P_1 is the same as in P_2, ..., the sum of m-th powers of elements of P_1 is the same as in P_2.

This problem has long history. In particular, Gauss and Euler studied some variations of this problem.

Prouhet came up with a solution. His solution may be interpreted in the following way. Let us consider an example. Take $m = 2$. Consider the word $p_2 \equiv abbabaab$, and produce the following table:

$$1\ 2\ 3\ 4\ 5\ 6\ 7\ 8$$
$$a\ b\ b\ a\ b\ a\ a\ b$$

© Springer International Publishing Switzerland 2014

M.V. Sapir, *Combinatorial Algebra: Syntax and Semantics*,
Springer Monographs in Mathematics, DOI 10.1007/978-3-319-08031-4_2

Now let P_a be the set of numbers from 1 to 8, that are above a in this table, P_b be the set of numbers, that are above b: $P_a = \{1, 4, 6, 7\}$, $P_b = \{2, 3, 5, 8\}$ (recall the connection between words and partitions from Example 1.2.2). Let us check the Prouhet condition:

$$1 + 4 + 6 + 7 = 18 = 2 + 3 + 5 + 8,$$

$$1^2 + 4^2 + 6^2 + 7^2 = 102 = 2^2 + 3^2 + 5^2 + 8^2.$$

If we want to construct the Prouhet decomposition for $m = 3$ we have to take this word $p_2 \equiv abbabaab$, change a by b and b by a (we'll get $baababba$), and concatenate these two words:

$$p_3 \equiv abbabaabbaababba.$$

By induction one can easily define the Prouhet word for every m: p_m is obtained by concatenating p_{m-1} and p'_{m-1} where p'_{m-1} is obtained from p_{m-1} by the substitution $a \mapsto b, b \mapsto a$.

Exercise 2.1.1. Prove that for every m the partition corresponding to the Prouhet word p_m satisfies the Prouhet condition for the sums of nth powers for every $n \leq m$.

The word p_m was rediscovered several times after Prouhet. Axel Thue rediscovered this word in 1906 [314] and he was the first to prove the following result:

Theorem 2.1.2. *The word p_m does not contain subwords of the form www where w is any nonempty word. Thus words p_m are cube-free.*

Arshon, Hedlund and Morse proved the same result in the late 1930s [14, 238]. Now it belongs to some collections of problems for high school students.

2.2 Proof of Thue's Theorem

Let us consider the following substitution:

$$\phi(a) \equiv ab, \quad \phi(b) \equiv ba.$$

Words ab and ba will be called *blocks*.
Let $t_1 = a$, ..., $t_n = \phi(t_{n-1})$.

Exercise 2.2.1. Prove that $t_n = p_{n-1}$ for every $n \geq 2$.

Lemma 2.2.2. *If w is cube-free, then $\phi(w)$ is also cube-free.*

Proof. Suppose that $\phi(w)$ contains a cube ppp.

Case 1. The length of p is even.

Case 1.1. The first occurrence of p starts with the first letter of a block. Since $\phi(w)$ is a product of blocks of length 2, and $|p|$ is even, p ends with the second letter of a block. Then the second and the third occurrence of p also start with the first letter of a block and end with the second letter of a block. Thus p is a product of blocks, so $p = \phi(q)$ for some word q. Now let us substitute every block in $\phi(w)$ by the corresponding letter. Then $\phi(w)$ will turn into w and $ppp = \phi(q)\phi(q)\phi(q)$ will turn into qqq. Therefore w contains a cube qqq.

Case 1.2. The first occurrence of p starts with the second letter of a block. Then it ends with the first letter of a block, and the same is true for the second and the third occurrence of p. Without loss of generality assume that p starts with a. This a is second letter of the block ba. Since the second occurrence of p also starts with a and this a is the second letter of a block, we can conclude that p ends with b. Therefore $p \equiv ap'b$. Then we have:

$$\phi(w) \equiv \ldots b\, ap'b\, ap'b\, ap'b \ldots .$$

Consider the word bap'. This word has even length, starts with the beginning of a block and repeats 3 times in $\phi(w)$, which is impossible by the previous case.

Case 2. The word p has odd length. If the first occurrence of p starts with the first letter of a block, then the second occurrence of p starts with the second letter of a block. If the first occurrence of p starts with the second letter of a block, then the second occurrence of p starts with the first letter of a block and the third occurrence of p starts with the second letter of a block.

In any case there are two consecutive occurrences of p such that the first one starts with the first letter of a block and the second one starts with the second letter of a block. Let us denote these copies of p by p' and p''.

It is clear that $|p| \geq 2$.

Suppose that p' starts with ab. Then p'' also starts with ab. This b is the first letter of a block. Therefore the third letter in p'' is a. Therefore the third letter in p' is also a. This a is the first letter of a block. The second letter of this block is b. Therefore the fourth letter of p' is b. Then the fourth letter of p'' is b, which is the first letter of a block, the fifth letter of p'' is a, same as the fifth letter in p', and so on. Every odd letter in p is a, every even letter in p is b. Since p' has odd number of letters, the last letter in p' is a. This a is the first letter of a block. The second letter of this block is the first letter of p'', which is a — a contradiction (we found a block aa). □

Exercise 2.2.3. Prove that t_n does not contain subwords of the form $qpqpq$ for any words p and q.

2.3 Square-Free Words

The words constructed in the previous section do not contain cubes. Now we
will consider words that do not contain squares, the *square-free* words.

A word is called *square-free* if it does not contain a subword of the form uu.

Exercise 2.3.1. Prove that every square-free word over an alphabet with 2
letters has length at most 3.

Theorem 2.3.2 (Thue, [314]). *There exist arbitrary long square-free words
over a 3-letter alphabet.*

Consider the following substitution:

$$\phi(a) \equiv abcab, \quad \phi(b) \equiv acabcb, \quad \phi(c) \equiv acbcacb.$$

Theorem 2.3.2 easily follows from the following

Lemma 2.3.3. *For every square-free word w, $\phi(w)$ is square-free.*

A substitution satisfying the condition of Lemma 2.3.3 will be called
square-free.

In turn, Lemma 2.3.3 will be a corollary of the following powerful theorem
of Thue.

Theorem 2.3.4. *Let M and N be alphabets and let ϕ be a substitution from
M to N^+. If*

(1) $\phi(w)$ *is square-free whenever w is a square-free word from M^+ of length
     no greater than 3,*
(2) $a = b$ *whenever $a, b \in M$ and $\phi(a)$ is a subword of $\phi(b)$,*

then ϕ is a square-free substitution.

Proof of Theorem 2.3.4. Let ϕ satisfy (1) and (2). First of all let us prove
the following "rigidity" statement:

Claim.   If $a, e_1, \ldots, e_n$ are letters from M, $E \equiv e_1 \ldots e_n$ is a square-free word,
       and $\phi(E) \equiv X\phi(a)Y$ then $a = e_j$, $X = \phi(e_1 \ldots e_{j-1})$, $Y = \phi(e_{j+1} \ldots e_n)$
       for some j.

Suppose this is not true. Since $\phi(e_i)$ cannot be a subword of $\phi(a)$ (by (2)),
$\phi(a)$ intersects with at most 2 factors $\phi(e_i)$. Since $\phi(a)$ cannot be a subword
of $\phi(e_i)$ (again by (2)), $\phi(a)$ intersects with exactly 2 factors, say $\phi(e_j e_{j+1})$.
Then $\phi(e_j) \equiv pq$, $\phi(e_{j+1}) \equiv rs$, and $\phi(a) \equiv qr$. Now $\phi(ae_j a) \equiv qrpqqr$ is not
square-free. By condition (1) the word $ae_j a$ is not square-free, thus $a = e_j$.
On the other hand: $\phi(ae_{j+1}a) \equiv qrrsqr$ also is not square-free. Thus $a = e_{j+1}$.
Therefore $e_j = e_{j+1}$, which contradicts the fact that $e_1 \ldots e_n$ is a square-free
word.

Now suppose that w is a square-free word from M^+ and $\phi(w) \equiv xyyz$ for some nonempty word y. Let $w \equiv e_0 \ldots e_n$. Let us denote $\phi(e_i)$ by E_i. We have

$$E_0 E_1 \ldots E_n \equiv xyyz.$$

If E_0 is contained in x or E_n is contained in z then we can shorten w (delete e_0 or e_n). Therefore we can suppose that

$$E_0 \equiv x E_0', \quad E_n = E_n'' z, \quad yy \equiv E_0' E_1 \ldots E_{n-1} E_n'.$$

By condition (0) we have that $n \geq 3$.

The word y is equal to $E_0' E_1 \ldots E_j' \equiv E_j'' E_{j+1} \ldots E_n'$ with $E_j' E_j'' \equiv E_j$. If $j = 0$ then $E_1 E_2$ must be a subword of E_0, which is impossible. Similarly, $j \neq n$.

Now by the rigidity statement,

$$E_0' \equiv E_j'', \quad E_1 \equiv E_{j+1}, \quad \ldots, E_j' \equiv E_n',$$

and, in particular, $n = 2j$. Therefore

$$\phi(e_0 e_j e_n) \equiv E_0 E_j E_n \equiv x E_0' E_j' E_j'' E_n'' z \equiv \\ x E_0' E_j' E_0' E_j' z.$$

By condition (1) either $e_0 = e_j$ or $e_j = e_n$. Without loss of generality let $e_0 = e_j$. We also know that $E_1 \equiv E_{j+1}, \ldots, E_{j-1} \equiv E_{2j-1}$. Condition (1) implies that ϕ is one-to-one. Therefore $e_0 = e_j, e_1 = e_{j+1}, \ldots$. Hence w is not square-free: it is equal to $e_0 e_1 \ldots e_{j-1} e_0 \ldots e_{j-1} e_n$. □

Theorem 2.3.4 implies Lemma 2.3.3 (check it!) and Theorem 2.3.2.

A complete algorithmic characterization of square-free substitutions was found first by Berstel [45, 46]. The best (in the computational sense) characterization was found by Crochemore.

Theorem 2.3.5 (Crochemore [75]). *Let ϕ be a substitution, M be the maximal size of a block, m be the minimal size of a block, k be the maximum of 3 and the number $1 + [(M-3)/m]$. Then ϕ is square-free if and only if for every square-free word w of length $\leq k$ $\phi(w)$ is square-free.*

Using Theorem 2.3.4 and a computer, one can establish a substitution from an infinite alphabet to $\{a, b, c\}^+$ that is square-free. We will present here a substitution from $\{x_1, x_2 \ldots\}$ to $\{a, b, c, d, e\}^+$. Let $w_1, w_2, \ldots$ be an infinite sequence of distinct square-free words on a three-letter alphabet $\{a, b, c\}$. Consider the following substitution from $x_1, x_2, \ldots$ to $\{a, b, c, d, e\}^+$:

$$x_i \to d w_i e w_i.$$

This substitution is square-free by Theorem 2.3.4. Indeed it is clear that a word $d w_i e w_i$ cannot be a subword of $d w_j e w_j$ because w_i and w_j do not contain d and e. Now if $dw_iew_idw_jew_jdw_kew_k$ contains a square uu then the numbers of d's and e's in uu must be even. Neither of them may be 0, because otherwise one of the words w_i, w_j, w_k would contain a square. So each of them is 2. Therefore each of the copies of u has one d and one e. The first d cannot participate, otherwise u must be equal to dw_iew_i, and $w_i \neq w_j$. Therefore u must start at the middle of the first w_i, and end in the middle of the first w_j. Then u must contain the subword ew_id. But this subword occurs in our product only once, so the second copy of u cannot contain it, a contradiction.

A square-free substitution from $\{a_1, a_2, a_3 \ldots\}$ to $\{a, b, c\}^+$ has been found by Bean, Ehrenfeucht and McNulty [28].

2.4 kth Power-Free Substitutions

For every natural $k \geq 2$, a substitution ϕ is called kth power free if the word $\phi(w)$ is k-power free whenever w is.

The following theorem, also proved by Bean–Ehrenfeucht–McNulty, gives a sufficient condition for a substitution to be kth power-free for $k > 2$.

Theorem 2.4.1. *Let M and N be alphabets and let ϕ be a substitution $M \to N^+$ that satisfies the following three conditions*

(1) $\phi(w)$ *is kth power-free whenever w is a k-power free word of length no greater than $k + 1$.*
(2) $a = b$ *whenever $\phi(a)$ is a subword of $\phi(b)$.*
(3) *If $a, b, c \in M$ and $x\phi(a)y \equiv \phi(b)\phi(c)$ then either x is empty and $a = b$ or y is empty and $a = c$.*

Then ϕ is kth power-free.

Exercise 2.4.2. Prove Theorem 2.4.1.

In particular, Theorem 2.4.1 implies existence of cube-free substitutions from the infinite set $\{a_1, a_2, \ldots, \}$ to $\{a, b\}^+$. An explicit substitution has been constructed in [28].

2.5 Avoidable Words

2.5.1 Examples and Simple Facts

We say that a word u *avoids* a word v if u does not contain the value of v under any substitution replacing letters by nonempty words (in that case we also call the word u v-*free*). A word v is called k-*avoidable* if there exists

an infinite word in a k-letter alphabet avoiding v (equivalently, if there are
infinitely many (finite) words in a k-letter alphabet avoiding v). If v is k-
avoidable for some k, then v is called *avoidable*. We already know that x^k,
$k \geq 2$, are avoidable words. It is clear that x^2yx and $xyxy$ are also avoidable
words because, in general,

Exercise 2.5.1. Prove that if u is a k-avoidable word and ϕ is any substitution
then any word containing $\phi(u)$ is also k-avoidable.

The following exercise is also useful.

Exercise 2.5.2. Prove that for every words u and w, if w avoids u, then every
subword of w avoids u.

On the other hand, the word xyx is *unavoidable*. Indeed, let k be any
natural number. Suppose that there exists an infinite set of words W in a
k-letter alphabet that avoid xyx. Then W must contain a word w of length
$> 2k$. Since we have only k letters in the alphabet, one of these letters, say
a, occurs twice in w, that is w contains a subword apa for some nonempty
word p. This word is of the form xyx ($\phi: x \mapsto a, y \mapsto p$).

Let us consider the following words Z_n, which will be called *Zimin words*:

$$Z_1 \equiv x_1, Z_2 \equiv x_1x_2x_1, \ldots, Z_n \equiv Z_{n-1}x_nZ_{n-1}. \qquad (2.5.1)$$

Notice that these words were studied before Zimin. They appear in [28] as
"maximal" unavoidable words. Before that they were considered by Coudrain
and Schützenberger [74], and even before that they were considered by Levitzki [201] also in connection to Burnside-type problems (for rings).

The next exercise shows that prehistoric humans discovered Zimin words
as soon as they learned how to count and divide by 2.

Exercise 2.5.3. Count the numbers i from 1 to $2^n - 1$ and write down the
maximal exponent of 2 dividing i. The word we obtain is 01020103.... Show
that this word is the value of Z_n under the substitution $x_j \mapsto j - 1$.

Zimin was first to understand the real role of Zimin words in the theory
of avoidable words and Burnside problems for semigroups.

Exercise 2.5.4 (Bean–Ehrenfeucht–McNulty [28], Zimin [337, 338]). Z_n is an
unavoidable word (for every n).

We shall show that the Zimin word Z_n is a "universal" unavoidable word
in an n-letter alphabet.

Let us list some properties of these words.

Exercise 2.5.5. (1) $|Z_n| = 2^n - 1$,

(2) Every odd letter in Z_n is x_1, every even letter in Z_n is x_i for $i > 1$,

(3) For every $k \leq n$, $Z_n \equiv Z_k(Z_{n-k+1}, x_{n-k+2}, \ldots, x_n)$,

(4) For every k, n, $Z_n(Z_{k+1}, x_{k+2}, \ldots, x_{k+n}) \equiv Z_{n+k}(x_1, \ldots, x_{n+k})$,

(5) For every k, n, $Z_k \cdot Z_n(x_{k+1}Z_k, \ldots, x_{k+n}Z_k) \equiv Z_{n+k}(x_1, \ldots, x_{n+k})$,

(6) If we delete the letter x_1 from Z_n then we get a word that differs from
Z_{n-1} only by names of the letters.

(7) Every word in the n-letter alphabet $\{x_1, \ldots, x_n\}$, which properly contains Z_n, contains a square, and so is avoidable.

2.5.2 Fusions, Free Sets and Free Deletions

Definition 2.5.6. Let u be a word in an alphabet X, let B and C be subsets of X. We call the pair (B, C) a *fusion* in u if for every two-letter subword xy in u

$$x \in B \text{ if and only if } y \in C.$$

We will call B and C *components* of the fusion.

For example the sets $\{x\}$ and $\{y\}$ form a fusion in the word $yxyzxytx$ and in the word $xyxy$. Sets $\{x, z\}$ and $\{z, t\}$ form a fusion in the word $xtxzztxz$.

Definition 2.5.7. If B, C is a fusion in u, then any subset $A \subseteq B \backslash C$ is called a *free set* of u.

For example, $\{x\}$ is a free set in $xtxzztxz$.

Exercise 2.5.8. Find all fusions and all free sets in Z_n.

Let u be a word, Y be a subset of letters of u. Consider the word obtained from u by deleting all letters that belong to Y. This word will be denoted by u_Y.

Definition 2.5.9. A deletion of a free set A in a word is called a *free deletion* and is denoted by σ_A.

Definition 2.5.10. A sequence of deletions $\sigma_Y, \sigma_Z, \ldots$ in a word u is called a *sequence of free deletions* if Y is a free set in u, Z is a free set in $u_Y \equiv \sigma_Y(u)$, etc.

Let us fix some notation. Let $u = x_1 x_2 \ldots x_n$ be a word, a be a letter. For $\alpha, \beta \in \{0, 1\}$ we denote by $[u, a]_\beta^\alpha$ the word u with letter a inserted between every two consecutive letters and possibly at the beginning and at the end. More precisely

$$[u, a]_\beta^\alpha = a^\alpha x_1 a x_2 a \ldots a x_n a^\beta \tag{2.5.2}$$

Definition 2.5.11. Let u be a word, B, C be a fusion in u, $A \subseteq B \backslash C$ and a be a letter not in $\text{cont}(u)$. For every substitution ϕ of the word $\sigma_A(u)$ we can define a substitution ϕ^* of u by the following rules:

$$\phi^*(x) = \begin{cases} a & \text{if } x \in A \\ [\phi(x), a]_0^0 & \text{if } x \in C \backslash B \\ [\phi(x), a]_1^0 & \text{if } x \in B \cap C \\ [\phi(x), a]_1^1 & \text{if } x \in B \backslash (C \cup A) \\ [\phi(x), a]_0^1 & \text{otherwise.} \end{cases} \tag{2.5.3}$$

We will call ϕ^* the substitution induced by ϕ relative to the triple (A, B, C).

For example, if $u \equiv xyzt$, $B = \{x, z\}$, $C = \{y, t\}$, $A = \{x\}$, and $\phi(x) \equiv x, \phi(y) \equiv y, \phi(z) \equiv z, \phi(t) \equiv t$, then $\phi^*(x) \equiv a$, $\phi(y) \equiv y$, $\phi(z) \equiv aza$, $\phi(t) \equiv t$. Therefore $\phi^*(u) \equiv ayazat$.

The following lemma contains the main property of free deletions.

Lemma 2.5.12. *For every word* u, $\phi^*(u) = [\phi(\sigma_A(u)), a]_\beta^\alpha$ *for some* α *and* β. *If* u *starts and ends with a letter from* $B\backslash C$ *then* $\alpha = \beta = 1$.

Exercise 2.5.13. Prove Lemma 2.5.12.

2.5.3 The Bean–Ehrenfeucht–McNulty and Zimin Theorem

2.5.3.1 The Formulation

Theorem 2.5.14 (Bean–Ehrenfeucht–McNulty, Zimin). *The following conditions are equivalent for every word* u:

(1) u *is unavoidable.*
(2) Z_n *contains a value of* u, *where* n *is the number of distinct letters in* u.
(3) *There exists a sequence of free deletions that reduces* u *to a 1-letter word.*

Exercise 2.5.15. Deduce the following two facts from Theorem 2.5.14

(1) Every word of length $\geq 2^n$ in a n-letter alphabet is avoidable.
(2) Every unavoidable word u contains a *linear letter*, i.e., a letter that occurs in u only once.

Exercise 2.5.16. Z_n has a sequence of free deletions of length $n - 1$ which reduces Z_n to x_n. **Hint:** The set $\{x_1\}$ is a free set in Z_n. Deleting x_1, we obtain a word that differs from Z_{n-1} only by names of letters (part (6) of Exercise 2.5.5).

Our proof of Theorem 2.5.14 is simpler than the original proofs of Bean–Ehrenfeucht–McNulty and Zimin. We shall follow [277]. It is also more general: the same ideas will be used later in a more complicated situation.

2.5.3.2 Proof of (3) → (2)

Exercise 2.5.17. Prove this implication. **Hint:** Suppose that a one letter word p is obtained from u by a sequence of free deletions $\sigma_1, \ldots, \sigma_k$: $u \to u_1 \to \ldots u_k \equiv p$. Note that $k \leq n-1$. Let ϕ be the identity substitution $p \mapsto p$. Then apply induced substitutions (2.5.3) and Lemma 2.5.12 k times using letters $x_k, x_{k-1}, \ldots, x_1$ and notice that $\phi^*(u_{k-1})$ is a subword of $x_k p x_k$, $\phi^{**}(u_{k-2})$ is a subword of $x_{k-1} x_k x_{k-1} p x_{k-1} x_k x_{k-1}$, etc.

2.5.3.3 Proof of (2) → (1)

This implication immediately follows from Exercises 2.5.1 and 2.5.2.

2.5.3.4 Proof of (1) → (3)

Suppose that there is no sequence of free deletions that reduces u to a 1-letter word. We shall prove that u is avoidable.

In fact we will construct a substitution γ such that for some letter a words $\gamma(a)$, $\gamma^2(a)$, ... are all different and avoid u.

The substitution γ is constructed as follows. Let r be a natural number. Let $\mathcal{A}$ denote the alphabet of r^2 letters a_{ij}, $1 \le i, j \le r$. Consider the $r^2 \times r$-matrix M, in which every odd column is equal to

$$(1, 1, \ldots, 1, 2, 2, \ldots, 2, \ldots, r, r, \ldots, r),$$

and each even column is equal to

$$(1, 2, \ldots, r, 1, 2, \ldots, r, \ldots, 1, 2, \ldots, r).$$

Replace every number i in the j-th column by the letter a_{ij}. The resulting matrix is denoted by M'. The rows of M' can be considered as words in the alphabet $\mathcal{A}$. Denote these words by $w_1, \ldots, w_{r^2}$ counting from top to bottom. Now define the substitution γ by the following rule:

$$\gamma(a_{ij}) = w_{(j-1)r+i}.$$

For every $j = 1, 2, \ldots$ let $\mathcal{A}_j$ be the set $\{a_{ij} \mid i = 1, 2, \ldots, r\}$.

Exercise 2.5.18. γ satisfies the following properties:

(1) The length of each block is r.
(2) No two different blocks have common 2-letter subwords.
(3) All letters in each block are different.
(4) The j-th letter in each block belongs to $\mathcal{A}_j$ (that is its second index is j).

Theorem 2.5.19. *Suppose that the word u cannot be reduced to a 1-letter word by a sequence of free deletions. Let $r > 6n + 1$ (where n is the number of letters in u). Then $\gamma^m(a_{11})$ avoids u for every m.*

Proof. By contradiction, suppose there exists m such that $\gamma^m(a_{11})$ contains $\delta(u)$ for some substitution δ. We may suppose that m is minimal possible. Clearly $m > 0$.

The idea of getting a contradiction is the following. First of all we will modify δ and find a substitution ϵ such that $\epsilon(u)$ has "long" sequences of free deletions. Then we prove that if a value of a word has "long" sequences of free deletions, then so does the word itself.

Let us call images of letters under γ *blocks*, and let us call products of blocks *integral words*.

Let w be an integral word. Let v be a subword of w. Then v is equal to a product of three words $p_1 p_2 p_3$ where p_1 is a suffix of a block, p_2 is a product of blocks or lies strictly inside a block, p_3 is a prefix of a block. Some of these words may be empty. The following property of integral words is important. It easily follows from Exercise 2.5.18.

Lemma 2.5.20. *The decomposition $v = p_1 p_2 p_3$ does not depend on the particular occurrence of v in an integral word: if we consider another occurrence of v, the decomposition $v = p_1 p_2 p_3$ will be the same.*

Now let $w = \gamma^m(a_{11})$. We have that $\delta(u)$ is a subword of w. For every letter $x \in \mathrm{cont}(u)$ let $\delta(x) = p_{x1} p_{x2} p_{x3}$ be the decomposition described above.

Now let us take any block B in w. This block can appear in many different places of w. Let us consider all words p_{xi} that are contained in all possible occurrences of B. There are no more than $3n$ such words p_{xi}. Each of them may occur in B at most once because all letters in B are different. Therefore we may consider B as an interval, which contains at most $3n$ other intervals. The length of B is at least $6n + 2$. Therefore there exists a 2-letter subword t_B of B that satisfies the following condition:

(T) For every subword p_{xi} inside B, the word t_B either is contained in p_{xi} or does not have common letters with it.

This easily follows from the trivial 1-dimensional geometry fact that k intervals on a line divide this line into at most $2k + 1$ subintervals two of which are infinite (why?).

Now let us replace the subword t_B in every block B in w by a new letter y_B. We shall get a new word w_1 in the alphabet $\mathcal{A} \cup \{ y_B \mid B = \gamma(a_{ij}) \}$. This word has the following form:

$$P_0 Q_1 y_1 P_1 Q_2 y_2 P_2 \ldots Q_k y_k P_k Q_{k+1},$$

where Q_i is a prefix of a block, P_i is a suffix of a block, P_i's and Q_i's do not overlap, and

$$\text{if } y_i = y_j \text{ then } Q_i = Q_j \text{ and } P_i = P_j. \tag{2.5.4}$$

Every word with these properties will be called a *quasi-integral* word.

Let us consider the substitution ϵ that is obtained from δ by the following procedure: take each $\delta(x)$ and replace each occurrence of t_B there by y_B. The property (T) of words t_B implies that occurrences of words t_B in $\delta(x)$ do not intersect, so our definition of ϵ is consistent. This property also implies that $\epsilon(u)$ is a subword of the quasi-integral word w_1.

Now we shall show that w_1 (and any other quasi-integral word) has "long" sequences of free deletions.

Exercise 2.5.21. In any quasi-integral word the sequence of deletions σ_{A_1}, $\sigma_{A_2} \ldots$, σ_{A_r} is a sequence of free deletions.

The result of this sequence of deletions is, of course, the deletion of all letters from $\mathcal{A}$. Now we have to understand how free deletions in $\epsilon(u)$ relate to the deletions in u.

First of all we have the following two simple observations.

Let θ be any substitution and let v be any word. Let D be a set of letters of $\theta(v)$. Let D' be a subset of $\mathrm{cont}(v)$ defined as follows:

$$D' = \{x \mid \mathrm{cont}(\theta(x)) \subseteq D\}.$$

Now let us define a substitution θ_D of $v_{D'}$:

$$\theta_D(x) = \theta(x)_D.$$

Exercise 2.5.22. $\theta_D(v_{D'}) = \theta(v)_D$.

The following exercise, while simple, is a yet another key property of free sets.

Exercise 2.5.23. If D is a free set in $\theta(v)$ then D' is a free set in v.

Let us apply Exercises 2.5.21, 2.5.22 and 2.5.23 to our situation: $\epsilon(u) \leq w_1$. We deduce that there exists a sequence of free deletions $\sigma_1 \ldots, \sigma_r$ in u. Let $u_{\mathcal{A}'}$ be the result of these deletions. Then by Exercise 2.5.22

$$\epsilon_{\mathcal{A}}(u_{\mathcal{A}'}) \leq (w_1)_{\mathcal{A}} = y_1 y_2 \ldots y_k.$$

Now, by definition, each y_B determines the block B. Let us consider the substitution α that takes each y_B to B. Then $\alpha((w_1)_{\mathcal{A}}) = w$. Therefore $\alpha\epsilon_{\mathcal{A}}(u_{\mathcal{A}'})$ is a subword of $w = \gamma^m(a_{11})$. But the image of each letter from $u_{\mathcal{A}'}$ under $\alpha\epsilon_{\mathcal{A}}$ is a product of blocks. Therefore we can apply γ^{-1} to $\alpha\epsilon_{\mathcal{A}}(u_{\mathcal{A}'})$. The result, $\gamma^{-1}\alpha\epsilon_{\mathcal{A}}(u_{\mathcal{A}'})$, is a subword of $\gamma^{m-1}(a_{11})$.

Now we can complete the proof. The word $u_{\mathcal{A}'}$ contains at most the same number of letters as u, and $m-1$ is strictly less than m. By the assumption (we assumed that m is minimal) there exists a sequence of free deletions, which reduces $u_{\mathcal{A}'}$ to a 1-letter word. If we combine this sequence of free deletions with the sequence that we used to get $u_{\mathcal{A}'}$ from u, we will get a sequence of free deletions that reduces u to a 1-letter word, which is impossible.

The theorem is proved.   □

2.5.4 Simultaneous Avoidability

Definition 2.5.24. A set of words W is said to be (k-)avoidable if there exists an infinite set of words in a finite alphabet (a k-letter alphabet), each of which avoids each word from W.

Exercise 2.5.25. Prove that a finite system of words W is avoidable if and only if each word in W is avoidable. **Hint:** Use Theorem 2.5.19. Alternatively, if X_w is the finite alphabet of an infinite word avoiding $w \in W$, find an infinite word in the alphabet $\prod_{w \in W} X_w$ (which is the Cartesian product of all X_w) avoiding all words from W simultaneously.

2.6 Further Reading and Open Problems

More on avoidable words, Prouhet–Morse–Thue words and related topics, applications to dynamics and arithmetics can be found in the books by Lothaire [206, 207] and N. Pytheas Fogg [103].

2.6.1 Square- and Cube-Free Words

The subject of k-power free words and certain nice generalizations of this concept is fast growing. Some of the recent papers are surveyed by Shur in [304].

One of the interesting problems was: how to decide, given a substitution $\phi \colon M \to M^+$ and $a \in M$ whether $\phi^n(a)$ is k-th power free for all n. Some sufficient conditions are in Thue's Theorem 2.3.4. Berstel [46] showed that over an alphabet with three letters, there is an algorithm for $k = 2$. Then Karhumäki [171] showed that over a binary alphabet, there is an algorithm for $k = 3$. The problem was solved in general by Mignosi and Séébold [231], who showed that there exists an algorithm for this problem for all alphabet sizes and all k.

There are several interesting generalizations of k-power free word studied, in particular, by Capri and Currie (see [63] and a survey [77]). For example, we say that a word u *does not contain Abelian k-th powers* if it does not contain products $w_1 w_2 \ldots w_k$ where all subwords w_i are permutations of each other. One of the typical results about such words is the following. Let $a(k)$ be the smallest alphabet size, on which Abelian k-powers are avoidable. Then $a(4) = 2, a(3) = 3, a(2) = 4$ (see [77]).

2.6.2 Avoidable Words

Open problems about avoidable words are discussed in McNulty's talks [227]. Since the slides of [227] are not generally available, we present a few open problems here. Some of these (and other) problems can be found in the survey by Currie [77] as well.

Problem 2.6.1. *For each natural number n find out how many unavoidable words there are in the n letter alphabet.*

If a word u is avoidable, then there is the smallest natural number $\mu(u)$ such that there is an infinite word on $\mu(u)$ letters, which avoids u. We can extend this function μ to all words by putting $\mu(u) = 1$ when u is unavoidable.

Problem 2.6.2. *Let u be an avoidable word with $\mu(u) = m$. Can every u-free word on m letters be extended to a maximal u-free word on m letters?*

The answer is yes if $m = 1$. The answer is not known even for $m = 2$.

Problem 2.6.3. *Is μ a recursive function? That is, is there an algorithm that computes μ?*

The proof of Theorem 2.5.14 in Chapter 2 gives an upper bound on μ as a quadratic polynomial in the number $c = c(u)$ of different letters in u. A linear upper bound was found by I. Mel'nichuk [229]: $\mu(u) \le 3c/2 + 3$. She also announced (in 1988) that $\mu(u) \le c + 6$ (unpublished). On the other hand we still do not know the answer to the following problem, which is open since [28, 337]

Problem 2.6.4. *Is $\mu(u)$ bounded from above by a constant?*

It is not even known if that constant is not equal to 5. By Exercise 2.3.1 and Theorem 2.3.4, $\mu(x^2) = 3$. In [17], Baker, McNulty and Taylot found the first word $u = abx_1bcx_2cax_3bax_4ac$, for which $\mu(u) = 4$. Clark [67] constructed a somewhat similar looking word $u = abx_1bax_2acx_3bcx_4cdax_5dcd$, for which $\mu(u) = 5$. This is all we know about the lower bound of μ. Note that the words from [17] and [67] contain linear letters, i.e., letters appearing only once.

Problem 2.6.5. *Suppose that u does not contain linear letters. Is it true that $\mu(u) \le 3$?*

One can also consider the following Abelian version of avoidability, which generalizes the notion of Abelian k-power free words from Section 2.6.1. Let ϕ be a substitution. We say that a word W *contains an Abelian image of a word w under ϕ* if it contains a word w' obtained from w by replacing each letter x by a permutation of the word $\phi(x)$. We say that w is *avoidable in the Abelian sense* if there are infinitely many words in a finite alphabet that do not contain Abelian images of w. The following nice problem from [77] was solved (in the affirmative) for $n = 1, 2, 3$ in Currie and Linek [78].

Problem 2.6.6. *Let w be a word in an alphabet with n letters. Is it true that w is avoidable in the Abelian sense if and only if the Zimin word Z_n does not contain Abelian images of w?*

Chapter 3
Semigroups

Semigroup presentations and identities provide "syntactic tools" to analyze the structure of semigroups. In this chapter, we start by presenting the "semantic tools", main building blocks of semigroups. We also present some results showing how a semigroup can be constructed from these building blocks.

Then we show how to use interaction between syntax and semantics to describe varieties of semigroups where nil-semigroups are locally finite and to describe finite inherently non-finitely based semigroups and inverse semigroups. One of the main tools used in the proofs here, uniformly recurrent words, come from symbolic dynamics. Another tool – Zimin words introduced in the previous chapter.

We give several examples of semigroups of intermediate growth, including a semigroup of matrices and a semigroup given by a Mealy automaton.

Finally we show that not only symbolic dynamics can help solve problems about semigroups, but also semigroups can help solve problems in symbolic dynamics: we present Trahtman's solution of the famous road coloring problem.

Theorems proved in this Chapter include

- Theorem 3.1.17 of Hedlund and Morse giving examples of infinite finitely generated nil-semigroups.
- My Theorem 3.3.4 characterizing varieties of semigroups where all nil-semigroups are locally finite.
- Theorem 3.4.1 of T.C. Brown showing how to construct locally finite semigroups from locally finite blocks.
- Theorem 3.6.16 by Rees and Sushkevich characterizing 0-simple periodic semigroups.
- Theorem 3.6.24 by Shevrin characterizing periodic semigroups without divisors A_2 and B_2.
- My Theorem 3.6.34 characterizing finite inherently non-finitely based semigroups.
- Theorem 3.7.4 of Chebyshev about prime numbers.

© Springer International Publishing Switzerland 2014
M.V. Sapir, *Combinatorial Algebra: Syntax and Semantics*,
Springer Monographs in Mathematics, DOI 10.1007/978-3-319-08031-4_3

- Theorem 3.7.11 by Nathanson and Theorem 3.7.14 by Bartholdi, Reznykov and Sushchansky giving examples of semigroups of intermediate growth.
- My Theorem 3.8.5 about inherently non-finitely based finite inverse semigroups.
- Trahtman's road coloring Theorem 3.9.11.
- Theorem 3.9.18 of Adler, Goodwyn and Weiss about the AGW equivalence of subshifts of finite type.

3.1 Structure of Semigroups

If a semigroup S does not contain an identity element or a zero, then we can formally add such an element to S and obtain a monoid or a semigroup with zero. These semigroups are denoted by S^1 and S^0 respectively. If S contains an identity element (resp. a zero) then by definition $S^1 = S$ (resp. $S^0 = S$). Note that the notation B_2^1 above agrees with this notation.

Exercise 3.1.1. Every semigroup has at most one identity element and at most one zero.

Recall that an element e of a semigroup is called an *idempotent* if $e^2 = e$. An identity element and a zero are of course idempotents. The set of idempotents of a semigroup S is denoted by $E(S)$. There is a natural *partial order* on the set $E(S)$: $e \leq f$ if and only if $ef = fe = e$.

Exercise 3.1.2. Prove that $\leq$ is indeed a partial order (i.e., it is reflexive, transitive and anti-symmetric).

A semigroup where every element is an idempotent, that is it satisfies the identity $x^2 = x$, is called a *band*.

A commutative band is called a *semilattice*.

Exercise 3.1.3. Let S be a set of subsets of a set X closed under taking intersections. Define a multiplication on S by $U \cdot V = U \cap V$.

(1) Show that S is a semilattice.
(2) Show that every semilattice is isomorphic to a semilattice constructed this way.
(3) Deduce from (2) that every finitely generated semilattice is finite.

Hint: To prove (2) let S be a semilattice. Consider a map ϕ from S to the semilattice of all subsets of S (with operation intersection) that takes e to eS. Show that ϕ is an injective homomorphism.

A band satisfying the identity $xy = x$ (resp., $xy = y$, $xyx = x$) is called a *left zero band* (resp., *right zero band*, *rectangular band*).

Exercise 3.1.4. Let I, J be two nonempty sets. Define a multiplication on $S = I \times J$ by $(i, j)(i', j') = (i, j')$.

(1) Show that S is a rectangular band, which is a left (right) zero band if and only if $|J| = 1$ (resp. $|I| = 1$).
(2) Show that every rectangular band is isomorphic to a band constructed this way.
(3) Deduce from (2) that every finitely generated rectangular band is finite.

Hint: To prove (2), let S be a rectangular band, $e \in S$. Let $I = Se, J = eS$. Show that I, J are subsemigroups of S, I is a left zero band, J is a right zero band, and S is isomorphic to the direct product $I \times J$. The isomorphism takes (xe, ey) to xey.

Definition 3.1.5. A semigroup S is called a *band of semigroups* S_α, $\alpha \in A$, if S is a disjoint union of S_α, and the corresponding partition is a congruence. The quotient semigroup of S over this congruence is a band. If this band is commutative (rectangular), then S is called a *semilattice* (resp., a *rectangular band*) of semigroups S_α.

Let S be any semigroup. Then we shall need four *Green's relations* on S:

- $a\mathcal{L}b \iff S^1 a = S^1 b$
- $a\mathcal{R}b \iff aS^1 = bS^1$
- $a\mathcal{J}b \iff S^1 a S^1 = S^1 b S^1$
- $a\mathcal{H}b \iff a\mathcal{L}b$ and $a\mathcal{R}b$.

Exercise 3.1.6. Prove that each of Green's relations is an equivalence relation.

Exercise 3.1.7. Describe Green's relations on the semigroup B_2^1.

Exercise 3.1.8. Prove that all four relations, $\mathcal{L}$, $\mathcal{R}$, $\mathcal{J}$ and $\mathcal{H}$, on the free semigroup are trivial.

Exercise 3.1.9. Let $M_n(\mathbb{R})$ be the multiplicative semigroup of $n \times n$-matrices with real entries. Prove that two matrices in M_n are $\mathcal{L}$-related if and only if they are row equivalent (that is one of them can be obtained from another one by a sequence of elementary row transformations); two matrices are $\mathcal{R}$-equivalent if and only if they are column-equivalent; two matrices are $\mathcal{J}$-equivalent if and only if they have the same rank.

3.1.1 Periodic Semigroups

A semigroup S is *periodic* if all its cyclic subsemigroups are finite, equivalently if for every element $x \in S$ there exist two positive integers m and n such that $x^m = x^{m+n}$.

Example 3.1.10. Every band is a periodic semigroup, every finite semigroup is periodic.

Exercise 3.1.11. Prove that if e is a minimal idempotent (with respect to the order $\leq$, see Exercise 3.1.2) of a periodic semigroup S, then eSe is a maximal subgroup of S with e as its identity element.

3.1.2 Periodic Semigroups with Exactly One Idempotent

A *nil-semigroup* is a semigroup with a zero element such that a power of every element is equal to zero. Clearly every nil-semigroup is periodic and contains exactly one idempotent, the zero.

Recall that a semigroup is *nilpotent of class k* if it has a zero and any product of k elements is zero. A nilpotent semigroup of class 2 is called a *semigroup with zero product*.

Exercise 3.1.12. Deduce from Exercise 1.8.7 that any finite cyclic semigroup contains exactly one idempotent. Prove that every periodic semigroup contains an idempotent.

Exercise 3.1.13. Show that a cyclic semigroup $\langle a \rangle$ is a group if and only if $a^n = a$ for some n. In this case a^{n-1} is the identity element of this group.

The next lemma gives a description of periodic semigroups with exactly one idempotent.

Lemma 3.1.14. *A periodic semigroup S contains exactly one idempotent if and only if S contains an ideal G, which is a subgroup, and the Rees quotient S/G is a nil-semigroup.*

Proof. If S has an ideal G that is a group, and S/G is nil, then S contains no idempotents outside G. There exists only one idempotent inside G, the identity element of G. Thus S contains exactly one idempotent.

Now let S be a periodic semigroup with exactly one idempotent e. Consider the set SeS. It is clear that SeS is an ideal of S.

For every $p \in S$, the cyclic subsemigroup $\langle pe \rangle$ contains the idempotent e by Exercise 3.1.12, so $(pe)^n = e$ for some n. Multiplying by pe on the left, gives us $(pe)^{n+1} = pe$. So $\langle pe \rangle$ is a group by Exercise 3.1.13. A group cannot contain two different idempotents, so e is the identity element for $\langle pe \rangle$. Therefore $epe = pe$ for every $p \in S$. Similarly $eqe = eq$ for every $q \in S$. Therefore $epeqe = peq$ for every $p, q \in S$, so e is the identity element in SeS. Since a power of any element from SeS is e, we conclude that SeS is a group.

In the quotient semigroup S/SeS the class SeS is a zero (Exercise 1.8.14). Since for every x in S, the subsemigroup $\langle x \rangle$ contains e, a power of any element of S/SeS is zero. Thus S/SeS is a nil-semigroup. □

Exercise 3.1.15. Let S be a periodic semigroup. Let E be the set of all idempotents of S. Prove that SES is an ideal of S and S/SES is a nil-semigroup.

3.1.3 Finite Nil-Semigroups

Lemma 3.1.16. *Every finite nil-semigroup is nilpotent.*

Proof. Let S be a nil-semigroup of order n. Consider the free semigroup S^+ with S as a set of free generators and the natural homomorphism from S^+ to S. We need to show that any word of sufficiently large length in S^+ represents 0 in S. Consider any word of length $\geq n+1$ from S^+. By the pigeon-hole principle there exist two different prefixes u and up of w that represent the same element in S. Thus $u = up$ in S. Multiplying this equality by p on the right we get $u = up = up^2 = \ldots = up^k$ in S for every $k \geq 1$. Since S is a nil-semigroup, one of the words p^k represents 0 in S. Therefore u represents 0 in S, and since u is a prefix of w, w represents 0 in S too. Therefore S is nilpotent of class $n + 1$.   □

It is easy to see that every finitely generated nilpotent semigroup is finite (prove it!). Thus by Lemma 3.1.16 a finitely generated nil-semigroup is finite if and only if it is nilpotent. The first example of an infinite finitely generated nil-semigroup was published by Morse and Hedlund [240], but the construction, which they used, is attributed to Dilworth.

Theorem 3.1.17. *There exist*

(a) *a 2-generated infinite semigroup with 0 that satisfies the identity $x^3 = 0$; therefore it satisfies the identity $x^3 = x^4$,*

(b) *a 3-generated infinite semigroup with 0 that satisfies the identity $x^2 = 0$; therefore it satisfies the identity $x^2 = x^3$.*

Proof. (**The Dilworth construction**) Let us take any set of (nonempty) words W closed under taking (nonempty) subwords. Let $S(W)$ be the set W with an additional symbol 0. Define an operation on $S(W)$ by the following rule: $u * v = uv$ if $uv \in W$, and $u * v = 0$ otherwise. The set $S(W)$ with this operation is a semigroup. It satisfies the identity $x^2 = 0$ (resp. $x^3 = 0$) provided W consists of square-free (cube-free) words. Indeed, since xx (resp. xxx) is not a word from W, $x * x$ (resp. $x * x * x$) must be equal to 0.   □

Using the Dilworth construction, it is easy to translate statements from the language of words into the language of semigroups. For example we can replace x^2 and x^3 by arbitrary words $u(x_1, \ldots, x_n)$, and ask the following question: When does $S(W)$ satisfy identically $u(x_1, \ldots, x_n) = 0$? The answer is almost clear:

Theorem 3.1.18. *$S(W)$ satisfies the identity $u = 0$ if and only if words from W are u-free.*

Exercise 3.1.19. Prove this theorem.

Translating the concept of avoidability from Chapter 2 into the language of semigroups we get

Theorem 3.1.20. *For every word u the following conditions are equivalent:*

(1) *u is k-avoidable;*

(2) *There exists an infinite semigroup S generated by k elements, which identically satisfies $u = 0$.*

Proof. (1) → (2). If u is k-avoidable, then the set W of all words in a k-letter alphabet that avoid u is infinite. It is clear that W is closed under taking subwords. Consider $S(W)$. This semigroup is infinite, k-generated, and, by Theorem 3.1.18 satisfies the identity $u = 0$.

(2) → (1). Let S be an infinite k-generated semigroup that satisfies the identity $u = 0$. Let $\{x_1, \ldots, x_k\}$ be the generators of S. Let $y_1, y_2, \ldots$ be all (infinitely many) non-zero elements of S. Since S is generated by x's, each y_i is represented by a word in x's (see Section 1.2.1). Take one such word w_i for each y_i.

Each of these words is written in the k-letter alphabet of x's. Suppose that one of these w_i does not avoid u. This means that w_i contains $\phi(u)$ for some substitution ϕ. Therefore $w_i \equiv p\phi(u)q$ where p, q are words in our k-letter alphabet.

For every word v in this alphabet let $\bar{v}$ be the element of S represented by this word. Then we have that $\bar{w}_i = y_i$. So we have

$$y_i = \bar{p}\phi\overline{(u)}\bar{q} = \bar{p}u(\phi(\bar{x}_1), \ldots, \phi(\bar{x}_k))\bar{q}.$$

But the middle term in the last product is 0 since S satisfies $u = 0$ identically. Therefore $y_i = 0$, a contradiction (we took y_i to be non-zero).   □

The proof of the following result is similar to Theorem 3.1.20.

Theorem 3.1.21. *A set of words W is k-avoidable if and only if there exists an infinite k-generated semigroup satisfying identities $w = 0$ for all $w \in W$.*

3.2 Free Semigroups and Varieties

3.2.1 Free Rees Factor-Semigroups

Let us take a set of words W, and an alphabet A, and consider the set $I(W)$ of all words from A^+ that *do not avoid* some words from W (i.e., $I(W)$ is the union of $I(\{w\})$ for all $w \in W$). The set $I(W)$ is an ideal in A^+ (prove it!).

Then we can consider the *Rees factor semigroup* $A^+/I(W)$ (see Exercise 1.8.13), which consists of the set of words $A^+\backslash I(W)$ (which is, of course, the set of all words that avoid W) and 0 with an obvious multiplication: $u \cdot v = uv$ if $uv \notin I(W)$ and 0 otherwise.

Theorem 3.2.1. *A Rees quotient semigroup A^+/I is relatively free if and only if $I = I(W)$ for some set of words W.*

Exercise 3.2.2. Prove Theorem 3.2.1. **Hint:** For the "if" part use Theorem 1.4.25. For the "only if part", assume that A^+/I is relatively free, then take $W = I$ and use Theorem 1.4.25 again.

Definition 3.2.3. The ideal $I(W)$ is called the *verbal* ideal defined by W.

3.2.2 A Description of Relatively Free Semigroups

Every semigroup is a quotient semigroup of a free semigroup A^+ over some congruence σ. So we have to describe congruences, which give us relatively free semigroups.

Theorem 3.2.4. *A quotient semigroup A^+/σ is relatively free if and only if σ is stable under all endomorphisms of A^+.*

Exercise 3.2.5. Prove this theorem. **Hint:** Use Theorem 1.4.25.

We will call congruences with the property from Theorem 3.2.4 *verbal* congruences.

We shall not distinguish between pairs of words (u,v) and identities. In particular if W is a set of pairs of words, then we will say that an identity $u = v$ belongs to W if $(u,v) \in W$.

Notice that if a pair (u,v) is from a verbal congruence σ then $u = v$ is always an identity of the relatively free semigroup A^+/σ. Conversely if $u = v$ is an identity of A^+/σ then $(u,v) \in \sigma$. In particular, if I is a verbal ideal of A^+ and $v \in I$ then A^+/I satisfies the identity $v = 0$.

Now, given a set of identities $W = \{u_i = v_i \mid i \in S\}$ and an alphabet A one can define a congruence $\sigma(W)$ as follows.

We say that a pair of words (u,v) *does not avoid* a pair of words (p,q) if $u \equiv s\phi(p)t, v \equiv s\phi(q)t$ for some substitution ϕ and some words s and t.

Take the set $I(W)$ of all pairs that do not avoid W and take the smallest equivalence relation containing $I(W)$.

Theorem 3.2.6. *A congruence σ on A^+ is verbal if and only if it is equal to $\sigma(W)$ for some set of identities W.*

Exercise 3.2.7. Prove this theorem.

3.2.2.1 Examples of Varieties of Semigroups

Exercise 3.2.8. Prove that the class of all semigroups with 0 satisfying identically $u = 0$ (u is a fixed word) is a variety. It is defined by two identities $uz = u$, $zu = u$ where z is a letter not in the content $\text{cont}(u)$. Prove that the verbal congruence on A^+, defined by these identities, coincides with the Rees ideal congruence corresponding to the ideal $I(\{u\})$.

Exercise 3.2.9. Prove that the class of all monoids satisfying identically $u = 1$ (u is a fixed word) is a variety consisting of groups. This variety is defined (as a variety of semigroups) by two identities $uz = z$, $zu = z$ where z is a letter not in cont(u). All groups of this variety satisfy the identity $x^n = 1$ for some n (n depends on u). What is the minimal n with this property? **Hint:** If S satisfies $u = 1$ and $s \in S$, substitute s for all letters from u and deduce $s^{|u|} = 1$, which implies that S is a group satisfying the (group) identity $x^{|u|} = 1$. P.

Exercise 3.2.10. The class of all commutative semigroups is a variety given by the identity $xy = yx$. The verbal congruence on A^+ corresponding to this identity consists of all pairs of words (u, v) such that v is a permutation of u.

Exercise 3.2.11. (a) Prove that the verbal congruence defined by $xyx = x^2 y$ consists of pairs of words (u, v) such that u and v start with the same letter, every letter occurs the same number of times in u and v, and for every two letters x, y, if the first occurrence of x precedes the first occurrence of y in u, then the same is true in v. **Hint:** Let $x_1, x_2, \ldots, x_k$ be the letters occurring in u listed in the order of their first occurrences in u, and x_i occurs k_i times in u, $i = 1, \ldots, k$. Prove that then $(u, x_1^{l_1} \ldots x_k^{l_k})$ belongs to the verbal congruence defined by $xyx = x^2 y$. Then show that different words of the form $x_1^{l_1} \ldots x_k^{l_k}$ (all x_i are distinct) are not in the same class of that verbal congruence.

(b) Describe the verbal congruence on A^+ defined by the identity $xyx = yx^2$.

(c) Describe the verbal congruence on A^+ defined by two identities $x^2 y = xy$ and $x^2 y^2 = y^2 x^2$.

3.3 The Burnside Problem for Varieties

Recall that we started with the problem of finding a finitely generated infinite periodic semigroup. We found such a semigroup that satisfies even the identity $x^2 = 0$. Then we described all words u such that there exists a finitely generated infinite periodic semigroup satisfying the identity $u = 0$. These are exactly the avoidable words. Then we described finite sets W of words such that there exists a finitely generated infinite periodic semigroup satisfying all identities $u = 0$ where $u \in W$.

Thus we have described all varieties given by finitely many identities of the form $u = 0$, which contain infinite finitely generated periodic semigroups. Actually we found a syntactic characterization of these varieties.

Now it is natural to ask

Problem 3.3.1. *What are all the varieties of semigroups that contain infinite periodic finitely generated semigroups?*

Exercise 3.3.2. Prove that every commutative finitely generated periodic semigroup is finite.

In general, Problem 3.3.1 is extremely difficult. Indeed, since the class of all groups satisfying the identity $x^n = 1$ is a variety of semigroups (see Exercise 3.2.9) this problem "contains" the problem of describing natural numbers n such that every finitely generated group with the identity $x^n = 1$ is finite. Already this problem seems "hopeless", and we still do not know if, say, 5 is such a number n (see Section 5.2.3).

Nevertheless the following similar problem turned out to be decidable. Notice that all infinite periodic semigroups that we constructed above turned out to be nil-semigroups. Thus we can ask

Problem 3.3.3. *What are all varieties of semigroups containing infinite finitely generated nil-semigroups?*

It is clear that every nil-semigroup is periodic, but not every periodic semigroup is nil. For example, a nontrivial periodic group can not be nil (it does not have 0). Thus if we consider nil-semigroups instead of arbitrary periodic semigroups, we avoid "bad" groups. It turned out that as soon as we do that, the situation becomes much more comfortable, and we have the following result.

Theorem 3.3.4 (Sapir, [277]). *Let V be a variety given by a set of identities Σ in at most n letters. Then the following conditions are equivalent:*

(1) *V does not contain an infinite finitely generated nil-semigroup.*
(2) *V does not contain an infinite finitely generated semigroup satisfying the identity $x^2 = 0$.*
(3) *V satisfies an identity $Z_{n+1} = W$ where W is a word distinct from Z_{n+1}.*
(4) *There exists an identity $u = v$ from Σ and a substitution ϕ such that Z_{n+1} contains $\phi(u)$ or $\phi(v)$ and $\phi(u) \not\equiv \phi(v)$.*

Note that the implication $(1) \to (2)$ is trivial.

The equivalence $(3) \leftrightarrow (4)$ is also easy to establish. In fact Z_{n+1} in these conditions could be replaced by an arbitrary word Z and they would still be equivalent. Indeed, if (4) holds, then $Z = s\phi(u)t$ and $s\phi(v)t \neq s\phi(u)t$ for some substitution ϕ, words s, t, and identity $u = v \in \Sigma$. Let $W = s\phi(v)t$. Then the pair (Z, W) does not avoid Σ. Therefore $(Z, W) \in \sigma(\Sigma)$. Hence Σ implies $Z = W$.

Conversely, if Σ implies $Z = W$ for some W distinct from Z then the pair (Z, W) belongs to the verbal congruence $\sigma(\Sigma)$, which is the smallest equivalence relation containing the set of pairs $I(\Sigma)$ that do not avoid Σ. Therefore there exists a chain of pairs

$$(Z, W_1), (W_1, W_2), \ldots, (W_k, W)$$

that do not avoid Σ or identities dual to identities from Σ ($u = v$ is dual to $v = u$). Since Z differs from W, we may suppose that W_1 differs from Z. Therefore the pair (Z, W_1) satisfies condition (4).

Remark 3.3.5. Condition (4) is effective: one needs to check only finitely many substitutions ϕ.

Remark 3.3.6. Let $u = v \in \Sigma$ where u is unavoidable but v is avoidable. Then condition (4) holds. Indeed, since u is unavoidable, Z_n contains a value $\phi(u)$ of u for some substitution ϕ. That is $Z_n = s\phi(u)t$ for some words s and t. Since v is avoidable, Z_n cannot contain values of v. In particular, $\phi(u) \neq \phi(v)$. Then the words $Z_n = s\phi(u)t$ and $W = s\phi(v)t$ are distinct. The pair (Z_n, W) does not avoid (u, v). Therefore the identity $Z_n = W$ follows from $u = v$. Hence $Z_{n+1} = W x_{n+1} Z_n$ follows from Σ. Therefore Σ satisfies condition (4).

Remark 3.3.7. Suppose that for every $u = v \in \Sigma$ both u and v are avoidable. Then condition (4) does not hold. Indeed, if (Z_{n+1}, W) does not avoid (u, v) or (v, u) then Z_{n+1} does not avoid u or v, and then u or v will be unavoidable.

Therefore the most complicated case is when Σ contains identities $u = v$ with both u and v unavoidable and does not contain identities $u = v$ where one of the words u or v is avoidable and another one isn't. The following example shows that even if both sides of an identity $u = v$ are unavoidable, $u = v$ can imply no identity of the form $Z_{n+1} = W$.

Exercise 3.3.8. Prove that both sides of the Mal'cev identity

$$axbybxazbxayaxb = bxayaxbzaxbybxa$$

(which in the case of groups defines the class of all *nilpotent groups* of class 3, see [217]) satisfies the following two conditions

- Both sides of this identity are unavoidable.
- This identity does not imply any identity $Z_6 = W$ where W differs from Z_6.

Hence by Theorem 3.3.4, there exists an infinite finitely generated semigroup satisfying the Mal'cev identity and $x^2 = 0$.

Thus in order to complete the proof of Theorem 3.3.4 it suffices to prove implications $(3) \to (1)$ and $(2) \to (4)$.

We will start with implication $(3) \to (1)$. This implication is a corollary of the following lemma.

Lemma 3.3.9. *Any finitely generated nil-semigroup S satisfying a nontrivial identity $Z_n = W$ is finite.*

We will return to this lemma later, after we get familiar with the following application of subshifts to semigroups.

3.3.1 Subshifts and Semigroups

Let $S = \langle A \rangle$ be a finitely generated semigroup. Then as we mentioned in Section 1.2.1, every element of S is represented by a word in the alphabet A.

For every element s in S take all shortest possible words representing s. These words are called *geodesic words*: they label *geodesic paths* on the Cayley graph of the semigroup, i.e., the shortest paths connecting given two points. Let W be the set of all geodesic words representing elements of S. Notice that W is closed under taking subwords (a subword of a geodesic word is a geodesic word itself).

Suppose S is infinite. Then W is infinite also. Now, in every word of W, mark a letter that is closest to the middle of this word. There must be an infinite subset $W_1 \subseteq W$ of words that have the same marked letters, an infinite subset $W_2 \subseteq W_1$ of words that have the same subwords of length 3 containing the marked letters in the middle, ..., an infinite subset $W_n \subseteq W_{n-1}$ of words that have the same subwords of length $2n-1$ containing the marked words in the middle, and so on. Therefore there is an infinite in both directions word α of marked letters from A such that every subword of α is a subword of a word from W. Thus every subword of α is a geodesic word. Bi-infinite words with this property will be called *bi-infinite geodesic words*. The set $D(S)$ of all bi-infinite geodesic words of S is a subshift (prove it!).

Remark 3.3.10. An arguments as above is usually called a *compactness argument*. It allows finding an infinite object with certain property given an infinitely many finite objects with that property.

Notice that if the semigroup S is finite, W is also finite and $D(S)$ is empty. Thus $D(S)$ is not empty if and only if S is infinite.

It is interesting that an arbitrary subshift is $D(S)$ for some S. Indeed, let $D \subseteq A^{\mathbb{Z}}$ be a subshift. Let (as in Section 1.6.2) $L(D)$ be the set of all words over A, which are not subwords of some words from D. Then $L(D)$ is an ideal in A^+ and the Rees factor-semigroup $S = A^+/L(D)$ will be denoted by $S(D)$.

Exercise 3.3.11. Prove that for every symbolic dynamical system D we have

$$D(S(D)) = D.$$

In order to translate properties of subshifts into properties of semigroups we shall need the following.

Lemma 3.3.12 (Furstenberg, [105]). *A bi-infinite word α of a subshift D is uniformly recurrent if and only if for every subword u of α there exists a number $N(u)$ such that every subword of α of length $N(u)$ contains u as a subword.*

Exercise 3.3.13. Prove this lemma.

Let α be a bi-infinite word from $A^{\mathbb{Z}}$. Let $D(\alpha)$ be the closure of the set of all shifts of α

$$\ldots, T^{-2}(\alpha), T^{-1}(\alpha), \alpha, T^1(\alpha), T^2(\alpha), \ldots$$

of α. This is a closed set and since T and T^{-1} are continuous maps, $D(\alpha)$ is stable under the shift and its inverse. Thus $D(\alpha)$ is a subshift. This is the minimal subshift containing α, thus we will call $D(\alpha)$ the subshift *generated by* α.

Lemma 3.3.14. *If $\beta \in D(\alpha)$ then every subword of β is a subword of α.*

Exercise 3.3.15. Prove this lemma.

Lemma 3.3.16. *Let α be a bi-infinite word in a finite alphabet X. Then there exist a letter $x \in X$ and an integer k such that for any n α has n consecutive occurrences of the letter x in α at positions $i_1 < i_2 < \ldots < i_n$ (depending on n) where $i_{j+1} - i_j \le k$, $j = 1, \ldots, n-1$.*

Proof. Indeed, consider the subshift generated by α. By Theorem 1.6.2 this subshift contains a uniformly recurrent bi-infinite word β. Every (finite) subword of β is a subword of α by Lemma 3.3.14. It remains to note that by Lemma 3.3.12 there exists k such that for every letter x occurring in β, every subword of length k of β contains x. $\square$

3.3.2 An Application of Subshifts to Semigroups

Theorem 1.6.2 and Lemma 3.3.12 imply

Lemma 3.3.17. *For every infinite finitely generated semigroup $S = \langle A \rangle$ there exists an infinite uniformly recurrent geodesic bi-infinite word.*

Now let us return to the proof of Lemma 3.3.9: if a finitely generated nil-semigroup S satisfies a nontrivial identity of the form $Z_n = W$ then it is finite. The following Lemma gives us a connection between uniformly recurrent bi-infinite words and Zimin words Z_n.

Lemma 3.3.18. *Let β be a uniformly recurrent bi-infinite word, $U_1 a U_2$ be an occurrence of letter a in β where U_1 is a word infinite to the left, U_2 is a word infinite to the right. Then for every natural number n there exists a substitution ϕ_n such that $U_3 \phi_n(Z_n) \equiv U_1 a$ for some word U_3 infinite to the left, $\phi_n(x_1) = a$, and $|\phi_n(Z_n)| \le A(n, \beta)$ where the number $A(n, \beta)$ depends only on β and n.*

Proof. Since β is uniformly recurrent, there exists a number $N = N(a)$ such that every subword of β of length N contains a. Therefore one can find another a at most $N+1$ letters to the left of our occurrence of a. Then we can set $\phi(x_1) = a$, and $\phi(x_2)$ equal to the word between our two occurrences of as. So we get a substitution of Z_2 that satisfies the required condition. Since β is uniformly recurrent, there exists a number $N_1 = N(\phi(Z_2))$ such that every subword of β of length N_1 contains $\phi(Z_2)$. So we can find another occurrence of $\phi(Z_2)$ to the left of the first occurrence, such that the distance between these two occurrences does not exceed $N_1 + 1$. Then we can define $\phi(x_3)$ to be equal to the word between these two occurrences of $\phi(Z_2)$. This gives us a substitution of Z_3 that satisfies the required condition. Now the proof is easy to complete by induction on n.   □

Suppose now that the finitely generated semigroup $S = \langle A \rangle$ is infinite. Then by Lemma 3.3.17 there exists a uniformly recurrent geodesic word in $D(S)$. Consider one of these uniformly recurrent geodesic words α.

Our goal is to get a contradiction with the following obvious fact.

Lemma 3.3.19. *None of the subwords of α is equal to 0 in S.*

If a word differs from Z_n only by the names of its letters, then we will denote it by Z'_n. In particular, words xyx, xzx, and Z_2 are denoted by Z'_2.

Recall that S satisfies the identity $Z_n = W$. First of all let us look at the word W.

Lemma 3.3.20. W *contains only letters* $x_1, \ldots, x_n$.

Proof. Indeed, suppose W contains an extra letter y. By Lemma 3.3.18 there exists a substitution ϕ such that $\phi(Z_n)$ is a subword of α . Since every letter of A is an element of S we can consider ϕ as a map into S. Let $\phi(y) = 0$. Then $\phi(W) = 0$ in S. Since S satisfies the identity $Z_n = W$, we have that $\phi(Z_n) = 0$ in S. Hence α has a subword that is equal to 0 in S — a contradiction with Lemma 3.3.19.   □

Exercise 3.3.21. Prove that W has one of the following 4 properties (see (2.5.2)).

(1) $W = [W_1, x_1]_1^1$ where $W_1 = W_{x_1}$.
(2) W contains x_1^2.
(3) W contains a subword $x_i x_j$ for some $i, j > 1$.
(4) W starts or ends not with x_1.

Hint: Less formally, the exercise says that every word W in $x_1, \ldots, x_n$ either has the form $x_1 * x_1 * \ldots * x_1$ where $*$ is any letter $\neq x_1$ or contains x_1^2 or contains a 2-letter subword without x_1 or does not start or end with x_1.

If u, v, p, q are words, then we will write

$$u = v \ (\text{mod} \ p = q)$$

if (u, v) belongs to $\sigma(p, q)$ (where that $\sigma(p, q)$ is the smallest equivalence relation containing the set of pairs that do not avoid (p, q)). Recall that in this case the identity $u = v$ follows from the identity $p = q$, so that if S satisfies $p = q$, then S satisfies $u = v$.

Notice that $Z_{n+1} = Z_n x_{n+1} W$, $Z_{n+1} = W x_{n+1} Z_n$ (mod $Z_n = W$). Now if W satisfies condition (4), then $Z_n x_{n+1} W$ or $W x_{n+1} Z_n$ satisfies condition (3). Thus we can assume that W satisfies one of the three conditions (1), (2), (3) of Exercise 3.3.21.

Suppose that W satisfies condition (2) or (3) of Exercise 3.3.21. Then the following statement holds.

Lemma 3.3.22. *Let β be an arbitrary uniformly recurrent bi-infinite word, a be a letter in β. Then β contains a subword u such that*

$$u = pa^2q \pmod{Z_n = W}$$

for some words p and q.

Proof. If W satisfies condition (2) of Exercise 3.3.21, the statement is a direct consequence of Lemma 3.3.18

Suppose W satisfies condition (3). Since β is uniformly recurrent, it can be represented in the form $\ldots p_{-2}ap_{-1}ap_0ap_1ap_2\ldots$ where lengths of the words p_i are smaller than $N(a)$. Let us introduce a new alphabet $B = \{a, p_i \mid i \in \mathbb{Z}\}$: we denote words p_i by letters, different words by different letters, equal words by equal letters. Since A is a finite alphabet and there are only finitely many words over A of any given length, B is also a finite alphabet. Let β_1 be the bi-infinite word that we get from β by replacing subwords p_i by the corresponding symbols. This bi-infinite word, β_1 may not be uniformly recurrent. But let us consider the subshift $D(\beta_1)$ generated by β_1. By Theorem 1.6.2 this subshift contains a uniformly recurrent bi-infinite word β_2. By Lemma 3.3.14 every subword of β_2 is a subword of β_1. Therefore β_2 has the form $\ldots p_{i_1}ap_{i_2}ap_{i_3}a\ldots$. Let p_1 be a letter from B occurring in β_2. By Lemma 3.3.18 there exists a substitution ϕ of the word Z_n such that $\phi(Z_n)$ is a subword of β_2 and $\phi(x_1) = p_1$. Then $\phi(x_i)$, $i = 2, 3, \ldots, n$, must start and end with a. Since W contains a subword x_ix_j for $i, j > 1$, we have that $\phi(W)$ contains a^2. The word $\phi(Z_n)$ is a subword of β_1. Let ψ be the substitution that takes a to a and the symbols p_i back to the words denoted by these symbols. Then $\psi(\phi(Z_n))$ is a subword of β and $\psi(\phi(W))$ contains a^2. The lemma is proved. $\square$

Lemma 3.3.23. *Assume that W satisfy condition (2) or (3) of Exercise 3.3.21. Then for every uniformly recurrent bi-infinite word β, every natural number k and every letter a occurring in β there exists a subword u of β such that*

$$u = sa^kt \pmod{Z_n = W}$$

for some words s and t.

Exercise 3.3.24. Prove this lemma. **Hint:** Use Lemma 3.3.22 as the base of induction ($k = 2$). To demonstrate the step of induction, let us show how to get from $k = 2$ to $k = 3$ (and in fact to $k = 4$). Replace every occurrence of the subword u (from Lemma 3.3.22) in β by $sa't$ (letter a' replaces a^2) and obtain a new bi-infinite word β'. Use Theorem 1.6.2 to find another uniformly recurrent word β_1 all of whose subwords are subwords of β'. Then use Lemma 3.3.22 again for the uniformly recurrent word β_1 and letter

a' (you need to observe that a' does occur in β_1 otherwise Lemma 3.3.22 does not apply). Afterward, replacing a' back by a^2, obtain the statement of Lemma 3.3.23 for $k = 4$.

Now we can finish the proof of Lemma 3.3.9 in the case when W satisfies one of the conditions (2) or (3) of Exercise 3.3.21. Indeed, let us apply Lemma 3.3.23 to α. Let a be a letter occurring in α. Since S is a nil-semigroup, $a^n = 0$ in S for some n. By Lemma 3.3.23 there exists a subword u in α such that

$$u = sa^n t \pmod{Z_n = W}.$$

Since S satisfies $Z_n = W$ we conclude that the word u is equal to $sa^n t$ in S. But $sa^n t = 0$ in S. This contradicts Lemma 3.3.19.

It remains to consider the case when W satisfies condition (1) of Exercise 3.3.21.

Lemma 3.3.25. *If $u = v \pmod{p = q}$ then $[u, a]_1^1 = [v, a]_1^1 \pmod{[p, a]_1^1 = [q, a]_1^1}$.*

Exercise 3.3.26. Prove this lemma.

Lemma 3.3.27. *Let W be an arbitrary word distinct from Z_n. Then for every uniformly recurrent bi-infinite word β there exists a subword $p \leq \beta$ such that for every natural number m there exists a subword u in β such that*

$$u = sp^m t \pmod{Z_n = W}$$

for some words s, t.

Proof. Induction on n. By Lemma 3.3.23 the statement is true in the case when W satisfies condition (2) or (3) of Exercise 3.3.21.

Suppose we have proved our lemma for $n-1$ (and arbitrary W) and that W satisfies the condition (1), that is $W = [W_1, x_1]_1^1$. We have $Z_n = [Z_{n-1}', x_1]_1^1$. Thus the identity $Z_{n-1}' = W_1$ is nontrivial. So we can suppose that the statement of our lemma holds for this identity. As in the proof of Lemma 3.3.22, let us represent β in the form $\ldots p_1 a p_2 a p_3 \ldots$, and replace each subword p_i by the corresponding symbol (different subwords are replaced by different symbols, equal subwords – by equal symbols). We get another bi-infinite word β_1. Let β_2 be the bi-infinite word obtained from β_1 by deleting a. Let β_3 be a uniformly recurrent bi-infinite word in $D(\beta_2)$. By the induction hypothesis there exists a subword p' in β_3 such that for every m there exists a subword u in β_3 such that $u = sp^m t \pmod{Z_{n-1}' = W_1}$. Then by Lemma 3.3.25 we have that $[u, a]_1^1 = [s, a]_1^0([p', a]_1^1)^m [t, a]_1^1 \pmod{Z_n = W}$. The word u is a subword of β_3. By Lemma 3.3.14 it is a subword of β_2. Then $[u, a]_1^1$ is a subword of β_1. Let ψ be the "return" substitution that takes a to a and every symbol p_i to the word that is denoted by this symbol. Then we have that

$$\psi([u, a]_1^1) = s_1(\psi([p', a]_1^0))^m t_1 \pmod{Z_n = W}$$

for some words s_1 and t_1. The word $\psi([u,a]_1^1)$ is a subword of β, and so $\psi([p',a]_1^0)$ is the desired word p. $\square$

Exercise 3.3.28. Complete the proof of Lemma 3.3.9.

3.3.3 The Completion of the Proof of Theorem 3.3.4

Lemma 3.3.9 gives us the implication (3) $\to$ (1). It remains to prove the implication (2) $\to$ (4).

Suppose that condition (4) of Theorem 3.3.4 does not hold. We have to prove that then there exists a finitely generated infinite semigroup S satisfying all identities of Σ and the identity $x^2 = 0$.

Definition 3.3.29. A word w is called an *isoterm* for an identity $u = v$ if for every substitution ϕ such that $\phi(u) \leq w$ (i.e., $\phi(u)$ is a subword in w) we have $\phi(v) \equiv \phi(u)$.

Remark 3.3.30. A word w may be an isoterm for $u = v$ but not for $v = u$. For example, xyx is an isoterm for $x^2 = x$ but not for $x = x^2$.

Remark 3.3.31. Condition (4) of Theorem 3.3.4 may be rewritten in the form:

(4') Z_{n+1} is not an isoterm for $u = v$ or $v = u$ for some identity $u = v$ in Σ.

Since we assume that condition (4') does not hold, Z_{n+1} is an isoterm for $u = v$ and $v = u$ for every $u = v \in \Sigma$.

In order to construct an infinite finitely generated semigroup satisfying Σ and $x^2 = 0$, we will employ the Dilworth construction again.

The following lemma is an analog of Theorem 3.1.18.

Lemma 3.3.32. *Let W be a set of words closed under taking subwords. The semigroup $S(W)$ satisfies an identity $u = v$ if and only if every word of W is an isoterm for $u = v$ and $v = u$.*

Exercise 3.3.33. Prove this lemma.

Thus, in order to find a finitely generated semigroup that satisfies Σ and $x^2 = 0$ it is enough to construct an infinite set W of square-free words over a finite alphabet, which are isoterms for $u = v$ and $v = u$ for every $u = v \in \Sigma$.

One can see that, again, we have translated a semantic question about semigroups into a syntactic question about words.

Let $\gamma = \gamma_r$, $r > 6n + 1$, be the substitution defined in Section 2.5.3.4. We will complete the proof of implication (2) $\to$ (4) if we prove the following result.

Lemma 3.3.34. *Let $u = v$ be an identity in n variables such that Z_{n+1} is an isoterm for $u = v$ and $v = u$. Then all words $\gamma^m(a_{11})$, $m \geq 1$ are isoterms for $u = v$ and $v = u$ whenever $r > 6n$.*

First of all we shall study, using fusions a free deletions, identities $u = v$ such that Z_m is an isoterm for $u = v$ and $v = u$. Instead of the infinite set of words Z_m, $m \geq 1$ we will consider one infinite word

$$Z_\infty = [\ldots [x_1, x_2]_1^1, x_3]_1^1 \ldots$$

which is a "limit" of Z_n'. Any word that differs from Z_∞ only by the names of letters, also will be denoted by Z_∞.

Exercise 3.3.35. Show that if Z_m is not an isoterm for $u = v$ then Z_{m+1} is not an isoterm for $u = v$.

Lemma 3.3.36. *Let* $|\mathrm{cont}(u)| = n$. *Suppose that a pair of sets of letters* B, C *is a fusion in* u *but not a fusion in* v. *Let* $A \subseteq B \backslash C$ *and suppose that* u_A *is unavoidable. Then* Z_∞ *is not an isoterm for* $u = v$.

Proof. By Theorem 2.5.14, there exists a substitution ϕ such that $\phi(u_A) \leq Z_\infty$. Consider the substitution ϕ^*. We have that $\phi^*(u) \prec Z_\infty$. Since B, C do not form a fusion in v, there exists a subword $xy \leq v$ such that either $x \notin B$, $y \in C$ or else $x \in B$, $y \notin C$. Suppose $x \notin B$, $y \in C$. Then $\phi^*(x)$ does not end with a, and $\phi^*(y)$ does not start with a (see the definition (2.5.3)). Therefore $\phi^*(v)$ contains a subword zt where $z \neq a \neq t$. It follows that $\phi^*(u) \neq \phi^*(v)$ since $\phi^*(u) \prec Z_\infty$. Suppose $x \in B$, $y \notin C$. Then $\phi^*(x)$ ends and $\phi^*(y)$ starts with a, so that $a^2 \leq \phi^*(v)$ and again $\phi^*(u) \neq \phi^*(v)$. Thus $\phi^*(u) \prec Z_\infty$ and $\phi^*(u) \not\equiv \phi^*(v)$, that is Z_∞ is not an isoterm for $u = v$. □

Lemma 3.3.37. *Let* A *be a free set in* u *such that* Z_m *is not an isoterm for* $u_A = v_A$. *Then* Z_{m+1} *is not an isoterm for* $u = v$.

Proof. Let ϕ be a substitution such that $\phi(u_A) \leq Z_m$ and $\phi(u_A) \not\equiv \phi(v_A)$. Then we can define ϕ^* as in (2.5.3). By Lemma 2.5.12, $\phi^*(u)$ is a subword of Z_{m+1}' and $\phi^*(u) \not\equiv \phi^*(v)$. □

Lemma 3.3.38. *Let* $u = v$ *be an identity of* n *variables such that* Z_{n+1} *is an isoterm for* $u = v$. *Then* Z_∞ *is an isoterm for* $u = v$.

Proof. Let $u = v$ be a counterexample to our statement and let the length $|uv|$ be minimal possible. Every subword of Z_∞ is contained in some Z_m'. Therefore, since Z_∞ is not an isoterm for $u = v$, there exists a minimal number m such that Z_m is not an isoterm for $u = v$. If $m \leq n + 1$ then we can apply Exercise 3.3.35. Thus we can assume that $m > n + 1$.

There exists a substitution ϕ such that $\phi(u) \leq Z_m$ and $\phi(u) \not\equiv \phi(v)$. Then u is an unavoidable word and by Theorem 2.5.14 there exists a substitution ψ such that $\psi(u) \leq Z_n$. Since $m > n + 1$, we have that $\psi(u) \equiv \psi(v)$ for every such substitution ψ. Therefore $\mathrm{cont}(v) \subseteq \mathrm{cont}(u)$ (indeed if a letter x is in $\mathrm{cont}(v)$ but not in $\mathrm{cont}(u)$, we can choose a long enough word for $\psi(x)$ so that $\psi(u) \not\equiv \psi(v)$).

Let us prove that $\phi(u)_{x_1} \equiv \phi(v)_{x_1}$. Suppose, by contradiction, that $u' \equiv \phi(u)_{x_1} \not\equiv \phi(v)_{x_1} \equiv v'$. Notice that $u' \leq Z'_{m-1}$. Let $A = \{x \in \text{cont}(u) \mid \phi(x) = x_1\}$.

If A is empty, then u' is a value of u under some substitution $\phi' = \phi_{x_1}$, and $v' \equiv \phi'(v)$ (see Exercise 2.5.22). Then $\phi'(u) \leq Z_{m-1}$ and $\phi'(u) \not\equiv \phi'(v)$ — a contradiction with the minimality of m.

Thus A is not empty. By Lemma 2.5.23 A is a free set in u. The word u' is a value of the word u_A under the substitution $\phi' = \phi_{x_1}$, and $v' \equiv \phi'(v_A)$. Therefore Z_{m-1} is not an isoterm for the identity $u_A = v_A$. Since $|u_A v_A| < |uv|$, we can conclude that Z_n is not an isoterm for $u_A = v_A$. By Lemma 3.3.37 we conclude that Z_{n+1} is not an isoterm for $u = v$, a contradiction.

Thus, indeed, $\phi(u)_{x_1} \equiv \phi(v)_{x_1}$. Therefore either one of the words $\phi(u)$ and $\phi(v)$ starts (ends) with x_1 and another one does not, or $\phi(v)$ contains a subword $x_p x_q$ for $p, q \neq 1$, or $\phi(v)$ contains x_1^2 (see Exercise 3.3.21). It is clear that these 2-letter subwords cannot occur in $\phi(x)$ for any letter x, otherwise $\phi(u)$ would contain such subwords also, which is impossible since $\phi(u) \leq Z'_m$. Therefore these subwords overlap with $\phi(x)$ and $\phi(y)$ for some letters x, y.

Let us define a substitution ψ^ϕ as follows. For every $x \in \text{cont}(u)$ we have $\phi(x) \equiv [\phi(x)_{x_1}, x_1]_{\epsilon_x}^{\delta_x}$. Then let $\psi^\phi(x) \equiv [\psi(x), a]_{\epsilon_x}^{\delta_x}$.

Now it is easy to see that if $\phi(x)$ starts (ends) with x_1 then $\psi^\phi(u)$ starts (ends) with a, and the same holds for v. Therefore if $\phi(u)$ (resp. $\phi(v)$) starts or ends with x_1, but $\phi(v)$ (resp. $\phi(u)$) does not then $\psi^\phi(u)$ (resp. $\psi^\phi(v)$) starts or ends with a, but $\psi^\phi(v)$ (resp. $\psi^\phi(u)$) does not, so that $\psi^\phi(u) \not\equiv \psi^\phi(v)$.

Also it is easy to see that $\psi^\phi(u)$ cannot contain a^2 and $x_i x_j$ for $i, j > 1$. Therefore (again by Exercise 3.3.21)

$$\psi^\phi(u) \equiv [\psi^\phi(u)_a, a]_\epsilon^\delta \equiv [\psi(u), a]_\epsilon^\delta$$

for some ϵ, δ, so $\psi^\phi(u) \leq Z_{n+1}$.

On the other hand, if v contains a 2-letter subword xy such that $\phi(x)\phi(y)$ contains x_1^2 or $x_i x_j$ for $i, j > 1$, and this word overlaps with $\phi(x)$ and $\phi(y)$, then we can conclude that $\psi^\phi(v) > \psi^\phi(xy)$ and so $\psi^\phi(v)$ contains either a^2 or a word $x_i x_j, i, j > 1$

In all cases $\psi^\phi(u) \leq Z_{n+1}$ and $\psi^\phi(u) \not\equiv \psi^\phi(v)$, which contradicts the minimality of m.   □

Recall that in the proof of Theorem 2.5.19, the crucial role was played by the quasi-integral words (see (2.5.4)). Now we need a more detailed analysis of such words.

Lemma 3.3.39. *Let*

$$u \equiv P_0 Q_1 y_1 P_1 Q_2 y_2 P_2 \dots Q_k y_k P_k Q_{k+1}$$

be a quasi-integral word, $v \equiv S_1 y_1 S_2 \dots y_k S_{k+1}$ where S_i are words in the alphabet $\mathcal{A}$. Let $T \equiv y_1 y_2 \dots y_k$ be an unavoidable word and $u \not\equiv v$. Then Z_∞ is not an isoterm for $u = v$.

Proof. By contradiction, suppose that Z_∞ is an isoterm for $u = v$.

Step 1. We can assume that u and v start and end with a letter from $Y = \{y_1, \dots, y_k\}$. Otherwise we can multiply this identity by a new letter from the left and by another new letter from the right, and then include these letters in L. All conditions of the lemma will be preserved. Thus we can assume that $P_0 Q_1, S_1, P_k Q_{k+1}, S_{k+1}$ are empty words.

Step 2. Since $u \not\equiv v$, there exists a number ℓ such that $S_\ell \not\equiv P_\ell Q_{\ell+1}$.

Step 3. Let, as before, σ_i be the deletion of letters from $\mathcal{A}_i$. We also denote the deletion of the letter a_{ij} by σ_{ij} and the deletion of all letters from $\mathcal{A}_j$ except a_{ij} by σ'_{ij}.

We know (Exercise 2.5.21) that the sequence $\sigma_2, \sigma_3, \dots, \sigma_r, \sigma_1$ is a sequence of free deletions in any quasi-integral word. As a result of these deletions we will get the word T that is unavoidable. By Theorem 2.5.14 of Bean–Ehrenfeucht–McNulty and Zimin we get that u is also unavoidable. Similarly every word that we can get from u by a sequence of deletions $\sigma_{i_1}, \dots, \sigma_{i_s}$ for $s \leq r$ is unavoidable.

Step 4. Take an arbitrary j, $1 \leq j \leq r$. Let $u_1 = \sigma_2 \sigma_3 \dots \sigma_r(u)$. In u_1, the deletion σ'_{j1} is a free deletion because a subset of a free set is free. Sets $B_1 = \{a_{j1}\}$ and $C_1 = \{y_t \mid Q_t \text{ contains } a_{j1}\}$ form a fusion in $\sigma'_{j1}(u_1)$ Therefore σ_{j1} is a free deletion in $\sigma'_{j1}(u_1)$. If B_1, C_1 do not form a fusion in $\sigma'_{j1}(v_1)$, then by Lemma 3.3.36 Z_∞ is not an isoterm for $u_1 = v_1$, and by Lemma 3.3.37 Z_∞ is not an isoterm for $u = v$, a contradiction.

Thus B_1 and C_1 form a fusion in $\sigma'_{j1}(v_1)$. This implies that S_ℓ contains a_{j1} if and only if $Q_{\ell+1}$ contains this letter, and that this letter appears in S_ℓ only once.

Step 5. Let now $1 \leq i, j \leq r$. Let u_2, v_2 be words obtained from u and v by the deletion of all letters that are not in $\mathcal{A}_j \cup \mathcal{A}_1 \cup Y$. We know that u_2 is obtained from u by a sequence of free deletions.

As above σ'_{ij} is a free deletion in u_2. In the word $\sigma'_{ij}(u_2)$ sets

$$B_2 = \{a_{ij}\} \cup \{y_t \mid P_t \text{ does not contain } a_{ij}\}$$

and

$$C_2 = \mathcal{A}_{r1} \cup \{y_t \mid Q_t \text{ either contains } a_{ij} \text{ or empty}\}$$

form a fusion. Therefore they form a fusion in $\sigma'_{ij}(v_2)$. This implies that S_ℓ can be represented in the form PQ where

- P contains the same letters as P_ℓ,
- Q contains the same letters as $Q_{\ell+1}$,
- if $Q_{\ell+1}$ is not empty, then Q and $Q_{\ell+1}$ start with the same letter from $\mathcal{A}_1$,
- each letter occurs in P (in Q) only once.

Therefore P is a permutation of the word P_ℓ, Q is a permutation of the word $Q_{\ell+1}$.

Step 6. Let $1 \le j \le r$. Let j' equal $j+1$ if $j < r$ and 1 if $j = r$. In the word u sets

$$B_3 = \mathcal{A}_j \cup \{y_t \mid P_t \text{ starts with a letter from } \mathcal{A}_{j'}\}$$

and

$$C_3 = \mathcal{A}_{j'} \cup \{y_t \mid Q_t \text{ ends with a letter from } \mathcal{A}_j\}$$

form a fusion. Since $\sigma_j(u)$ is unavoidable, these sets form a fusion in v. Therefore, in S_ℓ after every letter of $\mathcal{A}_j$ there is a letter from $\mathcal{A}_{j'}$. Since we have proved in Step 5 that words P and Q are permutations of P_ℓ and $Q_{\ell+1}$, and that Q and $Q_{\ell+1}$ have a common first letter, we can conclude that $P = P_\ell$, $Q = Q_{\ell+1}$, so that $S_\ell = P_\ell Q_{\ell+1}$, which contradicts the assumption in Step 2.   □

We will also need the following

Exercise 3.3.40. Let u be a word, and let ϕ be a substitution that takes every letter x to a product of distinct letters $x_1 x_2 \ldots x_{k_x}$ (x_i are different for different x). Then u can be obtained from $\phi(u)$ by a sequence of free deletions and by renaming letters.

Now we are ready to finish the proof of our Lemma 3.3.34.

Fix some numbers n and $r > 6n + 1$. We assume that Lemma 3.3.34 is false and that m is the smallest number such that $\gamma^m(a_{11})$ is not an isoterm for an identity $u = v$ in n letters, but Z_{n+1} is an isoterm for $u = v$ and $v = u$, and that n is the minimal number of letters for which such an m exists.

Since Z_{n+1} is an isoterm for $u = v$ and $v = u$, by Lemma 3.3.38 Z_∞ is an isoterm for $u = v$ and $v = u$.

By assumption there exists a substitution ϕ such that $\phi(u) \not\equiv \phi(v)$ and $\phi(u) \le \gamma^m(a_{11})$. We have met this situation before, in the proof of Theorem 2.5.14 and we have shown that there exists a substitution ϵ and a substitution ϕ_1 such that

- ϵ takes every letter to a product of at most three different letters. These letters are different for different letters of $\text{cont}(u)$.
- $\phi_1(x)$ is either a product of γ-blocks or a subword of a block.
- $\phi = \phi_1 \epsilon$.

We also know (Section 2.5.3.4) that in each γ-block one can find a two-letter subword p such that if p overlaps with one of $\phi_1(x)$ then p is a subword of $\phi_1(x)$.

Now we can define the following substitution η for every x in $\text{cont}(u)$:

$$\eta(x) = \begin{cases} x & \text{if } \phi_1(x) \text{ contains one of these two-letter subwords } p \\ \phi_1(x) & \text{otherwise.} \end{cases}$$

The word $\bar{u} = \eta\epsilon(u)$ is quasi-integral.

The word $\bar{v} = \eta\epsilon(v)$ has the form $S_0 x_1 S_1 \ldots S_f x_f S_{f+1}$ where $x_i \in \text{cont}(\epsilon(u))$ and $\text{cont}(S_i) \subseteq \mathcal{A}$.

Since $\bar{u}$ is quasi-integral $T_u = \bar{u}_{\mathcal{A}}$ is obtained from $\bar{u}$ by a sequence of free deletions (see Exercise 2.5.21). The word T_u is obtained from $\epsilon(u)$ by a deletion of some letters. Therefore by Exercise 2.5.23 T_u can be obtained from $\epsilon(u)$ by a sequence of free deletions. Therefore $T_u \equiv \epsilon_1(u_1)$ for some substitution ϵ_1 and some word u_1 which is obtained from u by a sequence of free deletions. Notice that for every letter x, $\epsilon_1(x)$ is obtained from $\epsilon(x)$ by a deletion of some letters.

Let us denote the word obtained from v by deleting letters from $\text{cont}(u) \setminus \text{cont}(u_1)$ by v_1. Then $\epsilon_1(v_1) \equiv T_v$.

Now define a substitution δ by $\delta(x_i) = Q_i x_i P_i$, $x_i \in \text{cont}(T_u)$. Then the word $\phi'\delta(x_i)$ is a product of γ-blocks where ϕ' is the substitution that takes every x_i to the two-letter word replaced by x_i. Therefore we can consider $\gamma^{-1}\phi'\delta(x)$. Since $\phi'(T_u)$ is a subword of $\gamma^m(a_{11})$, we have that $\gamma^{-1}\phi'\delta(T_u)$ is a subword of $\gamma^{m-1}(a_{11})$. Since $T_u \equiv \epsilon_1(u_1)$ we have that $\gamma^{m-1}(a_{11})$ contains a value of u_1. By Theorems 2.5.14 and 2.5.19 the word u_1 is unavoidable. Since u_1 is obtained from u by a series of free deletions and Z_∞ is an isoterm for $u = v$ we can use Lemma 3.3.37 and conclude that Z_∞ is an isoterm for $u_1 = v_1$.

Suppose that Z_∞ is not an isoterm for $v_1 = u_1$. Then Z_∞ contains a value of v_1, which differs from the corresponding value of u_1. In particular, since Z_∞ contains a value of v_1, this word is unavoidable. By Lemma 3.3.36 v_1 is obtained from v by a sequence of free deletions (the same sequence was used to get u_1 from u). This contradicts Lemma 3.3.37. Thus Z_∞ is an isoterm for $u_1 = v_1$ and $v_1 = u_1$.

From the minimality of m we can deduce that $\gamma^{m-1}(a_{11})$ is an isoterm for $u_1 = v_1$ and $v_1 = u_1$. Since $\gamma^{m-1}(a_{11})$ contains a value $\gamma^{-1}\phi'\delta\epsilon_1(u_1)$ of u_1, this value of u_1 must coincide with the corresponding value of v_1. Therefore, in particular, $\gamma^{m-1}(a_{11})$ contains a value of v_1.

As above, this implies that v_1 is unavoidable. As we know v_1 is obtained from v by a series of free deletions. Since Z_∞ is an isoterm for $v = u$, we can conclude that it is an isoterm for $v_1 = u_1$.

We have already proved that

$$\gamma^{-1}\phi'\delta\epsilon_1(u_1) \equiv \gamma^{-1}\phi'\delta\epsilon_1(v_1).$$

Therefore

$$\phi'\delta\epsilon_1(u_1) \equiv \phi'\delta\epsilon_1(v_1).$$

Let us denote $P_0\delta\epsilon_1(v_1)Q_{f+1}$ by w. We have

$$\phi'(w) \equiv P_0\phi'\delta\epsilon_1(v_1)Q_{f+1} \equiv \phi(u) \not\equiv \phi(v) \equiv \phi'(\bar{v}).$$

Thus, in particular, $w \not\equiv \bar{v}$. By definition, w is a quasi-integral word and T_v is equal to $w_{\mathcal{A}}$. Recall that T_v is a value of an unavoidable word v_1, and the corresponding substitution ϵ_1 takes every letter to a product of (at most 3) different letters. By Exercise 3.3.40, T_v is unavoidable. Thus all conditions of Lemma 3.3.39 hold and we can conclude that Z_∞ is not an isoterm for $w = \bar{v}$.

Therefore there exists a substitution θ such that $\theta(w) \preceq Z_\infty$ and $\theta(w) \neq \theta(\bar{v})$.

We have proved that Z_∞ is an isoterm for $v_1 = u_1$. Since

$$w \equiv P_0 \delta \epsilon_1(v_1) Q_{f+1}$$

and

$$\bar{u} \equiv P_0 \delta \epsilon_1(u_1) Q_{f+1},$$

we have that Z_∞ is an isoterm for $w = \bar{u}$. Hence

$$\theta(\bar{u}) \equiv \theta(w) \not\equiv \theta(\bar{v}).$$

This means that Z_∞ is not an isoterm for $\bar{u} = \bar{v}$. But

$$\bar{u} \equiv \eta \epsilon(u), \quad \bar{v} \equiv \eta \epsilon(v).$$

Therefore Z_∞ is not an isoterm for $u = v$, a contradiction.

Theorem 3.3.4 is proved.

3.4 Brown's Theorem and Uniformly Recurrent Words

The following important and general result of T.C. Brown is much less trivial than its group theoretic counterpart (and corollary): every extension of a locally finite group by a locally finite group is locally finite. There are at least four different proofs of this theorem. We present here our proof using uniformly recurrent bi-infinite words.

Theorem 3.4.1 (T.C. Brown [55]). *Let S be a semigroup and σ be a congruence on S such that S/σ is localy finite and every equivalence class of σ that is a subsemigroup is locally finite. Then S is locally finite.*

Proof. We can assume that S is finitely generated, hence $M = S/\sigma$ is finite. Let X be a finite generating set of S, $M = S/\sigma$, ϕ be the natural map $S \to M$. Then $\phi(X)$ generates M. We shall use the induction on the pair of numbers $(|M|, |X|)$ ordered lexicographically. Thus assume that S is a counterexample, $|M|$ is as small as possible (among all counterexamples), $|X|$ is the smallest possible among all counterexamples with the same $|M|$. Then by Theorem 1.6.2 there exists a geodesic uniformly recurrent bi-infinite word α for S. By Lemma 3.3.12 there exists a number k such that every letter in

α occurs in every subword of α of length k. By the minimality assumption, we can assume that every generator of S occurs in α.

For every generator x of S consider the set U of all subwords of the form xu of α where u does not contain x (u may be empty). Since $|u| < k$, U is finite. The set U generates a subsemigroup S' of S and we can rewrite α into a bi-infinite word $\beta \equiv \dots xu_{i_1} xu_{i_2} \dots$ in the alphabet U. Note that then β is a geodesic word for S'. Hence S' is infinite. The images $\phi(xu)$, $xu \in U$ are in the subsemigroup $\phi(x)M$ of M. By the minimality of M, we can assume that $\phi(x)M = M$ for every generator x. Similarly $M\phi(x) = M$ for every generator x of S. Therefore M is a group (prove that!). Now consider the following bi-infinite word γ of elements of M. For every $i \in \mathbb{Z}$, $\gamma(i,i)$ is $\phi(\alpha(0,i))$ if $i \geq 0$ and $\phi(\alpha(i,0))$ if $i < 0$. By Lemma 3.3.16 there exist an element $x \in S/\sigma$ and an integer k' such that for any n there are integers $i_1 < i_2 < \dots < i_n$ with $x = \gamma(i_j, i_j)$, $j = 1, \dots, n$ and $i_{j+1} - i_j \leq k'$, $j = 1, \dots, n-1$. Without loss of generality assume that all $i_j > 0$. Therefore the elements $\phi(\alpha(0, i_j))$ are all equal. Hence $\phi(\alpha(i_j + 1, i_{j+1}))$, $j = 1, \dots, n-1$ represent elements of $\phi^{-1}(1)$. Since the number of possible elements $\phi(\alpha(i_j + 1, i_{j+1}))$ is finite and the subsemigroup $\phi^{-1}(1)$ is locally finite, the subsemigroup of S generated by these elements is finite. Thus for a sufficiently large n, there must exist $j' > j$ such that $\alpha(i_1 + 1, i_j) = \alpha(i_1 + 1, i_{j'})$ in S, which contradicts the assumption that α is geodesic.　　　　　　　□

As an immediate corollary we get

Corollary 3.4.2 (Schmidt [285]). *If G has a locally finite normal subgroup N and G/N is locally finite, then G is locally finite.*

Applying Theorem 3.4.1 and Part (3) of Exercise 3.1.3 we get

Corollary 3.4.3 (Shevrin [293]). *Every semilattice of locally finite semigroups is locally finite. If a semigroup S has a locally finite ideal I and S/I is locally finite, then S is locally finite.*

This corollary will be used later in Section 3.6.2.

Remark 3.4.4. In fact we can deduce from Theorem 3.4.1 that every band of locally finite semigroups is locally finite (which was proved in [293]). Indeed every band is locally finite by Theorem 3.3.4 applied to $\Sigma = \{x = x^2\}$ (that was first proved by Green and Rees [116]).

3.5 Burnside Problems and the Finite Basis Property

All varieties of semigroups that we met before were finitely based. Now we will use Theorem 3.3.4 to construct non-finitely based varieties.

We can rewrite this theorem in the following way.

Theorem 3.5.1 (Sapir [277]). *Let $\mathcal{V}$ be a variety of semigroups that satisfies the following two properties:*

(1) *every finitely generated nil-semigroup in $\mathcal{V}$ is finite;*
(2) *$\mathcal{V}$ does not satisfy any nontrivial identity of the form $Z_n = W$.*

Then $\mathcal{V}$ cannot be defined by a finite set of identities.

Here is an easy way to construct a variety that satisfies both conditions 1 and 2.

Let us take all subwords of Z_∞ and construct semigroup $S(Z_\infty)$ using the Dilworth construction.

Exercise 3.5.2. Using Theorem 1.4.31 prove that the variety generated by $S(Z_\infty)$ is locally finite. Estimate the function $n(m)$ from Theorem 1.4.31 for the semigroup $S(Z_\infty)$.

Theorem 3.5.3. *The variety $\operatorname{var} S(Z_\infty)$ is inherently non-finitely based.*

Proof. By Exercise 3.5.2, the variety $\operatorname{var} S(Z_\infty)$ is locally finite. By Theorem 3.5.1, it is enough to show that $S(Z_\infty)$ does not satisfy any nontrivial identity of the form $Z_n = W$. By Lemma 3.3.32 it is enough to show that for every n some subwords of Z_∞ are not isoterms for $Z_n = W$. But Z_n itself is a subword of Z_∞, and Z_n is, of course, not an isoterm for $Z_n = W$. □

3.6 Inherently Non-finitely Based Finite Semigroups

3.6.1 Some Advanced Semigroup Theory

3.6.1.1 Ideals and 0-Simple Semigroups

Exercise 3.6.1. Let S be a semigroup.

(1) Show that the set of ideals of S is closed under any unions and finite intersections.
(2) Show that for every element a of S the set $S^1 a S^1 = SaS \cup Sa \cup aS \cup \{a\}$ is an ideal of S. This ideal is called the *principal ideal generated by* a. Elements of $S^1 a S^1$ are precisely the elements of S *divisible by* a.

A semigroup S is called *0-simple* if it does not contain any ideals distinct from zero and the whole S. Note that a 0-simple semigroup does not necessarily contain a zero. If a 0-simple semigroup does not contain 0, it is called *simple*.

Example 3.6.2. Every group is obviously a simple semigroup. The 2-element semigroup with zero product is 0-simple. This semigroup is an exception and in some books on semigroup theory this semigroup is not included into the class of 0-simple semigroups.

Exercise 3.6.3. If a semigroup S is 0-simple, then S^0 is also 0-simple.

Exercise 3.6.4. Prove that every *rectangular band* of groups (see Definition 3.1.5) is simple.

Exercise 3.6.5. Prove that the Brandt semigroup B_2 is 0-simple.

Exercise 3.6.6. Let a be an element of a semigroup S. Let M be the union of all ideals of S contained in the principal ideal S^1aS^1 and not containing a. Prove that S^1aS^1/M is 0-simple. This semigroup is called a *principal factor* of S.

Exercise 3.6.7 (Rees–Sushkevich construction). Let I, J be sets, G be a group and P be a function $P : I \times J \to G^0$ (it is convenient to consider P as a $I \times J$-matrix over G^0). Suppose that P is *regular* that is it does not contain zero rows or zero columns. Consider the set $M^0(G, I, J, P) = I \times G \times J \cup \{0\}$ and define an operation on this set by the following formula:

$$(i, a, j)(i', a', j') = \begin{cases} (i, aP(i', j)a', j') & \text{if } P(i', j) \neq 0; \\ 0 & \text{if } P(i', j) = 0. \end{cases}$$

Prove that $M^0(G, I, J, P)$ is a 0-simple semigroup. Prove also that if $P: I \times J \to G$ then the set $M(G, I, J, P) = M^0(G, I, J, P) \backslash \{0\}$ is a subsemigroup of $M^0(G, I, J, P)$ that is simple. Prove that $M^0(G, I, J, P)$ is periodic if and only if G is periodic. The semigroups $M^0(G, I, J, P)$ and $M(G, I, J, P)$ are called the *Rees–Sushkevich semigroups over the group G with the sandwich matrix P*.

We shall always assume that sandwich matrices are regular.

Semigroups of the form $M^0(G, I, J, P)$ (respectively, $M(G, I, J, P)$) are called *completely 0-simple* (respectively, *completely simple*). The next three exercises describe idempotents and maximal subgroups of completely 0-simple semigroups.

Exercise 3.6.8. Let $M = M^0(G, I, J, P)$. Prove that idempotents of M are the elements of the form $(i, P(i, j)^{-1}, j)$ where $P(i, j) \neq 0$, and 0.

Exercise 3.6.9. Let $M = M(G, I, J, P)$. Prove that for every two idempotents $e, f \in M$ there exist idempotents $x, y \in M$ such that $ex = e = ye, xf = f = fy, xe = x = fx, ey = y = yf$.

Exercise 3.6.10. Let $M = M^0(G, I, J, P)$. Prove that two elements (i, g, j) and (i', g', j') are $\mathcal{L}$-related if and only if $j = j'$. Prove that these elements are $\mathcal{R}$-related if and only if $i = i'$. Prove that all maximal subgroups of M are isomorphic to G and have the form $\{i\} \times G \times \{j\}$ where $i \in I, j \in J$ and $P(i, j) \neq 0$.

Exercise 3.6.11. Prove that $M = M(G, I, J, K)$ is a rectangular band of groups. More precisely, for every $i \in I, j \in J$ the set $\{i\} \times G \times \{j\}$ is a subgroup. These subgroups cover M and form a partition of M. This partition is a congruence, and the quotient semigroup is a rectangular band isomorphic to the set $I \times J$ with operation $(i, j)(i', j') = (i, j')$ (see Exercise 3.1.4).

Exercise 3.6.12. If a and b belong to the same subgroup of the Rees–Sushkevich semigroup $S = M(G, I, J, K)$ and $a \neq b$, then for any x in S we have $ax \neq bx$ and $xa \neq xb$.

We shall now show that every periodic 0-simple semigroup either is the 2-element semigroup with zero product or is isomorphic to a Rees–Sushkevich semigroup.

Lemma 3.6.13. *Let S be a 0-simple periodic semigroup, $a, p \in S$, $ap \neq 0$. Then there exists an element q in S such that $apq = a$. Similarly if $pa \neq 0$ then there exists an element $q \in S$ such that $qpa = a$.*

Proof. It is enough to prove the first statement of the lemma. Since $S^1 a p S^1$ is an ideal of S containing ap and $ap \neq 0$, $S^1 a p S^1 = S$. Therefore $uapv = a$ for some $u, v \in S^1$. Multiplying by u, pv on the left and on the right $n-1$ times, we deduce $u^n a (pv)^n = a$ for every $n \geq 1$. Since S is periodic, there exists n such that $u^n = e$ is an idempotent (Exercise 3.1.12). Then $ea(pv)^n = a$. Multiplying by e on the left and using $e = e^2$, we get $ea = a$. Hence $a(pv)^n = a$. This means $ap[(vp)^{n-1}v] = a$. So we can take $q = (vp)^{n-1}v$.   □

Lemma 3.6.14. *Let S be a 0-simple periodic semigroup, $a \in S$, $a \neq 0$. Then either S is a 2-element semigroup with zero product or S contains two idempotents e, f such that $ea = a, af = a$.*

Proof. If S is a 0-simple semigroup with zero product, then S cannot contain more than two elements, because every subset of S containing zero is an ideal of S. So we can assume that S is not a semigroup with zero product. Therefore there exist two elements $x, y \in S$ such that $xy \neq 0$. The ideal $S^1 xy S^1$ coincides with S. So $a = uxyv$ for some $u, v \in S^1$. Since $S^1 a S^1 = S$, we can find elements $b, c \in S^1$ such that $y = bac$. Therefore $a = uxyv = uxbacv$. Notice that $uxb \in S$. As in the proof of Lemma 3.6.13, multiplying by uxb on the left and by cv on the right $n - 1$ times, we get $a = (uxb)^n a (cv)^n$. Since S is periodic, there exists an n such that $(uxb)^n = e$ is an idempotent in S. Multiplying by e on the left, we get $ea = a$. Similarly we can find an idempotent f such that $af = a$.   □

Lemma 3.6.15. *Let S be a 0-simple periodic semigroup. Then for every idempotent $e \in S$, the set eSe is a group with zero formally added.*

Proof. It is clear that eSe is a subsemigroup and that e is the identity element of e. If $e = 0$ then $eSe = \{0\}$ and there is nothing to prove. Let $e \neq 0$ and let eae be a non-zero element of S. By Lemma 3.6.13 there exists an element $p \in S$ such that $eaep = e$. Multiplying by e on the right gives us: $eaepe = e$. So $(eae)(epe) = e$. Similarly we can find an element q such that $(eqe)(eae) = e$. Therefore every element of eSe has a left inverse and a right inverse elements, so eSe is a group (see Exercise 1.8.1).   □

Theorem 3.6.16. *Every periodic 0-simple semigroup either is a 2-element semigroup with zero product, or is isomorphic to a Rees–Sushkevich semigroup with a regular sandwich matrix.*

Proof. Assume that S is not isomorphic to the 2-element semigroup with zero product. By Exercise 3.6.3 we can assume that S contains zero.

Then by Lemma 3.6.14 there exists a non-zero idempotent $e \in S$. By Lemma 3.6.15 the subset $G^0 = eSe$ is a subgroup of S with zero formally added.

Consider the set of all $\mathcal{R}$-classes of the semigroup S. Choose one representative of the form ae from each class containing an element of this form. Let I be the set of all such representatives.

Consider the set of all $\mathcal{L}$-classes of S. Choose one representative of the form ea from each class containing an element of this form. Let J be the set of all such representatives.

It is clear that I and J are not empty because there exist an $\mathcal{L}$-class and an $\mathcal{R}$-class containing $e = ee$.

For every $i = ae \in I$ and $j = eb \in J$ the element $ji = ebae$ is in G^0. This gives us the matrix $P : I \times J \to G^0$, $P(i,j) = ji$.

We will prove that S is isomorphic to $M^0(G, I, J, P)$.

Consider the map $\phi : M^0(G, I, J, P) \to S$ that takes 0 to 0 and any triple (i, g, j) to the element $igj \in S$ (recall that I, J, G are subsets of S). It is clear that this map is well defined.

Since

$$(igj)(i'g'j') = i(gji'g')j' = i(gP(i',j)g')j',$$

ϕ is a homomorphism.

Let us prove that ϕ is injective. Suppose that $igj = i'g'j'$ where $i = ae, g = ese, j = eb, i' = a'e, g' = es'e, j' = b'e$. Hence

$$ae \cdot ese \cdot eb = a'e \cdot es'e \cdot eb'$$

in S. Then by Lemma 3.6.13 there exist two elements q_1 and q_2 such that

$$ae = a'e \cdot es'e \cdot eb'q_1, \ ae \cdot es'e \cdot ebq_2 = a'e.$$

This implies that $ae\mathcal{R}a'e$, so $ae = a'e$. Similarly $eb = eb'$. Now we can apply Lemma 3.6.13 to $e \cdot eb$ and to $ae \cdot e$ and obtain that $ese = es'e$. Thus $i = i', j = j'$ and $g = g'$. So ϕ is injective.

Finally we need to prove that ϕ is surjective. Let $a \in S \backslash \{0\}$. We need to show that $a = igj$ for some $i \in I, j \in J, g \in G$. Since SeS is an ideal of S, we have $SeS = S$. Then $a = xey = xe \cdot ey$ for some $x, y \in S$. There exist an element $i = ue \in I$ and an element $j = ev$ in J such that $xe\mathcal{R}ue, ey\mathcal{L}ev$. Then $xe = ues$, $ey = tev$ for some $s, t \in S^1$. Since $e = e^2$, we get

$$xe = ue \cdot es, \ ey = te \cdot ev.$$

Thus

$$a = xe \cdot ey = ue \cdot este \cdot ev = igj$$

where $g = este \in G = eSe \setminus \{0\}$. This proves that ϕ is surjective.

It remains to show that the matrix P is regular. Suppose P has a zero column, that is $P(i, j_0) = 0$ for some $j_0 \in J$ and every $i \in I$. Then take any $i_0 \in I$ and $g_0 \in G$. Let $a = (i_0, g_0, j_0)$. By the definition of $M(G, I, J, P)$ the product $(i_0, g_0, j_0)(i', g', j')$ is 0 for every $i' \in I, g \in G, j' \in J$. Thus $ax = 0$ for every $x \in S$. This contradicts Lemma 3.6.14. We get a similar contradiction if P has a zero row.   □

Exercise 3.6.17. Prove that $B_2 = M^0(1, \mathbf{2}, \mathbf{2}, I_2)$ where 1 is the trivial group, $\mathbf{2} = \{1, 2\}$, and I_2 is the 2×2-identity matrix.

Definition 3.6.18. Let G be any group, I_n be the identity $n \times n$-matrix over G^0. Then the semigroup $M^0(G, \mathbf{n}, \mathbf{n}, I_n)$ is called the $(n \times n)$-*Brandt semigroup over the group* G (here and below $\mathbf{n}$ denotes the set $\{1, \ldots, n\}$).

Exercise 3.6.19. Prove that $M^0(G, \mathbf{n}, \mathbf{n}, P)$ is isomorphic to the $n \times n$-Brandt semigroup over G if and only if P has exactly one non-zero element in every row and in every column.

Corollary 3.6.20. *A periodic semigroup is simple if and only if it is a rectangular band of groups.*

Proof. This corollary immediately follows from Exercise 3.6.11 and Theorem 3.6.16.   □

The next exercise shows that the local finiteness question for completely 0-simple semigroups reduces to the question for groups.

Exercise 3.6.21. Prove that if all maximal subgroups of a completely 0-simple semigroup S are locally finite, then S is locally finite.

3.6.1.2 Semigroups Without Divisors Isomorphic to A_2 and B_2

A *divisor* of a semigroup S is a quotient semigroup of a subsemigroup of S (it is a specialization of the definition from Section 1.4.7).

Lemma 3.6.22. *Let A_2 be the 0-simple semigroup $M^0(1, \mathbf{2}, \mathbf{2}, P)$ where*

$$P = \begin{pmatrix} 0 & 1 \\ 1 & 1 \end{pmatrix}.$$

Then the Brandt semigroup B_2 is a divisor of the semigroup $A_2 \times A_2$.

Proof. Let W be a subset of $A_2 \times A_2$ consisting of all pairs with one coordinate equal 0. Then W is clearly an ideal of $A_2 \times A_2$. It is easy to check that

$S = (A_2 \times A_2)/W$ is isomorphic to $M^0(1, 4, 4, P \otimes P)$ where $P \otimes P$ is the *tensor square* of P,

$$P \otimes P = \begin{pmatrix} 0\ 0\ 0\ 1 \\ 0\ 0\ 1\ 1 \\ 0\ 1\ 0\ 1 \\ 1\ 1\ 1\ 1 \end{pmatrix}$$

(check it!). Then the elements $(2, 1, 2), (3, 1, 2), (2, 1, 3), (3, 1, 3), 0$ from S form a subsemigroup isomorphic to B_2 (prove it!). $\qquad\square$

We shall show that the Brandt semigroup B_2 plays a very special role in the theory of semigroups because a periodic semigroup without divisors isomorphic to B_2 and its close relative A_2 has a very nice structure. We start with 0-simple semigroups.

Lemma 3.6.23. *Let $S = M^0(G, I, J, P)$ be a 0-simple semigroup and the sandwich-matrix P contains zero entries. Then S contains a copy of A_2 or B_2.*

Proof. Let $P(i, j) = 0$. Since P is regular, there exist $m \in I, n \in J$ such that $P(m, j) \neq 0$ and $P(i, n) \neq 0$.
  Case 1. Let $P(m, n) \neq 0$. Then the elements

$$(i, P(i, n)^{-1}P(m, n)P(m, j)^{-1}, j), (i, P(i, n)^{-1}n),$$

$$(m, P(m, j)^{-1}, j), (m, P(m, n)^{-1}, n),$$

and 0 form a subsemigroup of S isomorphic to A_2.
  Case 2. Let $P(m, n) = 0$. Then the elements

$$(i, P(i, n)^{-1}n), (i, 1, j), (m, 1, n), (m, P(m, j)^{-1}, j),$$

and 0 form a subsemigroup of S isomorphic to B_2. $\qquad\square$

The next theorem (due to Shevrin [294, 295]) describes the structure of arbitrary periodic semigroups that do not have divisors isomorphic to A_2 and B_2. The proof of this theorem shows how a combination of algebraic methods and manipulation with words can elucidate the structure of a semigroup.

Theorem 3.6.24. *Every periodic semigroup that does not have divisors isomorphic to A_2 or B_2 is a semilattice of semigroups each one of which is an ideal extension of a Rees–Sushkevich semigroup of the form $M(G, I, J, P)$ by a nil-semigroup.*

Proof. Let S be a periodic semigroup without divisors isomorphic to A_2 or B_2.
  For every element a in a periodic semigroup let a^0 denote the idempotent in $\langle a \rangle$. Consider the following relation $\mho$ on S:

$$a \mho b \iff Sa^0S = Sb^0S.$$

We claim that $\mho$ is a congruence, the quotient semigroup $S/\mho$ is a semilattice, and each class of $\mho$ is a subsemigroup that is an ideal extension of a simple semigroup by a nil-semigroup.

The fact that $\mho$ is an equivalence relation is obvious.

Let us prove that $\mho$ is a congruence. We shall divide the proof into a number of steps.

Step 1. For every $a, b \in S$ we have $ab \mho ba$. Indeed $(ab)^n = a(ba)^{n-1}b$. So $(ab)^0$ and $(ba)^0$ are divisible by each other.

Step 2. If $a \mho b$ then $a \mho a^m b^n$ for every $m, n \geq 1$ and there exist idempotents $x, y \in S$ such that $a^0 x = a^0 = ya^0, xb^0 = b^0 = b^0 y, xa^0 = x = b^0 x, a^0 y = y = yb^0$. Indeed, we have that $Sa^0 S = Sb^0 S$. Consider the principal factor $K = Sa^0 S/W$ as in Exercise 3.6.6. We know that it is a 0-simple semigroup. By Theorem 3.6.16 K is isomorphic to $M^0(G, I, J, P)$ for some G, I, J, P. By Lemma 3.6.23, the matrix P does not have zero entries. By Exercise 3.6.7 $K \setminus \{0\}$ is a simple semigroup. Notice that a^m and b^n are in $K \setminus \{0\}$. Therefore $a^m b^n$ and $(a^m b^n)^0$ are in $K \setminus \{0\}$ too. Since $K \setminus \{0\}$ is simple, a^0 and $(a^m b^n)^0$ are divisible by each other. So $a \mho a^m b^n$. The second part of the statement follows from Exercise 3.6.9.

Step 3. For every $a, c \in S$, we have $ac \mho a^0 c^0$. By Step 1, $ac \mho ca$. Using Step 2 and Step 1 several times, we get:

$$ac \mho ac^2 a \mho a^2 c^2 \mho c^2 a^2 \mho a^2 c^3 a \mho a^3 c^3 \ldots \mho a^n c^n$$

for every n. Since S is periodic, there exists an n such that $a^n = a^0$ and $c^n = c^0$. Therefore $ac \mho a^0 c^0$.

Step 4. If $a \mho b$ then for every $c \in C$ we have $ac \mho bc$ and $ca \mho cb$. Indeed, by Step 2 there exists an idempotent x such that $a^0 x = a^0, xa^0 = x = b^0 x, xb^0 = b^0$. By Step 3,

$$a^0 c^0 = a^0 x c^0 \mho a^0 (x c^0)^0.$$

On the other hand $xc^0 \mho xa^0 c^0 \mho x(a^0 c^0)^0$. Therefore $(xc^0)^0$ and $(a^0 c^0)^0$ are divisible by each other. Hence $xc \mho xc^0 \mho a^0 c^0 \mho ac$. Similarly one can prove that $cx \mho cb$. By Step 1, we can conclude that $ac \mho bc$ and $ca \mho cb$. Thus $\mho$ is a congruence.

Clearly $a \mho a^0$ for every $a \in S$. So $S/\mho$ is a band (i.e., every element of $S/\mho$ is an idempotent). Since by Step 1 $ab \mho ba$ for every $a, b \in S$, $S/\mho$ is commutative. Thus $S/\mho$ is a semilattice.

Finally let S_α be an equivalence class of $\mho$. Let E be the set of idempotents in S_α and let $W = S_\alpha E S_\alpha$ be the ideal generated by E. As we have seen on Step 2, for every two idempotents $e, f \in S_\alpha$ there exists an idempotent $x \in S_\alpha$ such that $e = ex = efx$. This implies that any two idempotents in W are divisible by each other inside W. Every element in W divides an idempotent because W is periodic, and every element in W is divisible by an idempotent by the definition of W. Therefore any two elements in W are divisible by each other. Thus W is a simple semigroup. By Theorem 3.6.16 and Exercise 3.6.7, W is isomorphic to a semigroup of the form $M(G, I, J, P)$.

The semigroup S_α/W has exactly one idempotent, 0. Therefore by Lemma 3.1.14, S_α/W is a nil-semigroup. □

The following exercise shows that a statement converse to Theorem 3.6.24 also holds.

Exercise 3.6.25. Let S be a periodic semigroup that is a semilattice of ideal extensions of simple semigroups by nil-semigroups. Prove that S satisfies the following law:

$$((ab)^0(ba)^0)^0(ab)^0 = (ab)^0$$

for every $a, b \in S$ (here as in the proof of Theorem 3.6.24 a^0 denotes the idempotent from $\langle a \rangle$). Prove that A_2 and B_2 do not satisfy this law. Deduce from it that S cannot have divisors isomorphic to A_2 or B_2.

Exercise 3.6.26. Prove that for a finite semigroup S the following conditions are equivalent.

(1) S has no divisors isomorphic to the semigroups A_2 and B_2.
(2) For any idempotent e of S and any element a dividing e the element eae belongs to the maximal subgroup of S with identity element e, denoted by S_e.
(3) For any idempotent e of S and any element a dividing e the element a^2 divides e.
(4) S possesses a series of ideals $I_1 < I_2 < \ldots < I_n = S$, with each quotient I_{k+1}/I_k either a completely simple semigroup, or a completely simple semigroup with an adjoined zero, or a semigroup with zero multiplication.

3.6.2 A Description of Inherently Non-finitely Based Finite Semigroups

First let us give an example of a finite inherently non-finitely based semigroup.

Let B_2^1 be the Brandt monoid from Exercise 1.8.4.

Lemma 3.6.27. B_2^1 *does not satisfy any nontrivial identity of the form* $Z_n = W$.

Proof. By contradiction, suppose B_2^1 satisfies a nontrivial identity $Z_n = W$. Since W differs from Z_n, by Exercise 3.3.21, W satisfies one of four conditions listed in this lemma. Suppose W contains x_1^2 or $x_i x_j$ where $i, j \neq 1$ or W starts or ends not with x_1. Consider the matrix representation of B_2^1 from Exercise 1.8.4 and let $x_1 = \begin{pmatrix} 0 & 1 \\ 0 & 0 \end{pmatrix}$, $x_i = \begin{pmatrix} 0 & 0 \\ 1 & 0 \end{pmatrix}$ for all $i \neq 1$. Then it is easy to

see that the value of W is not equal to the value of Z_n (which is $\begin{pmatrix} 0 & 1 \\ 0 & 0 \end{pmatrix}$).
This contradicts the assumption that $Z_n = W$ is an identity of B_2^1.

If $W = [W_1, x_1]_1^1$, then $W_1 \neq (Z_n)_{x_1}$. Now let $x_1 = \begin{pmatrix} 0 & 1 \\ 0 & 0 \end{pmatrix}$ and all other x_i
be arbitrary elements of B_2^1. Since B_2^1 satisfies $Z_n = W$, we will have that
$(Z_n)_{x_1} = W_1$ holds for an arbitrary choice of $x_i \in B_2^1$. Therefore B_2^1 satisfies
the nontrivial identity $(Z_n)_{x_1} = W_1$. But this identity is of the form $Z'_{n-1} = W'$
and we can finish the proof by induction on n.   □

Now the following theorem can be deduced in the same way as
Theorem 3.5.3.

Theorem 3.6.28 (Sapir [277]). *The semigroup B_2^1 is inherently non-finitely
based.*

Therefore every finite semigroup having B_2^1 as a quotient is inherently
non-finitely based. In particular, by the definition of B_2^1, the multiplicative
semigroup of all matrices of size > 1 over an arbitrary finite ring with unit
contains B_2^1. Thus it is inherently non-finitely based. So is the semigroup of
all transformations of a more than 2-element set.

It is interesting that if we consider the set of all matrices over a finite associative ring as a ring, then it generates a finitely based variety (as any finite
associative ring) — by a theorem of Kruse–L'vov [208] (see Theorem 1.4.33).

Remark 3.6.29. Note that if a group G satisfies the identity $x^d = 1$, then in
every identity $u = v$ or $u = 1$ we can replace every occurrence of a letter x^{-1}
by x^{d-1} and get an equivalent monoid identity (modulo $x^d = 1$). Therefore
every periodic variety of groups can be defined by monoid identities. Note
also that every group identity $u = 1$ is equivalent to the conjunction of two
semigroup identities $uy = y, yu = y$ where y is a letter that does not occur in
u. Thus every periodic variety of groups is a variety of semigroups.

In order to describe all finite inherently non-finitely based semigroups, we
will need the following two exercises.

Exercise 3.6.30. Prove that for every semigroup variety $\mathcal{V}$ satisfying the identity $x^m = x^{m+d}$ with $d \geq 1$, the class of all groups in $\mathcal{V}$ is a semigroup variety
$\mathcal{V}_0$, which is equal to $\mathcal{V} \cap \text{var}\{x^d y = yx^d = y\}$. **Hint:** Use Remark 3.6.29.

Exercise 3.6.31. If S is a finite semigroup, $\mathcal{V} = \text{var}\, S$, then $\mathcal{V}_0$ is generated
by the maximal subgroups of S. **Hint:** By (1.4.1) every group from $\mathcal{V}$ is a
homomorphic image of a subsemigroup of a Cartesian product of copies of
S. Show that if a group G is a homomorphic image of a periodic semigroup
U, then G is a homomorphic image of a subgroup of U (what is the identity
element of that subgroup?) and that a subgroup of a Cartesian product of
semigroups U_i is a subgroup of a Cartesian product of subgroups of U_i.

The proof of the next lemma employs uniformly recurrent bi-infinite words again.

Lemma 3.6.32. *Let E be the set of all idempotents of a semigroup S. If each subsemigroup eSe, $e \in E$, is locally finite, then the ideal SES is locally finite.*

Proof. Suppose that the ideal SES is not locally finite, so there exists a finite number of elements $s_i e_i s_i'$, $e_i = e_i^2$, $s_i, s_i' \in S$, $i = 1, \ldots, l$, which generate an infinite subsemigroup S_1. Then there exists a bi-infinite uniformly recurrent geodesic word α in the letters $s_i e_i s_i'$, $i = 1, \ldots, l$ for S_1. Without loss of generality we can assume that $s_1 e_1 s_1'$ appear in α. Then α can be represented as $\ldots s_1 e_1 s_1' u_1 s_1 e_1 s_1' u_2 \ldots$ where the words $u_j, j \in \mathbb{Z}$ are of bounded length. Hence the set of elements $e_1 s_1' u_j s_1 e_1$ is finite. That set is contained in eSe, and hence generates a finite subsemigroup of eSe. Hence there exists k such that $(e_1 s_1' u_1 s_1 e_1)(e_1 s_1' u_2 s_1 e_1) \cdots (e_1 s_1' u_k s_1 e_1)$ is equal to a shorter product of words $e_1 s_1' u_j s_1 e_1$, which contradicts the assumption that α is geodesic.   □

Lemma 3.6.33. *Let $\mathcal{V}$ be a locally finite semigroup variety such that $\mathcal{V}_0$ (see Exercise 3.6.30) is finitely based. Then $\mathcal{V}$ is inherently non-finitely based if and only if it does not satisfy any nontrivial identity of the form $Z_n = W$.*

Proof. If $\mathcal{V}$ is a locally finite variety that does not satisfy nontrivial identities of the form $Z_n = W$, then any bigger variety cannot satisfy such identities either, and so by Theorem 3.5.1 $\mathcal{V}$ is not contained in a finitely based locally finite variety, that is $\mathcal{V}$ is inherently non-finitely based.

Suppose that $\mathcal{V}$ satisfies a nontrivial identity of the form $Z_n = W$. Since $\mathcal{V}$ is locally finite, it does not contain the infinite cyclic semigroup, so it satisfies some nontrivial identity of the form $x^m = x^{m+n}$ (why?) Let $v_1(x_1, \ldots, x_k) = 1, \ldots, v_\ell(x_1, \ldots, x_k) = 1$ be a basis of identities of $\mathcal{V}_0$. We may assume that the words $v_1, \ldots, v_\ell$ are positive (not containing inverses of letters) because of the identity $x^{-1} = x^{n-1}$ that is true in every group from $\mathcal{V}$. Consider the free semigroup F of $\mathcal{V}$ on k generators $x_1, \ldots, x_k$. Since F is a finite semigroup, it has a minimal idempotent. Let $v(x_1, \ldots, x_k)$ be a minimal idempotent in F. Then, vFv is a group by Exercise 3.1.11. Consequently, F satisfies the equalities $v_i(vx_1 v, \ldots, vx_k v) = v_i(vx_1 v, \ldots, vx_k v)^2$, $i = 1, \ldots, m$. These equalities are identities in $\mathcal{V}$ by Exercise 1.4.26. Consider the variety $\mathcal{V}_1$ defined by these identities, and the identities $x^m = x^{m+n}$ and $Z_n = W$. By construction $\mathcal{V} \subseteq \mathcal{V}_1$. By Theorem 3.3.4, all nil-semigroups in $\mathcal{V}_1$ are locally finite. The groups in $\mathcal{V}_1$ satisfy the identities $v_i = 1, i = 1, \ldots, m$ and therefore belong to $\mathcal{V}_0 \subset \mathcal{V}$. It follows that all groups in $\mathcal{V}_1$ are locally finite. Since A_2^1 and B_2^1 do not satisfy the identity $Z_n = W$ by Lemmas 3.6.22 and 3.6.27, every semigroup Y of $\mathcal{V}_1$ is a periodic semigroup without divisors isomorphic to A_2^1 or B_2^1. Then for every idempotent $e \in Y$, the monoid eYe does not have divisors isomorphic to A_2 or B_2. By Theorem 3.6.24, eYe is a semilattice of semigroups Y_α, $\alpha \in A$ each one of which is an ideal extension of a semigroup of the form $M(G_\alpha, I_\alpha, J_\alpha, P_\alpha)$ by a nil-semigroup. All these groups G_α belong to

$\mathcal{V}_0$ and are locally finite. By Exercise 3.6.21 all $M(G_\alpha, I_\alpha, J_\alpha, P_\alpha)$ are locally finite. Since all nil-semigroups in $\mathcal{V}_1$ are locally finite, by Corollary 3.4.3 applied several times (to ideal extensions and semilattices of semigroups), we conclude that each monoid eYe is locally finite. By Lemma 3.6.32, YEY is locally finite where E is the set of all idempotents of Y. Since Rees quotient Y/YEY is a nil-semigroup by Lemma 3.1.14 (all idempotents of Y are identified in this Rees quotient semigroup with zero), and belongs to $\mathcal{V}_1$, it is locally finite by Theorem 3.5.1. Applying Corollary 3.4.3 again, we conclude that Y is locally finite. Hence $\mathcal{V}_1$ is a finitely based locally finite variety containing $\mathcal{V}$, and $\mathcal{V}$ is not inherently non-finitely based.   □

Now we are in a position to give an algorithmic description of finite inherently non-finitely based semigroups. The following theorem gives in fact two algorithmic descriptions (parts (1) and (2) give one description, and Part (3) gives another one). The first one is easier to verify, the second one is easier to formulate. Recall that the *center* $Z(G)$ of a group G consists of all $x \in G$ such that $xy = yx$ for every $y \in G$. The series of normal subgroups $G_1 \le G_2 \le \dots$ of G is called the *upper central series* if $G_1 = Z(G), G_{i+1}/G_i = Z(G/G_i), i \ge 1$. The *upper hypercenter* of G is the union of all G_i in the upper central series of G. A group G is called *nilpotent of class n* if $G_n = G$.

Theorem 3.6.34 (Sapir [278]). *Let S be a finite semigroup.*

(1) *If S is inherently non-finitely based, then it contains an inherently non-finitely based subsemigroup of the form eSe, $e^2 = e$.*

(2) *Suppose S is a finite monoid and d is the period of S (that is S satisfies the identity $x^m = x^{m+d}$ for some m). The semigroup S is inherently non-finitely based if and only if the following property holds*

    (H) *for some idempotent e of S and some element a dividing e the elements eae and $ea^{d+1}e$ do not lie in the same coset of the maximal subgroup S_e with respect to the upper hypercenter $\Gamma(S_e)$.*

(3) *S is inherently non-finitely based if and only if S does not satisfy any nontrivial identity of the form $Z_n = W$ with $n \le |S|^3$.*

For each $n > 3$ we define a word

$$T_n(x, y, x_2, \dots, x_n)$$

by replacing in $Z_n(x, x_2, \dots, x_n)$ the occurrences number $2, 3, \dots, n+1$ of x by the letter y. For each identity $A = B$ we denote by $\alpha_n(A = B)$ the identity

$$Z_n(A, x'_2, \dots, x'_n) = T_n(A, B, x'_2, \dots, x'_n),$$

(the variables x'_i do not occur in A or B).

Lemma 3.6.35. *Suppose S is a semigroup and I is an ideal of S.uppose that I is a semigroup with zero multiplication. Suppose also that S/I satisfies an identity $A = B$. Then S satisfies the identity $\alpha_n(A = B)$ for any $n > 3$.*

Proof. Consider an arbitrary substitution λ of elements of S for the letters of the identity $\alpha_n(A = B)$. Put $a = \lambda(A)$ and $b = \lambda(B)$. If a or b does not belong to I, then $a = b$ by the assumption that S/I satisfies the identity $A = B$. Therefore, by definition of the word T_n, we have the equality $\lambda\alpha_n(A = B)$ (this expression means that if we apply λ to the left and right sides of the identity $\alpha_n(A = B)$, we obtain a true equality). If, however, $a, b \in I$, then the equality $\lambda\alpha_n(A = B)$ is true because I is a semigroup with zero multiplication (both sides of this equality are equal to zero in I). $\square$

We assume up to Lemma 3.6.41 that S is a finite semigroup and I is an ideal of S that is a completely simple semigroup $M(G, L, R, P)$ for some G, L, R, P.

Lemma 3.6.36. *If $a, b \in S$ and $eae = ebe$ for every idempotent e of I, then $xay = xby$ for every $x, y \in I$.*

Proof. For every x in I we denote by x^0 the identity element of a subgroup containing x. Suppose $xay \neq xby$ for some x, y in I. Since $x = xx^0$ and $y = y^0 y$, we have $x^0 a y^0 \neq x^0 b y^0$. Using the representation of I as a set of triples $L \times G \times R$, by Exercise 3.6.11, the elements $x^0 a y^0$ and $x^0 b y^0$ lie in a subgroup of I of the form $\{l\} \times G \times \{r\}$ (l and r are some elements of L and R respectively). By Exercise 3.6.12, $y^0 x^0 a y^0 x^0 \neq y^0 x^0 b y^0 x^0$. Therefore $(y^0 x^0)^0 a (y^0 x^0)^0 \neq (y^0 x^0)^0 b (y^0 x^0)^0$. Taking $e = (y^0 x^0)^0$, we obtain $eae \neq ebe$, which contradicts an assumption of the lemma. $\square$

Lemma 3.6.37. *Suppose that for some natural number p and any $a \in S$ and $x, y \in I$ we have $xay = xa^p y$. Suppose also that S/I satisfies an identity $A(x_1, \ldots, x_n) = B(x_1, \ldots, x_n)$ and B is obtained from A by replacing some occurrences of x_1 by x_1^p. Then S satisfies the identity $\alpha_k(A = B)$ for every $k > 3$.*

Proof. Consider a substitution λ of elements of S for the letters of the identity $\alpha_k(A = B)$. We represent the right side of the identity $\alpha_k(A = B)$ arbitrarily in the form $ux_1^p v$, where u and v are words. Recall that the word T_k is obtained from $Z_k(x, x_2, \ldots, x_k)$ by replacing k occurrences of x, beginning with the second one, by y. It follows that A is an initial segment of u and a terminal segment of v. We may assume $\lambda(A) \neq \lambda(B)$; hence $\lambda(A) \in I$. Consequently, $\lambda(u), \lambda(v) \in I$. Therefore, by hypothesis, $\lambda(ux_1^p v) = \lambda(ux_1 v)$. Continuing in this way, we can replace all occurrences of $\lambda(x_1^p)$ in the right side of the equality $\lambda\alpha_k(A = B)$ by $\lambda(x_1)$ so that each occurrence of $\lambda(B)$ is replaced by $\lambda(A)$. As a result, the right side of this equality becomes graphically equal to the left. $\square$

Lemma 3.6.38. *Suppose N is the upper hypercenter of G, k is divisible by the exponent of G (i.e., the minimal number ℓ such that $x^\ell = 1$ for every $x \in G$), and m is greater than the length of the upper central series of G. Then G satisfies the implication*

$$x = y \pmod{N} \to \alpha_k^m(x = y),$$

where $\alpha_k^m(x = y) = \alpha_k(\alpha_k(\ldots \alpha_k(x = y)\ldots))$.

Proof. Let C be the center of G. If $C = 1$, the lemma follows from the definition of α_k. Suppose $C \neq 1$. Using induction on the length of the upper central series of G, we may assume that in G satisfies the implication

$$x = y \pmod{N} \to \alpha_k^{m-1}(x = y) \pmod{C} \tag{3.6.1}$$

Let us denote the left and right sides of the identity $\alpha_k^{m-1}(x = y)$ by A and B. We replace the letters of the identity $\alpha_k(A = B)$ (or, what is the same, the letters of the identity $\alpha_k^m(x = y)$) by elements of G by means of the substitution λ. In view of (3.6.1) if $\lambda(x) = \lambda(y) \pmod{N}$, then $\lambda(A) = \lambda(B) \pmod{C}$. Then $\lambda(A) = \lambda(B)c$ for some $c \in C$, and

$$\lambda(T_k(A, B, x_2', \ldots, x_k')) = T_k(\lambda(A), \lambda(B), \ldots) = c^k T_k(\lambda(A), \lambda(A), \ldots)$$
$$= T_k(\lambda(A), \lambda(A), \ldots) = Z_k(\lambda(A), \lambda(x_2'), \ldots, \lambda(x_k')) = \lambda(Z_k(A, x_2', \ldots, x_k')),$$

as desired.   □

Up to Lemma 3.6.41 we fix a number k that is a multiple of the period of the semigroup S, and a number m that is greater than the length of the upper central series of G. Let N be a normal subgroup of G. We define a binary relation σ_N on S by setting

1. $(x, y) \in \sigma_N \leftrightarrow x = y$ or $x, y \in I$ and $x = (l, g, r), y = (l, h, r)$ and $g = h \pmod{N}$.

Exercise 3.6.39. Prove that σ_N is a congruence on S.

Lemma 3.6.40. *Let $N = \Gamma(G)$. If S/σ_N satisfies the identity $A = B$, then S satisfies the identity $\alpha_k^m(A = B)$.*

Proof. We replace each letter t of the identity $\alpha_k^m(A = B)$ by some element $\lambda(t)$ of S. Put $a = \lambda(A)$ and $b = \lambda(B)$. It suffices to consider the case where $a \neq b$. In this case $a, b \in I$ and $(a, b) \in \sigma$. Consequently, $a = (l, g, r)$ and $b = (l, h, r)$ for some l, r, g, h with $g = h \pmod{N}$. We denote the idempotent (l, p_{lr}^{-1}, r) by e (see Exercise 3.6.8). It is an identity element for a and b. Consider the substitution λ_1 that coincides with λ on the letters of the words A and B, and sends every other letter t to $e\lambda(t)e$. Since in the words Z_k the letter x_1 appears in all odd positions (Exercise 2.5.5) and since e is an identity element for a and b, the left (right) side of the equality $\lambda_1\alpha_k^m(A = B)$ is equal to the left (right) side of the equality $\lambda\alpha_k^m(A = B)$. It is easy to see that $\lambda_1(t) \in I_e = \{l\} \times G \times \{r\}$ for every letter t (prove it!). Since the group I_e is isomorphic to G and $g, h \in G$, $g = h \pmod{N}$, we can use Lemma 3.6.38 to conclude that the equality $\lambda_1 \alpha_k^m (A = B)$ is true. Consequently, the equality $\lambda \alpha_k^m (A = B)$ is also true. $\qquad\qquad\qquad\qquad\qquad\qquad\qquad\qquad\qquad\qquad\qquad\square$

Lemma 3.6.41. *Suppose S is a finite semigroup of period d, and suppose that for every idempotent e of S and every element a dividing e the elements eae and $ea^{d+1}e$ lie in the same coset of the upper hypercenter $\Gamma(S_e)$ in the group S_e. Then there exists a number $m_0 \leq |S|^2$, such that for any $m > m_0$ and any k that is a multiple of d, $k > 3$, the identity $\alpha_k^m (x = x^{d+1})$ is true in S.*

Proof. By Exercise 3.6.26, S possesses a chain of ideals $I_1 < I_2 < \cdots < I_n = S$, such that each of the quotients I_j / I_{j-1} is either a semigroup with zero multiplication, or a completely simple semigroup, or a completely simple semigroup with adjoined zero. Using induction on n, we may assume that we have already found a number $m_1 \leq |S/I_1|^2$ such that for any $m > m_1$ and any $k > 3$ that is a multiple of d, the identity $\alpha_k^m (x = x^{d+1})$ is true in S/I_1. If I_1 is a semigroup with zero multiplication, then, by Lemma 3.6.35, the identity $\alpha_k^{m+1} (x = x^{d+1})$ is true in S for any $m > m_1$, and as the desired number m_0 we can take, say, $m_1 + 1 \leq |S|^2$. Suppose I_1 is a completely simple semigroup. Then each element of S divides each element of I_1. Let N be the upper hypercenter of G. Consider the congruence $\sigma = \sigma_N$ defined before Exercise 3.6.39. Put $\bar{S} = S/\sigma$ and $\bar{I} = I_1/\sigma$. It is clear from the definition of σ that $\bar{S}/\bar{I} \cong S/I_1$. By hypothesis, for any idempotent $e \in \bar{I}$ and any $a \in \bar{S}$ we have $eae = ea^{d+1}e$. Then, by Lemma 3.6.36, $xay = xa^{d+1}y$ for every $x, y \in I$. Since the right side of the identity $\alpha_k^m (x = x^{d+1})$ is obtained from the left by replacing several occurrences of x by x^{d+1}, we can apply Lemma 3.6.37. It implies that $\bar{S}$ satisfies the identity $\alpha_k^m (x = x^{d+1})$. Therefore, in view of Lemma 3.6.40, S satisfies the identity $\alpha_k^{m+1+l} (x = x^{d+1})$, where l is any number greater than the length of the upper central series of a maximal subgroup of I_1. Thus, as the desired m_0 we can take, say, $m_1 + |S| \leq (|S| - 1)^2 + |S| \leq |S|^2$.

Now suppose that I_1 is a simple semigroup with an adjoined zero. Take $e = e^2 \in I_1$. Let J denote the set of elements of S not dividing e. It is clear that J is an ideal of S and that $I_1 \cap J = \{0\}$. Since $S < S/I_1 \times S/J$ (check it!), we may assume that $J = \{0\}$ (otherwise we use the induction assumption for S/I_1 and S/J). We will prove that S contains no *zero divisors*, i.e., non-zero elements a, b such that $ab = 0$. Suppose $a, b \in S \setminus \{0\}$. We have $e = u_1 a v_1 = u_2 b v_2$ for some u_1, v_1, u_2, v_2 in S^1. Consequently $e = u_2 b v_2 u_1 a v_1$, and hence $bv_2 u_1 a \neq 0$. By Part (3) of Exercise 3.6.26, $(bv_2 u_1 a)^2 = bv_2 u_1 (ab) v_2 u_1 a \neq 0$ hence $ab \neq 0$. Thus $S \setminus \{0\}$ is a subsemigroup. By the induction assumption, the identity $\alpha_k^m (x = x^{d+1})$ is true in $S \setminus \{0\}$ for any m greater than some $m_0 \leq (|S| - 1)^2$. The left and right sides of this identity consist of the same letters. Therefore it is true on S. Consequently this m_0 is as desired. $\qquad\qquad\square$

Remark 3.6.42. Note that by definition, the left side of the identity $\alpha_k(x = x^{d+1})$ is equal to $Z_k(x, x_2', \ldots, x_k')$. In view of Exercise 2.5.5, the left side of the identity $\alpha_k^m(x = x^{d+1})$ differs from some word Z_n only in the names of the variables. It is also obvious that this identity is nontrivial.

Exercise 3.6.43. Suppose that G is a group, $a, b, c, d \in G$, and $axb = cxd$ for all $x \in G$. Then ac^{-1}, bd^{-1} belong to the center of G. **Hint:** Take $x = 1$ to show that $c^{-1}a = db^{-1}$. Then for every x, $c^{-1}ax = xdb^{-1} = xc^{-1}a$. Hence $c^{-1}a = db^{-1}$ is in the center of G. Finally since ac^{-1} is a conjugate of $c^{-1}a$, we get $ac^{-1} = c^{-1}a$.

Lemma 3.6.44. *Suppose $I = M(G, L, R, P)$ is a finite completely simple semigroup and the center of the group G is trivial. Suppose also that $a, b, c, d \in I$ and $axb = cxd$ for every $x \in I$. Then $ax = cx$ and $xb = xd$ for every $x \in I$.*

Proof. The equality $axb = cxd$ and the definition of the Rees–Sushkevich semigroup $M(G, L, R, P)$ imply that ax and cx belong to the same maximal subgroup I_e of I, and xb and xd belong to another maximal subgroup, say I_f. Fix x and z. Suppose $ax = (i, g_1, j)$, $cx = (i, g_2, j)$, $zb = (s, h_1, q)$, and $zd = (s, h_2, q)$. Take any $y = (n, t, m) \in I$. We have $axyzb = cxyzd$, i.e., $(i, g_1 p_{nj} t p_{sm} h_1, q) = (i, g_2 p_{nj} t p_{sm} h_2, q)$. Hence

$$g_1 p_{nj} t p_{sm} h_1 = g_2 p_{nj} t p_{sm} h_2.$$

Therefore by Exercise 3.6.43, $g_1 g_2^{-1}$ and $h_1 h_2^{-1}$ belong to the center of G. But, by our assumption, the center of G is trivial; hence $g_1 = g_2$ and $h_1 = h_2$. Consequently $ax = cx, zb = zd$, as required.   □

Lemma 3.6.45. *Suppose that S is a periodic semigroup, $E = E(S)$ is the set of all idempotents of S, and for every e in E the semigroup eSe satisfies an identity $A(x_i, \ldots, x_n) = B(x_i, \ldots, x_n)$. Then the semigroup SES satisfies the identity*

$$xA(x_1 x, \ldots, x_n x)x = xB(x_1 x, \ldots, x_n x).$$

Proof. Indeed

$$uevA(x_1 uev, \ldots, x_n uev) = uA(evx_1 ue, \ldots, evx_n ue)v$$
$$= uB(evx_1 ue, \ldots, evx_n ue)v = uevB(x_1 uev, \ldots, x_n uev)$$

for every $e \in E, u, v, x_1, \ldots, x_n \in S$.   □

Proof of Theorem 3.6.34. We start with Part (2).

Let S be a finite monoid, d be the period of S. Suppose that S is not inherently non-finitely based and $|S|$ is the smallest among counterexamples to property (H) from Part (2) of the theorem. Then by Lemma 3.6.33 S satisfies a nontrivial identity of the form

$$Z_k(x_1, \ldots, x_k) = W(x_1, \ldots, x_k). \tag{3.6.2}$$

Since S is a monoid, removing x_1 from this identity gives another (shorter) identity of S of this form. So we can assume that removing x_1 from W produces $Z_{k-1}(x_2, \ldots, x_k)$.

By Lemmas 3.6.22 and 3.6.27 S does not have divisors of the form A_2 or B_2 (for if, say, B_2 was a divisor of the monoid S, then B_2^1 would be a divisor too – check it!). By Exercise 3.6.26, S possesses a series of ideals $I_1 < I_2 < \ldots < I_m = S$ each of whose quotients I_{j+1}/I_j is either a semigroup with zero multiplication, or a completely simple semigroup or a completely simple semigroup with zero adjoined. Take an idempotent $e \in S$ and an element a dividing e, that is $e \in SaS$, but

$$eae \neq ea^{d+1}e \pmod{\Gamma(S_e)}. \qquad (3.6.3)$$

By the minimality of $|S|$, we may assume that $e \in I_1$ and S does not contain 0: otherwise we take S/I_1 or $S \smallsetminus \{0\}$ instead of S. We may also assume that the center of S_e is trivial: otherwise we pass from S to S/σ_N where $N = \Gamma(S_e)$ (the congruence σ_N was defined before Exercise 3.6.39). Indeed, S/σ_N is not inherently non-finitely based as a quotient of S, and eae is not equal to $ea^{d+1}e$ modulo $\Gamma(G/N) = \{1\}$, but the number of elements in S/σ_N would be smaller than in S.

Take the smallest number k with the following property: there exists a word $U(x_1, \ldots, x_k)$ different from $Z_k(x_1, \ldots, x_k)$ and turning into $Z_{k-1}(x_2, \ldots, x_k)$ after deleting x_1, such that

$$pZ_k(x_1, \ldots, x_k)q = pU(x_1, \ldots, x_k)q \qquad (3.6.4)$$

for all $p, q \in I_1$ and $x_1, \ldots, x_k \in S$. Such a k exists in view of (3.6.2) and our assumptions about W (indeed, W satisfies all the needed conditions except having the smallest number of distinct letters).

Assume that $k > 1$. Then x_k occurs only once in U and U can be represented as $U_1 x_k U_2$ where U_1, U_2 do not contain x_k. By (3.6.4) we have

$$pZ_{k-1}x_k Z_{k-1}q = pU_1 x_k U_2 q$$

for all $p, q, x_1 \in I_1$ and $x_1, \ldots, x_{k-1} \in S$. By Lemma 3.6.44 we have

$$pZ_{k-1}q = pU_1 q, pZ_{k-1}q = pU_2 q.$$

Since U differs from Z_k, one of U_1, U_2 differs from Z_{k-1}, which contradicts the minimality of k.

Hence $k = 1$. Thus for any $p, q \in I_1$ and any $x \in S$ we have $pxq = px^l q$ where l is some natural number greater than 1 and not depending on p, q, x. Since S is finite of period d, it satisfies the identity $x^m = x^{m+d}$ for some m. Hence

$$eae = ea^l e = \ldots = ea^{l^m}e = ea^{l^m + d}e = ea^{l^m}(a^d e) = ea(a^d e) = ea^{d+1}e.$$

Here we used the fact that if $a^m = a^{m+d}$, then $a^{m'} = a^{m'+d}$ for every $m' > m$ (prove this property of finite cyclic semigroups!). We get a contradiction with (3.6.3).

This proves that if S is inherently non-finitely based, then property (H) holds.

Suppose that property (H) holds for a finite semigroup S (not necessarily a monoid). Then by Lemma 3.6.41, S satisfies the identity $\alpha_k^m(x = x^{d+1})$ for some k, m. This identity is nontrivial and its left-hand side differs from some Z_n only by names of variables by Remark 3.6.42. Hence S is not inherently non-finitely based by Lemma 3.6.33. This proves Part (2) of Theorem 3.6.34.

Now let us prove Part (1). Suppose that all subsemigroups of the form eSe are not inherently non-finitely based. We must show that S itself is not inherently non-finitely based. By the property (H), every submonoid eSe of S satisfies the assumptions of Lemma 3.6.41. By Lemma 3.6.41, eSe satisfies $\alpha_k^m(x = x^{d+1})$ for some k, m. We can of course choose k, m (big enough) independent of e. Therefore all submonoids eSe satisfy the same nontrivial identity of the form $Z_n = W$ by Remark 3.6.42. Then by Lemma 3.6.45, the identity

$$x Z_n(x_1 x, \ldots, x_n x) = x W(x_1 x, \ldots, x_n x)$$

is true in SES where E is the set of idempotents of S. The left-hand side of this identity is equal to $Z_{n+1}(x, x_1, \ldots, x_n)$ by Exercise 2.5.5 and differs from the right-hand side. Since S/SES is nilpotent by Lemma 3.1.16, it satisfies the identity $Z_k = 0$ for some k (say, $k = |S|$ is enough). Therefore S satisfies a nontrivial identity,

$$Z_{n+1}(Z_k, x_{k+1}, \ldots, x_{k+n}) = Z_k W(x_{k+1} Z_k, \ldots, x_{k+n} Z_k),$$

which completes the proof of Part (1).

Finally let us prove Part (3). If S satisfies a nontrivial identity of the form $Z_k = W$, then S is not inherently non-finitely based by Lemma 3.6.33. Suppose that S is not inherently non-finitely based. Then all its submonoids eSe are not inherently non-finitely based. By Lemma 3.6.41 each eSe satisfies $\alpha_k^m(x = x^{d+1})$, and we can take $k, m \leq |S|$. By Remark 3.6.42 the left-hand side of this identity differs from some Z_r, $r \leq |S|^2$, only by the names of the letters. Hence eSe satisfies a nontrivial identity of the form $Z_{|S|^2} = W$ (independent of e). By Lemma 3.6.45 then SES satisfies the nontrivial identity $Z_{|S|^2+1} = W'$. Since S/SES is nilpotent of class at most $|S|$, we obtain that S satisfies a nontrivial identity of the form $Z_{|S|^2+1+|S|} = W''$. It remains to note that $|S|^2 + 1 + |S| \leq |S|^3$ if $|S| > 1$. If $|S| = 1$, then S satisfies the identity $x = x^2$, and the condition of Part (1) holds also.   □

Remark 3.6.46. Note that by using a little bit more care we could lower the estimate in Property (3) of Theorem 3.6.34 from $|S|^3$ to $|S|^2$.

Remark 3.6.47. By Theorem 3.6.28 B_2^1 is inherently non-finitely based, and so every finite semigroup having B_2^1 as a divisor is inherently non-finitely based.

Note, though, that not every inherently non-finitely based finite semigroup S has B_2^1 in var S, examples have been found in [278]. Jackson [159] described all minimal finite inherently non-finitely based semigroups T. In other words he found a list of finite inherently non-finitely based semigroups so that for every inherently non-finitely based finite semigroup S one of the semigroups from that list is in var S. It turned out that any such list is infinite.

Nevertheless we have the following result, which we shall give here as an exercise. It follows from Theorem 3.6.34.

Exercise 3.6.48 (Sapir [278]). If all subgroups of a finite semigroup S are nilpotent, then S is inherently non-finitely based if and only if $B_2^1 \in$ var S (in fact if and only if B_2^1 is a divisor of $S \times S$).

3.7 Growth Functions of Semigroups

3.7.1 The Definition

Let S be a semigroup generated by a finite set A, and let f_A be the *growth function* of S with respect to A, that is $f_A(n)$ is the number of elements of S represented by words in A of length at most n (see Section 1.5). Clearly, $f_A(n)$ is bounded from above by the exponential function $|A|^n$. We say that the growth of S is *polynomial* if $f_A(n)$ is bounded from above by n^k for some $k < \infty$. The infimum of all such k is called the *Gelfand–Kirillov dimension* of S [107]. We say that growth is *exponential* if f_A is bounded from below by an exponential function $e^{\alpha n}$, $\alpha > 0$. In this case the number $\limsup_{n\to\infty} \sqrt[n]{f_A(n)}$ is called the *exponential growth rate of S* with respect to A. We say that the growth is *intermediate* if it is neither polynomial nor exponential. Since the growth functions of a semigroup relative to two different finite generating sets are equivalent (see Section 1.5), having polynomial (exponential, intermediate) growth does not depend on the choice of a finite generating set. Moreover in the case of polynomial growth, the Gelfand–Kirillov dimension does not depend on the choice of a finite generating set.

Example 3.7.1. Let S be the free commutative monoid with free generating set $A = \{a_1, \ldots, a_k\}$ (see Exercise 1.8.19). Then every element s of S is uniquely represented by a word of the form

$$a_1^{r_1} \ldots a_k^{r_k}, \tag{3.7.1}$$

$r_i \geq 0$. The sum $r_1 + r_2 + \ldots + r_k$ is the minimal length of a word representing s in S. Hence the number $f(n, k)$ of elements of length n satisfies $f(0, k) = 1$ for every $k \geq 1$, $f(n, 1) = n + 1$ and the relation

$$f(n, k) = f(n - 1, k) + f(n, k - 1)$$

for every $n \geq 1, k \geq 2$ (indeed, the first summand in the right-hand side counts words (3.7.1) with $r_1 > 0$ and the second summand counts words with $r_1 = 0$). These conditions completely determine the number $f(n, k)$ for all n, k. Note that the binomial coefficient $\binom{n+k}{k}$ satisfies the same conditions (prove it using Exercise 1.1.4). Hence $f(n, k) = \binom{n+k}{k}$ for all n, k. The right-hand side of this equality is a polynomial in n with the highest term $\frac{n^k}{k!}$ (see Exercise 1.1.1), so the growth function is polynomial and the Gelfand–Kirillov dimension of S is k.

Example 3.7.2. Let S be the free k-generated monoid with free generating set A, then all words of length $\leq n$ in A represent different elements of S, so

$$f_A(n) = \sum_{m=0}^{n} k^m = \frac{k^{n+1} - 1}{k - 1},$$

so the growth is exponential and the exponential growth rate is k.

There exists a natural *partial order* on growth functions, and in general on all non-decreasing functions $\mathbb{N} \to \mathbb{N}$:

$$f_1 \prec f_2 \iff \text{there exists } C > 1 \text{ such that for all } n \; f_1(n) \leq C f_2(Cn) + C.$$

So f_1 and f_2 are equivalent (in the sense of Section 1.5) if and only if $f_1 \prec f_2$ and $f_2 \prec f_1$ (prove it!).

Exercise 3.7.3. Prove that if a semigroup S_1 is a subsemigroup (homomorphic image) of a semigroup S_2 and both semigroups are finitely generated, then the growth function of S_1 does not exceed the growth function of S_2. **Hint:** If $S_1 = \langle X_1 \rangle$ is a subsemigroup of $S_2 = \langle X_2 \rangle$, show that the growth function of S_2 with respect to the generating set $X_1 \cup X_2$ cannot be smaller than the growth function of S_1 with respect to X_1, then use Exercise 1.5.1. If S_1 is a homomorphic image of $S_2 = \langle X_2 \rangle$, then use the fact that the image X_1 of X_2 generates S_1 and if an element $s \in S_2$ is represented by a word of length $\leq n$ in the alphabet X_2, then the image of s in S_1 is represented by a word of length $\leq n$ in X_1. Moreover every element of S_1 that is represented by a word of length $\leq n$ in X_1 is a homomorphic image of an element of S_2 that can be represented by a word of length $\leq n$ in X_2.

Hence every subsemigroup of a semigroup with polynomial growth has polynomial growth (and the Gelfand–Kirillov dimension of a subsemigroup does not exceed the Gelfand–Kirillov dimension of the ambient semigroup), and every semigroup containing a subsemigroup of exponential growth has exponential growth itself.

3.7.2 Chebyshev, Hardy–Ramanujan and Semigroups of Intermediate Growth

Although it is not obvious that semigroups of intermediate growth exist, here we present relatively easy constructions of such semigroups. For this we need two classical results from outside of Combinatorial Algebra.

3.7.2.1 The Chebyshev Theorem About Primes

For every natural number n let $\pi(n)$ denotes the number of primes $\leq n$. Say, $\pi(1) = 0$, $\pi(2) = 1$, $\pi(10) = 4$, etc. The next theorem was proved by Chebyshev in 1850. We present a proof based on some ideas of Erdös (he used these ideas to prove another Chebyshev theorem that for every natural n there is a prime between n and $2n$).

Theorem 3.7.4. *There exists a constant $c > 0$ such that for every natural number n we have*

$$\frac{cn}{\ln n} \leq \pi(n) \leq n - 1.$$

Remark 3.7.5. In fact as was proved by Hadamard and de la Vallee-Poussin in 1890 [145], the limit $\lim_{n \to \infty} \frac{\pi(n)}{n/\ln n}$ is equal to 1.

Proof of Theorem 3.7.4. The inequality $\pi(n) \leq n - 1$ is obvious, so we should only prove the inequality $\frac{cn}{\ln n} \leq \pi(n)$. For every natural number m consider the middle binomial coefficient $\binom{2m}{m}$ in the Newton expansion of $(1 + 1)^{2m} = 4^m$.

Exercise 3.7.6. Prove that $\binom{2m}{m} \geq \frac{4^m}{2m+1}$. **Hint:** Use the Pascal triangle formula (see Exercise 1.1.4) to prove that the middle binomial coefficient is the biggest one. Then use the fact that the sum of the binomial coefficients is 4^m (by Exercise 1.1.1), and the number of them is $2m + 1$.

For a prime p and a number m we define $o_p(m)$ as the largest exponent of p that divides m. Note that $o_p(m_1 m_2) = o_p(m_1) + o_p(m_2)$ and if m_1 divides m_2, then $o_p(\frac{m_1}{m_2}) = o_p(m_1) - o_p(m_2)$ (why?).

Exercise 3.7.7. For every natural number m and prime p we have

$$o_p(m!) = \sum_{i=1}^{r} \left\lfloor \frac{m}{p^i} \right\rfloor$$

where r is such that $p^r \leq m \leq p^{r+1}$. **Hint:** There are $\left\lfloor \frac{m}{p} \right\rfloor$ numbers $\leq m$ that are divisible by p, $\left\lfloor \frac{m}{p^2} \right\rfloor$ numbers $\leq m$ that are divisible by p^2, etc.

Lemma 3.7.8. *If* $k = o_p\left(\binom{2m}{m}\right) > 0$, *i.e.,* p *divides* $\binom{2m}{m}$, *then* $p^k \leq 2m$.

Proof. Let r be the natural number such that $p^r \leq 2m < p^{r+1}$. Then by Exercise 3.7.7 $k = o_p\left(\binom{2m}{m}\right) = o_p((2m)!) - 2o_p(m!) = \sum_{i=1}^{r}\left\lfloor\frac{2m}{p^i}\right\rfloor - 2\sum_{i=1}^{r}\left\lfloor\frac{m}{p^i}\right\rfloor = \sum_{i=1}^{r}\left(\left\lfloor\frac{2m}{p^i}\right\rfloor - 2\left\lfloor\frac{m}{p}\right\rfloor\right) \leq r$ because for every positive real number x we have $\lfloor 2x \rfloor - 2\lfloor x \rfloor \leq 1$ (why?). Therefore $p^k \leq p^r \leq 2m$. $\square$

Now let us deduce the theorem from this lemma. Since $\pi(n+1) \leq \pi(n)+1$ for every n, we can assume that $n = 2m$ is even. Let $k = \pi(2m)$. Every prime that divides $\binom{2m}{m}$ does not exceed $2m$ (why?). Therefore $\binom{2m}{m} = \prod_{i=1}^{\ell} p_i^{a_i}$, where all primes p_i are different and $p_i \leq 2m$, so $\ell \leq k$. Hence

$$\ln\binom{2m}{m} = \sum a_i \ln p_i.$$

By Lemma 3.7.8, each $a_i \ln p_i$ does not exceed $\ln 2m$. Hence by Exercise 3.7.6

$$\ell \geq \frac{\ln\binom{2m}{m}}{\ln(2m)} \geq \frac{\ln\frac{4^m}{2m+1}}{\ln(2m)} \geq c\frac{2m}{\ln(2m)}$$

for some $c > 0$. Thus $\pi(2m) \geq c\frac{2m}{\ln(2m)}$. $\square$

3.7.2.2 The Hardy–Ramanujan Theorem About Partitions

Let n be a natural number. A representation of n as a sum $k_1 + k_2 + \ldots + k_m$ where $k_1 \geq k_2 \ldots \geq k_m$ is called a *partition* of n. The number $p(n)$ of different partitions of n is the Hardy–Ramanujan partition function [144].

The famous Hardy–Ramanujan partition formula [308] gives that

$$p(n) \sim \frac{1}{4n\sqrt{3}}e^{\pi\sqrt{\frac{2n}{3}}},$$

i.e., the quotient of $p(n)$ by the right-hand side of this formula tends to 1 as $n \to \infty$. We shall use the following much easier inequality.

Theorem 3.7.9. *For every natural* n,

$$p(n) \leq 2n^{3\sqrt{n}}. \tag{3.7.2}$$

Proof. First consider the partitions where every summand does not exceed $\sqrt{n}$. The number of summands is at most n, so the number of such partitions does not exceed $n^{\sqrt{n}}$. Now consider the partitions with a summand $> \sqrt{n}$. The number of possible maximal summans of such a partition does not exceed n. If we remove a biggest summand a, we get a partition of $n - a$. Therefore

$$p(n) \leq np(n - \lfloor\sqrt{n}\rfloor) + n^{\sqrt{n}}.$$

Exercise 3.7.10. Complete the proof of the theorem.

□

3.7.2.3 A Semigroup of Matrices

The semigroup from the next theorem was found by Okniński, the proof is due to Nathanson [244].

Theorem 3.7.11. *Let S be the semigroup of 2×2-matrices (under multiplication) generated by the set $A = \{a, b\}$ where*

$$a = \begin{pmatrix} 1 & 1 \\ 0 & 1 \end{pmatrix}, b = \begin{pmatrix} 1 & 0 \\ 1 & 0 \end{pmatrix}.$$

Then S has intermediate growth.

Proof. Let us observe that S satisfies the following relations:

$$b^2 = b, ba^k b = (k + 1)b.$$

Therefore for any positive integers $k_1, \ldots, k_r$ we have

$$ba^{k_1} ba^{k_2} b \cdots ba^{k_r} b = \prod_{i=1}^{r} ba^{k_i} b = \left(\prod_{i=1}^{r}(k_i + 1)\right) b$$

Let $p_1, \ldots, p_r$ be distinct prime numbers not exceeding $\sqrt{n}$ for some $n \geq 1$. Then

$$ba^{p_1-1} ba^{p_2-1} b \cdots ba^{p_r-1} b = \left(\prod_{i=1}^{r} p_i\right) b. \qquad (3.7.3)$$

The length of that element does not exceed

$$\sum_{i=1}^{r} p_i + 1 \leq r\sqrt{n} + 1 \leq \sqrt{n}\pi(\sqrt{n}) + 1$$

where $\pi(m)$ is the number of primes not exceeding m. By Theorem 3.7.4

$$\frac{cn}{\ln n} \leq \pi(n) \leq n - 1$$

for some constant $c > 0$. The length of the element in (3.7.3) does not exceed

$$\sqrt{n}(\sqrt{n} - 1) + 1 \leq n$$

for every $n \geq 1$. All the elements (3.7.3) are distinct since every integer uniquely decomposes as a product of primes [145], and so every subset of primes not exceeding $\sqrt{n}$ gives a different element of S of length at most n. This gives the superpolynomial lower bound for the growth function:

$$f_A(n) \geq 2^{\pi(\sqrt{n})} \geq 2^{\frac{2c\sqrt{n}}{\ln n}}.$$

To compute the upper bound let $s \in S$ and the length of the shortest word in a, b representing s be at most n. Since we are finding a super-polynomial upper bound for the number of such elements, we can ignore a few (bounded number) of sets with polynomial number of elements. Thus we can ignore elements of the form a^k, $k = 1, \ldots, n$ and $a^u b a^v$ with $u + v \leq n - 1$. Hence we can assume that s has the form $a^u s' a^v$, where

$$s' = ba^{k_1} ba^{k_2} \cdots ba^{k_r} b$$

where $u+v \leq n$, $r, k_1, \ldots, k_r$ are positive integers such that $k_1 + \ldots + k_r + r + 1 \leq n$. By (3.7.3), every permutation of k_i's gives the same element, hence we can assume that $k_1 \geq \ldots \geq k_r \geq 1$. With every such element s' we associate the following *partition* of the natural number n:

$$n = (k_1 + 1) + \ldots + (k_r + 1) + 1 + \ldots + 1$$

where the number of 1's at the end is

$$n - \sum_{i=1}^{r} k_i - r \geq 0.$$

Thus we have a surjective map from the set of all partitions of n to the set of elements s' as above.

Theorem 3.7.9 gives $n^2 e^{\sqrt{n} \ln n}$ as an upper bound for the growth function up to the equivalence. Thus S has intermediate growth.   □

Here is a syntactic representation of S. Consider the semigroup S' defined by the following presentation

$$\mathrm{sg}\langle a, b \mid b^2 = b, \ ba^m ba^n b = ba^{mn+m+n} b, \ m, n \in \mathbb{N} \rangle.$$

The identity map $a \mapsto a$, $b \mapsto b$ extends to a homomorphism from S' to S (since the relations of S' obviously hold in S). It is easy to see that the rewriting system $b^2 \to b$, $ba^m ba^n b \to ba^{mn+m+n} b$, $m, n \in \mathbb{N}$ is Church–Rosser (prove it!). The canonical words have one of the forms b, a^u $(u \geq 1)$, $a^u ba^v$ $(u + v \geq 1)$, $a^u ba^v ba^w$ $(v \geq 1)$. It is easy to see that these words represent different elements in S (prove it!). Hence the homomorphism $S' \to S$ is an isomorphism.

Note that a similar semigroup

$$S'' = \mathrm{sg}\langle a, b \mid b^2 = b, \ ba^m ba^n b = ba^n ba^m b, \ m, n \in \mathbb{N} \rangle$$

was considered by Belyaev, Sesekin and Trofimov in [41]. They proved that S'' has intermediate growth also. The semigroup S is clearly a quotient of S'', and the estimation of growth from the above for S'' is similar to the estimation from the proof of Theorem 3.7.11. An exact estimate (up to equivalence) of the growth functions of S and S'' was found in [193]. It turned out that the growth function of S is equivalent to $2^{\sqrt{n/\ln n}}$ and the growth function of S'' is equivalent to $2^{\sqrt{n}}$.

Okniński and others (see [64, 253, 254] and the references therein) studied other finitely generated matrix semigroups over fields. In fact Okniński [253] found the first matrix semigroup of intermediate growth. Cedó and Okniński formulated the following

Conjecture 3.7.12. *A finitely generated semigroup of matrices over a field has exponential growth if and only if it contains a free subsemigroup of rank at least 2.*

The "if" part of the conjecture is obvious, but the "only if" part is still open. They proved the conjecture in the case when the sizes of matrices are at most 4. Note that Okniński and Salwa proved the following related result:

Theorem 3.7.13 (Okniński, Salwa [255]). *A (not necessarily finitely generated!) semigroup of matrices over a finitely generated field contains no free subsemigroups of rank at least 2 if and only if S satisfies a nontrivial semigroup identity.*

3.7.2.4 A Semigroup of Automatic Transformations

Another example of a semigroup of intermediate growth was found in [24]. It seems quite different but is in fact very similar.

Theorem 3.7.14 (Bartholdi, Reznykov, Sushchansky, [24]). *The semigroup of automatic transformations defined by the Mealy automaton on Figure 1.2 has growth function equivalent to $2^{\sqrt{n}}$.*

Proof. It turns out that the semigroup S defined by this Mealy automaton has the following syntactic description (presentation):

$$\mathcal{P} = \mathrm{mn}\langle\, x, y \mid x^2 = 1,\ y(xy)^p(yx)^p y^2 = y(xy)^p(yx)^p,\ p \geq 0 \,\rangle.$$

The proof is contained in the following four exercises.

Exercise 3.7.15. Prove that the semigroup S satisfies the relations from $\mathcal{P}$ if we replace x by M_{q_0} and y by M_{q_1}.

Exercise 3.7.16. Prove that the presentation $\mathcal{P}$ is Church–Rosser.

Exercise 3.7.17. Prove that the canonical words of the rewriting system $\mathcal{P}$ (i.e., all words that do not contain the left-hand sides of the relations from $\mathcal{P}$) are the empty word, the word x, and all words of the form

$$x^{\epsilon_1} y(xy)^{p_1} y(xy)^{p_2} y \cdots (xy)^{p_k} y(xy)^{p_{k+1}} x^{\epsilon_2} \qquad (3.7.4)$$

where $\epsilon_1, \epsilon_2 \in \{0,1\}$, $0 \le p_1 < p_2 < \ldots < p_k$, $p_{k+1} \ge 0$. Deduce that for every word w in x, y, its canonical form is not longer than w.

Exercise 3.7.18. If u, v are canonical words of the rewriting system $\mathcal{P}$, and M_u, M_v are the automatic transformations corresponding to these words (where x is replaced by M_{q_0}, y is replaced by M_{q_1}), then M_u, M_v are different. Thus there exists a word w such that $M_u(w) \ne M_v(w)$.

Now we see (using the second part of Exercise 3.7.17) that elements of S of length at most $n > 1$ are precisely the empty word, the word x, and the words (3.7.4) with $\epsilon_1 + \epsilon_2 + k + 2(p_1 + \ldots + p_{k+1}) \le n$. As in the proof of Theorem 3.7.11, this gives the desired estimates of the growth function.   □

3.8 Inverse Semigroups

3.8.1 Basic Facts About Inverse Semigroups

Recall that a semigroup with unary operation $^{-1}$ is called an inverse semigroup if it satisfies the identities $(x^{-1})^{-1} = x$, $xx^{-1}x = x$, $xx^{-1}yy^{-1} = yy^{-1}xx^{-1}$. The basic properties of inverse semigroups are formulated below as an exercise.

Exercise 3.8.1 (Basic properties of inverse semigroups). Show that the following properties of inverse semigroups hold.

(1) Every inverse semigroup contains an idempotent, idempotents form a commutative subsemigroup (i.e., a semilattice).
(2) A semigroup is inverse if and only if for every x there exists y such that $xyx = x$ (i.e., the semigroup is *regular*) and its idempotents commute.
(3) The monoid $B = \mathrm{mn}\langle a, b \mid ab = 1 \rangle$ is an inverse semigroup (it is called the *bicyclic semigroup*). The presentation of B is Church–Rosser. The normal forms are $b^m a^n$, $m, n \ge 0$. An element of B is an idempotent if and only if it is equal to $b^n a^n$ for some n. Every homomorphic image of B is isomorphic to either B itself or to a cyclic group. Show that B is the syntactic monoid of the Dyck language (see Exercise 1.8.53).
(4) Every semilattice of groups (in particular, every semilattice) is an inverse semigroup.
(5) A completely 0-simple semigroup $M = M^0(G, I, J, P)$ is inverse if and only if P is a matrix of 0's and 1's where every row and every column contains exactly one 1, i.e., P is obtained from the identity matrix by

permuting the rows and permuting the columns (in this case M is called a *Brandt semigroup over the group* G).

(6) If $S_1, S_2, \ldots$ are inverse semigroups with zero 0, so that $S_i \cap S_j = \{0\}$ for every $i \neq j$, then define an operation on the union $S = \bigcup S_i$: $a \cdot b = ab$, if a, b are in one of the S_i, and $a \cdot b = 0$ otherwise. Then S is an inverse semigroup (it is called the 0-*direct sum* of $S_1, S_2, \ldots$).

Here are three more advanced (but still not that difficult) exercises, see [269]. We say that an inverse semigroup S has *height* h if every chain of idempotents $e_1 < e_2 < \ldots$ (in the natural partial order, see Section 3.1) has at most h elements.

Exercise 3.8.2. Prove that an inverse semigroup is of height 1 if and only if it is a group. Show that an inverse semigroup of height > 1 has an ideal that is of height 1.

Exercise 3.8.3. Prove that an inverse semigroup with zero has height 2 if and only if it is a zero-direct sum of Brandt semigroups. Prove that every inverse semigroup S of height $h \geq 2$ with zero has an ideal I, which is of height 2, and such that S/I has height $h - 1$.

Exercise 3.8.4. Prove that an inverse semigroup does not contain the bi-cyclic semigroup as a subsemigroup if and only if every $\mathcal{J}$-class is $\{0\}$ or a Brandt semigroup with zero removed (a semigroup where all $\mathcal{J}$-classes are $\{0\}$ or completely 0-simple semigroups with zeroes removed is called *completely semisimple*).

3.8.2 Identities of Finite Inverse Semigroups, Zimin Words, and Subshifts

Here we shall prove

Theorem 3.8.5 (Sapir [280]). *Every finite inverse semigroup belongs to a locally finite finitely based variety of inverse semigroups, so it is not inherently non-finitely based.*

Lemma 3.8.6. *The Brandt semigroup B_2 considered as an inverse semigroup satisfies the following identity*

$$xyx = xyxyy^{-1} \tag{3.8.1}$$

Therefore every 0-direct sum of Brandt semigroups of arbitrary sizes over arbitrary groups satisfies this identity.

Exercise 3.8.7. Prove this lemma.

Let $\hat{Z}_n$ be the Zimin word Z_n without the last letter (which is x_1).

Lemma 3.8.8. *Every inverse semigroup S of height h satisfies the identity*

$$\hat{Z}_{h+1} = \hat{Z}_{h+1}x_1x_1^{-1} \tag{3.8.2}$$

Proof. Let us use induction on height h. Suppose first that S does not have a zero. If $h = 1$, then S is a group by Exercise 3.8.3 and (3.8.2) clearly holds. Thus we may assume that $h \geq 2$. Then by Exercise 3.8.3 S has an ideal I of height 1 that is a nontrivial group and S/I is of height at most h. Consider two homomorphisms ϕ, ψ of S. The homomorphism ϕ is the natural homomorphism $S \to S/I$. The homomorphism ψ is from S to I, it takes every $s \in S$ to ese where e is the identity element of I (show that ψ is a homomorphism). Then the homomorphism $\gamma: S \to S/I \times I$ that takes every $s \in S$ to $(\phi(s), \psi(s))$ is injective (prove that!). Therefore S is a subsemigroup of $S/I \times I$. The group I satisfies the identity (3.8.2). Thus it is enough to show that S/I satisfies this identity.

Since S/I has a zero, it is suffices to assume that S has a zero and height ≥ 2.

Then by Exercise 3.8.3 S has an ideal I of height 2 such that S/I has height at most $h - 1$.

By Exercise 3.8.3, I is a 0-direct sum of Brandt semigroups. By the induction hypothesis S/I satisfies $\hat{Z}_h = \hat{Z}_h x_1 x_1^{-1}$. Consider the word $\hat{Z}_{h+1}$. By the definition of Z_n abd the definition of an inverse semigroup we have

$$\hat{Z}_{h+1} \equiv Z_h x_{h+1} \hat{Z}_h = \hat{Z}_h x_1 x_{h+1} \hat{Z}_h = \hat{Z}_h x_1 (\hat{Z}_h x_1)^{-1}(\hat{Z}_h x_1) x_{h+1} \hat{Z}_h.$$

Let us take any homomorphism ϕ of the free inverse semigroup into S and let $a = \phi(\hat{Z}_h), b = \phi(x_1), c = \phi((\hat{Z}_h x_1)^{-1}(\hat{Z}_h x_1) x_{h+1})$. We have to prove

$$abca = abcabb^{-1}.$$

Suppose that $a \notin I$. Since the identity $\hat{Z}_h = \hat{Z}_h x_1 x_1^{-1}$ holds in S/I, we have $a = abb^{-1}$ hence

$$abca = abcabb^{-1}.$$

Suppose that $a \in I$. Then $bc \in I$ since $bc \in SaS$. By Lemma 3.8.6, I satisfies the identity $xyx = xyxyy^{-1}$. Therefore we have

$$abca = abca(bc)(bc)^{-1} = abcabcc^{-1}b^{-1} = abcabcc^{-1}b^{-1}(bb^{-1})$$
$$= abcabc(bc)^{-1}(bb^{-1}) = abcabb^{-1}.$$

Thus in both cases we obtained the desired equality $abca = abcabb^{-1}$. □

Now let us take a finitely generated inverse semigroup S with all subgroups locally finite, which satisfies identity (3.8.2). Suppose that S is infinite. Then there exists a geodesic uniformly recurrent bi-infinite word U. Let $A = A(n, U)$ be the number from Lemma 3.3.18.

Lemma 3.8.9. *Let u, v, w be consecutive subwords of U such that $|u| \geq A, |v| \geq A$, w may be empty. Then (3.8.2) implies the identity $uvwu = uvwuvv^{-1}$.*

Proof. Let p be the longest prefix of v such that the identity $uvwu = uvwupp^{-1}$ follows from (3.8.2). If $v = p$ we are done, so suppose that $v \neq p$. Let $v = paq$ for some letter a and word q. We have $uvwup = upaqwup$. Since upa is a subword in U and $|upa| \geq A$ by Lemma 3.3.18 there exists an endomorphism ϕ of the free semigroup such that $upa = u_1\phi(Z_n)$ and $\phi(x_1) = a$. Therefore $up = u_1\phi(\hat{Z}_n)$. Thus identity (3.8.2) implies the identity $up = upaa^{-1}$. Therefore we have identities $uvwu = uvwupp^{-1} = uvupaa^{-1}p^{-1} = uvwupa(pa)^{-1}$, which contradicts the maximality of the prefix p.   □

Remark 3.8.10. Consider the (finite) set of all subwords of U of length $2A$. Since U is uniformly recurrent, there exists a number B such that every subword of U of length B contains every subword in this set. Let u and v be any two subwords of U of lengths $\geq 2B$. Let a, b be the corresponding elements of S.

Lemma 3.8.11. *The elements a and b are $\mathcal{J}$-related, i.e., $a = c_1bd_1, b = c_2ad_2$ for some elements c_1, c_2, d_1, d_2 from S^1.*

Proof. It is enough to prove that (3.8.2) implies an identity of the form $u = svt$ for some words s, t. We may assume that the length of v is a multiple of A. Let us represent v as a product $v_1 \dots v_k$ of words of length A. For every $i = 1, \dots, k$ let $v(i) = v_1 \dots v_i$. Since u contains every subword of U of length $2A$, it may be represented as a product $u_1v_1v_2u_2$ and $|u_1| \geq B$. Let $i \geq 2$ be the biggest number such that the identity $u = u_1v(i)v(i)^{-1}v_1v_2u_2$ follows from (3.8.2). If $i = k$, then $u = u_1v(v^{-1}v_1v_2u_2)$ and we are done. Suppose $i < k$.

Since $|u_1| \geq B$, u_1 contains the subword v_iv_{i+1}. Therefore $u_1 = u_3v_iv_{i+1}u_4$ for some u_3, u_4 and

$$u_1v(i)v(i)^{-1}v_1v_2u_2 = u_3v_iv_{i+1}u_4v(i)v(i)^{-1}v_1v_2u_2.$$

By Lemma 3.8.9 the identity (3.8.2) implies the identity

$$v_iv_{i+1}u_4v(i) = v_iv_{i+1}u_4v(i)v_{i+1}v_{i+1}^{-1}.$$

Hence we can deduce identities

$$u = u_3v_iv_{i+1}u_4v(i)v_{i+1}v_{i+1}^{-1}v(i)^{-1}v_1v_2u_2 = u_1v(i+1)v(i+1)^{-1}v_1v_2u_2,$$

which contradicts the maximality of i.   □

Exercise 3.8.12. Prove that the identity (3.8.2) does not hold in the bicyclic semigroup.

Lemma 3.8.13. *Let S be an inverse semigroup with all subgroups locally finite. If S satisfies identity (3.8.2), then S is locally finite.*

Proof. By Exercise 3.8.12, S does not contain the bicyclic semigroup. Therefore S is completely semisimple, so every $\mathcal{J}$-class of S is a Brandt semigroup with zero removed. Suppose S is infinite and U is an irreducible bi-infinite uniformly recurrent word for S. Let number B be as in Remark 3.8.10. By Lemma 3.8.11 all subwords of U of length $\geq 2B$ represent elements of the same $\mathcal{J}$-class D of S. Let $R \subset S$ be the set of these words, I be the ideal generated by D, J be the (unique) maximal ideal in I that does not contain D. Then elements represented by words in R are in $I \smallsetminus J$. Then I/J is a Brandt semigroup whose maximal subgroup is a maximal subgroup of S. Hence I/J is locally finite by Exercise 3.6.21. Therefore words from R represent only finitely many elements of I/J. Hence they represent only finitely many elements in S. This contradicts the fact that U is irreducible.   □

Now we are ready to **prove Theorem 3.8.5.**

Proof. Let T be a finite inverse semigroup. Then all groups in the variety of inverse semigroups generated by S form a subvariety $\mathcal{V}$ generated by the maximal subgroups of S (see Exercise 3.6.31), so $\mathcal{V}$ is locally finite and by the Oates–Powell Theorem 1.4.33 is given by a finite number of identities $v_1 = 1, v_2 = 1, \ldots, v_m = 1$ in the class of all groups, where $v_i = v_i(x_1, \ldots, x_k)$ is a *group word* (that is a word over the alphabet $\{x_1, \ldots, x_k\}^{\pm 1}$). Consider the relatively free semigroup F of rank k in $\mathrm{var}\, T$. By Lemma 3.8.13, F is finite. So F has a minimal idempotent e (see Exercise 3.1.2). Let u be a word in the alphabet $x_1, \ldots, x_k$ that represents e in F. Then T satisfies the identity

$$u = u^2. \tag{3.8.3}$$

By Exercise 3.1.11, eFe is a subgroup of F. Hence T satisfies the identities

$$v_i(ux_1u, \ldots, ux_ku) = u, i = 1, \ldots, m \tag{3.8.4}$$

Every group satisfying identities (3.8.3), (3.8.4) satisfies all identities $v_i = 1$, $i = 1, \ldots, m$, and so belongs to $\mathcal{V}$ and is locally finite. By Lemma 3.8.8, T satisfies the identity (3.8.2). Thus T belongs to the variety $\mathcal{W}$ given by the identities (3.8.2), (3.8.3), (3.8.4). Now by Lemma 3.8.13 $\mathcal{W}$ is locally finite and Theorem 3.8.5 is proved.   □

3.8.3 Inverse Semigroups of Bi-rooted Inverse Automata

Here are some further nice examples of important inverse semigroups (see Margolis and Meakin [220]).

Let G be a group generated by a finite set X, and let Γ be the corresponding Cayley graph. We build an inverse monoid from Γ in the following way. Let $M_X(G)$ be the set of all pairs (Ψ, g) where Ψ is a finite connected subgraph of Γ, $g \in G$, and Ψ contains the vertices 1 and g. With every such Ψ and every $h \in G$, we can define the translation of Ψ by h, the sub-graph $h\Psi$ in the natural way (just multiply all vertices by h, and use the fact that Γ is the right Cayley graph, so if $g \to gx$ is an edge, then $hg \to hgx$ is also an edge). We can view each pair (Ψ, g) from $M_X(G)$ as a rooted inverse automaton with input vertex 1 and output vertex g. Since (Ψ, g) has just one input vertex and one output vertex, it is called a *bi-rooted automaton*. We define the multiplication operation on $M_X(G)$ as follows:

$$(\Psi_1, g_1) \cdot (\Psi_2, g_2) = (\Psi_1 \cup g_1 \cdot \Psi_2, g_1 g_2),$$

where $\Gamma \cup g \cdot \Delta$ simply denotes the union of the corresponding subgraphs of Γ.

We also define the inverse by $(\Psi, g)^{-1} = (g^{-1}\Psi, g^{-1})$.

Exercise 3.8.14. Show that $M_X(G)$ with these operations is an inverse monoid. Show that the idempotents of $M_X(G)$ are precisely all the pairs $(\Psi, 1)$, i.e., the bi-rooted automata were input and output vertices coincide.

Exercise 3.8.15. Show that the map $(\Psi, g) \mapsto g$ is a homomorphism from $M_X(G)$ onto G such that the preimage of 1 is the set of all idempotents of $M_X(G)$. Thus $M_X(G)$ is an *E-unitary cover* of G.

Exercise 3.8.16 (Munn, [241]). Let $F = \mathrm{gp}\langle X \rangle$ be a free group. Show that then $M_X(F)$ is the *free inverse monoid* whose free generators are the pairs (e, x), $x \in X$, where e is an edge labeled by x with tail 1. **Hint:** First show that $M_X(F)$ is indeed generated by the pairs (e, x) as above, the inverse of (e, x) is $(e^{-1}x)$. Let S be an inverse semigroup generated by the set X. For every bi-rooted automaton (Ψ, g) let p be any path in Ψ that starts at 1, ends at g and passes through every edge in at least one direction at least once. Let $\mathrm{Lab}(p)$ be the label of that path and $\psi(p)$ be the element of S represented by $\mathrm{Lab}(p)$ in S. Show (using the fact that the underlying graph of Ψ is a tree) that $\psi(p)$ does not depend on the choice of p, and depends only on (Ψ, g). Then show that ψ coincides with ϕ on the bi-rooted automata (e, x), and is a homomorphism from $M_X(F)$ to S.

For more on *E*-unitary covers and semigroups of bi-rooted automata see the survey by Margolis, Meakin, Sapir [221] and the references therein.

3.9 Subshifts and Automata. The Road Coloring Problem

We have seen that subshifts are useful for studying semigroups (Section 3.3.2). Here we show that semigroups (more precisely, semigroup acts) are useful in studying subshifts.

3.9.1 Synchronizing Automata

A complete automaton $\mathcal{A} = (Q, A)$, from the algebraic viewpoint, is an algebra
with $|A|$ unary operations defined on the set Q since we can view $\mathcal{A}$ as the
Cayley graph of the free monoid A^* acting on Q. This means that we may
apply basic algebraic notions presented in Section 1.4 to complete automata.

Let (Q, A) be the Cayley graph of a free semigroup A^+ acting on a finite
set Q. Then every word w in A^* corresponds to a function $q \mapsto q \cdot w$ from
Q to Q. The automaton (Q, A) is called *synchronizing* if there exists a word
$w \in A^+$ such that the corresponding function is constant, that is all paths
labeled by w in (Q, A) have the same terminal vertex. Any word w with this
property is said to be a *reset* word for the automaton.

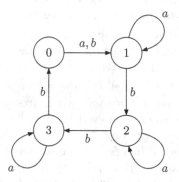

Figure 3.1 The automaton $\mathcal{C}$

Exercise 3.9.1. Figure 3.1 shows a synchronizing automaton with 4 vertices[1]
denoted by $\mathcal{C}$. Prove that the word ab^3ab^3a resets the automaton, the corresponding transformation takes every vertex to 1. Prove that ab^3ab^3a is the
shortest reset word for $\mathcal{C}$.

The notion of a synchronizing automaton came from real life applications
of automata in E. Moore [236] and Ginsburg [111]. In these applications,
finite automata served as a mathematical model of discrete devices such as
computers or relay control systems. The problem they tried to solve was:
how can we return the device to an appropriate state if we do not know
its current state and cannot observe outputs produced by the device under
various control actions (think of a satellite that circles around the Moon
and can be controlled from the Earth, but the result of executing control
commands cannot be observed while the satellite is "behind" the Moon.)?

---

[1] Here and below we adopt the convention that edges having multiple labels represent
bunches of parallel edges. In particular, the edge $0 \xrightarrow{a,b} 1$ in Figure 3.1 represents the
two parallel edges $0 \xrightarrow{a} 1$ and $0 \xrightarrow{b} 1$.

3.9.2 The Road Coloring Problem

Recall that for every complete automaton (Q, A), in the underlying graph Γ, each vertex has the same out-degree $k = |A|$.

Given a graph $\Gamma = (V, E)$ where all vertices have the same out-degree k, one can ask whether there exists an action of $mn\langle A \rangle$ on V such that Γ is the underlying graph of the corresponding complete automaton, and the automaton satisfies some nice properties. The *road coloring problem* is certainly the most famous question within this framework. It asks under which conditions on Γ, one of the automata (Q, A) with underlying graph Γ admits a reset word.

The name of the problem suggested in [6] comes from the following interpretation. Imagine a city with many streets and street intersections (exactly two streets meet at every intersection). We would like to help a traveler who is lost in the city find his/her way to a prescribed destination (the City Hall or a hospital). For this, we color the blocks of the streets between intersections in certain colors. The result is an automaton whose underlying graph is the map of the city and the alphabet - the colors of the street blocks. Now if there was a reset word, we could put it on the map, so that the traveler can follow the word and arrive at a prescribed destination starting at any point in the city. Moreover, if there are several travelers at different street corners in the city and they start following the reset word at the same time walking with the same speed, then all of them will meet at the same time at the hospital (or City Hall). Of course that problem is not very practical because, for example, the reset word can be very long, and paths following the reset word could visit the same street intersection many times before the traveler arrives at the hospital. The problem actually originated in a serious study of subshifts by Adler, Goodwyn and Weiss [6]; in an implicit form it was present already in an earlier book by Adler and Weiss [8]. As Adler, Goodwyn and Weiss, we shall consider only strongly connected graphs (this is not a very restrictive assumption, see Volkov [327].)

Example 3.9.2. The graph in Figure 1.1 admits a synchronizing labeling—in fact, each of the two of its labelings shown in Figure 1.1 is synchronizing. Moreover, it can be shown that **every** complete automaton with that underlying graph is synchronizing.

Exercise 3.9.3. In contrast to Example 3.9.2, the graph shown in Figure 3.2 admits both synchronizing and non-synchronizing labelings. Find such labelings, and for the synchronizing one, construct a reset word of minimum length that leads to the vertex 0.

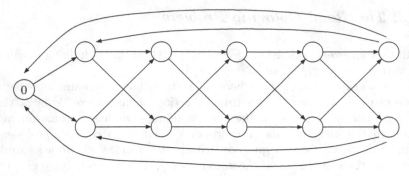

Figure 3.2 Graph admitting a synchronizing labeling

Exercise 3.9.4. (1) Find a synchronizing labeling of the Cayley graph of the group of all permutations of $\{1,2,3\}$ with respect to the generating set consisting of permutations

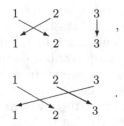

(2) Do the same for the Cayley graph of the Dihedral group D_{2n} for odd n with respect to its generating set introduced in Exercise 1.8.22.

(3) Is there a synchronizing labeling for the Cayley graph of D_{2n} (with respect to the generating set from Exercise 1.8.22) if n is even?

The following necessary condition was found in [6]:

Proposition 3.9.5. *If a strongly connected graph $\Gamma = (V, E)$ admits a synchronizing labeling, then the greatest common divisor of lengths of all cycles in Γ is equal to* 1.

Proof. By contradiction, let $k > 1$ be a common divisor of lengths of the cycles in Γ. Take a vertex $v_0 \in V$ and, for $i = 0, 1, \ldots, k-1$, let

$$V_i = \{v \in V \mid \text{there exists a path from } v_0 \text{ to } v \text{ of length } i \pmod{k}\}.$$

Clearly, $V = \bigcup_{i=0}^{k-1} V_i$. We claim that $V_i \cap V_j = \varnothing$ if $i \neq j$.

Let $v \in V_i \cap V_j$ where $i \neq j$. This means that in Γ there are two paths from v_0 to v: of length $\ell \equiv i \pmod{k}$ and of length $m \equiv j \pmod{k}$. Since Γ is strongly connected, there exists also a path from v to v_0. Let n be its length. Combining that path with the two paths above we get a cycle of length $\ell + n$

and a cycle of length $m + n$. Since k divides the length of any cycle in Γ, we have $\ell + n \equiv i + n \equiv 0 \pmod{k}$ and $m + n \equiv j + n \equiv 0 \pmod{k}$, hence $i \equiv j \pmod{k}$, a contradiction.

Thus, V is a disjoint union of $V_0, V_1, \ldots, V_{k-1}$, and by the definition each edge in Γ leads from V_i to $V_{i+1 \pmod{k}}$. Then Γ definitely cannot be converted into a synchronizing automaton by any labeling of its edges: no paths of the same length ℓ originating in V_0 and V_1 can terminate in the same vertex because they end in $V_{\ell \pmod{k}}$ and in $V_{\ell+1 \pmod{k}}$ respectively.   □

Graphs satisfying the conclusion of Proposition 3.9.5 are called *primitive*. Adler, Goodwyn and Weiss [6] conjectured that primitivity not only is necessary for a graph to have a synchronizing labeling but also is sufficient. In other word, they suggested the following

Conjecture 3.9.6 (Road coloring conjecture). *Every strongly connected primitive graph with constant out-degree admits a synchronizing labeling.*

The road coloring conjecture was finally proved by Trahtman [317]. But the first step was done by Culik, Karhumäki and Kari [76]. They defined the *confluence relation* $\sim$ on an automaton Q as follows:

$$q \sim q' \iff \text{for all } u \in A^* \text{ there exists } v \in A^* \text{ such that } q \cdot uv = q' \cdot uv.$$

Any pair of vertices (q, q') such that $q \neq q'$ and $q \sim q'$ is called *confluent*.[2]

Exercise 3.9.7. Prove that the confluence relation on an automaton (Q, A) is an equivalence relation, and a congruence relation on the automaton considered as an algebra with $|A|$ unary operations corresponding to the labels of edges.

Proposition 3.9.8 (Culik, Karhumäki and Kari [76]). *If every strongly connected primitive graph with constant out-degree and more than one vertex has a labeling with a confluent pair of vertices, then the road coloring conjecture is true.*

Proof. Let Γ be a strongly connected primitive graph with constant out-degree. We show that Γ has a synchronizing labeling by induction on the number of vertices in Γ. If Γ has only one vertex, there is nothing to prove. If Γ has more than one vertex, then, by the assumption, it admits a labeling with a confluent pair of vertices by the letters of some alphabet A. Let $\mathcal{A}$ be the automaton resulting from this labeling. By Exercise 3.9.7, the confluence relation is a congruence of $\mathcal{A}$. Since the congruence is nontrivial, the quotient automaton $\mathcal{A}/\sim$ has fewer vertices than $\mathcal{A}$. Since the quotient graph of every strongly connected graph is again strongly connected (check it!), the underlying graph $\Gamma/\sim$ of $\mathcal{A}/\sim$ is strongly connected. Moreover, since each cycle in Γ induces a cycle of the same length in $\Gamma/\sim$, the latter graph is primitive

---

[2] It is somewhat similar to the notion of confluence for rewriting systems, see 1.7.2.

as well. Therefore, by the induction assumption, the graph $\Gamma/\!\sim$ admits a synchronizing labeling.

We lift this labeling to a labeling of Γ in the following natural way. If $p \to q$ is an edge in Γ, let $[p], [q]$ denote the $\sim$-congruence classes of p, q. If in the new labeling of the quotient graph, the edge $[p] \to [q]$ has label a', then we relabel $p \to q$ by a' as well. An important feature of this relabeling procedure is that it is consistent with the confluence relation $\sim$ in the following sense. Suppose $p \xrightarrow{a} q$ and $p' \xrightarrow{a} q'$ are two transitions with the same label in $\mathcal{A}$ such that $p \sim p'$ and $q \sim q'$. Then $[p] = [p']$, $[q] = [q']$ and the two transitions induce the same transition $[p] \xrightarrow{a} [q]$ in $\mathcal{A}/\!\sim$. If it is being recolored to $[p] \xrightarrow{a'} [q]$ for some $a' \in A$, then the two transitions are being changed in the same way such that the resulting transitions $p \xrightarrow{a'} q$ and $p' \xrightarrow{a'} q'$ still have a common label. This implies, in particular, that if two vertices are in the same equivalence class of $\sim$ in the old labeling, then they are in the same class of $\sim$ in the new labeling (why?).

Let $\mathcal{A}'$ be the automaton resulting from the lifted labeling; we will show that $\mathcal{A}'$ is synchronizing.

Let w be a reset word for the synchronizing labeling of $\Gamma/\!\sim$. Then w maps the set of all vertices to a set S that is contained in a single congruence class of $\sim$. Let us order the vertices in S arbitrarily: $S = \{x_1, x_2, \ldots, x_n\}$. Since $x_1 \sim x_2$, there exists a word v_1 such that $x_1 \cdot v_1 = x_2 \cdot v_1$. Then $S \cdot v_1$ is again inside a congruence class of $\sim$ and there exists a word v_2 such that $(x_2 \cdot v_1) \cdot v_2 = (x_3 \cdot v_1) \cdot v_2$. Then consider $S \cdot v_1 v_2$ and the pair of vertices $x_3 \cdot v_1 v_2$, $x_4 \cdot v_1 v_2$. Continuing in that manner, we will find words $v_1, \ldots, v_{n-1}$ such that $S \cdot v_1 v_2 \ldots v_{n-1}$ consists of one vertex (see Figure 3.3). Therefore the word $w v_1 v_2 \cdots v_{n-1}$ is a reset word for $\mathcal{A}'$.

□

Using Proposition 3.9.8, Kari proved [172] the road coloring conjecture for complete automata with the same in-degree and out-degree of every vertex. For Trahtman's proof of the complete road coloring conjecture we need to concentrate on the action of the cyclic semigroup generated by a single letter. For this, we need some auxiliary notions and results.

Let $\mathcal{A} = (Q, A)$ be a complete automaton. A pair (p, q) of distinct vertices is *compressible* if $p \cdot w = q \cdot w$ for some $w \in A^*$; otherwise it is *incompressible*. A subset $P \subseteq Q$ is said to be *compressible* if P contains a compressible pair and to be *incompressible* if every pair of distinct vertices in P is incompressible. Clearly, if P is incompressible, then for every word $u \in A^*$, the set $P \cdot u = \{p \cdot u \mid p \in P\}$ also is incompressible and $|P| = |P \cdot u|$.

Lemma 3.9.9. *Let P be an incompressible set of vertices of maximum size in a complete automaton $\mathcal{A} = (Q, A)$. Suppose that there exist a word $w \in A^*$ and a vertex $q \in P$ such that $q \cdot w \neq q$ but $p \cdot w = p$ for all $p \in P' = P \smallsetminus \{q\}$. Then the pair $(q, q \cdot w)$ is confluent.*

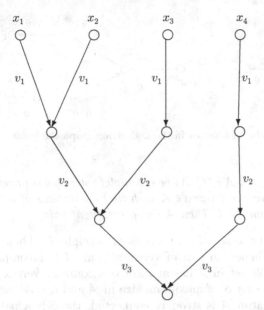

Figure 3.3 Synchronizing a ~-class

Proof. Let $q' = q \cdot w$. Take an arbitrary word $u \in A^*$; we have to show that $q \cdot uv = q' \cdot uv$ for a suitable word $v \in A^*$. Clearly, we may assume that $q \cdot u \neq q' \cdot u$. Since the set $P \cdot wu$ is incompressible, the vertex $q' \cdot u = q \cdot wu$ forms an incompressible pair with every vertex in $P' \cdot u = P' \cdot wu$. Similarly, since the set $P \cdot u$ is incompressible, the vertex $q \cdot u$ also forms an incompressible pair with every vertex in $P' \cdot u$, and of course every pair of distinct vertices in $P' \cdot u$ is incompressible too. Now $P' \cdot u \cup \{q \cdot u, q' \cdot u\}$ has more than $|P|$ elements so it must be compressible, and the above analysis shows that the only pair in $P' \cdot u \cup \{q \cdot u, q' \cdot u\}$ that may be compressible is the pair $(q \cdot u, q' \cdot u)$. Thus, there is a word $v \in A^*$ such that $q \cdot uv = q' \cdot uv$.   □

Suppose that $\mathcal{A} = (Q, A)$ is a complete automaton. Fix a letter $a \in A$ and remove all edges of $\mathcal{A}$ except those labeled by a (in other words, consider the act induced by the cyclic monoid $\{a\}^*$ and its Cayley graph). The remaining graph is called the *underlying graph of a* or simply the *a-graph*. Thus, in the a-graph every vertex is the tail of exactly one edge. Since the function on Q induced by the action of a generates a finite cyclic semigroup, we can apply Exercise 1.8.7 and conclude that for every vertex $q \in Q$ there exists a nonnegative integer ℓ and some integer $m > \ell$ with $q \cdot a^\ell = q \cdot a^m$, so we have a path as on Figure 3.4. The least nonnegative integer ℓ such that $q \cdot a^\ell = q \cdot a^m$ for some $m > \ell$ is called the *a-height* of the vertex q and the vertex $q \cdot a^\ell$ is called the *bud* of q. The cycles of the a-graph are referred to as *a-cycles*.

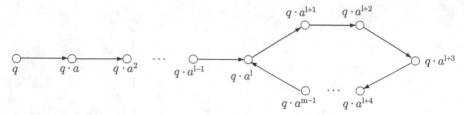

Figure 3.4 The orbit of a vertex in the underlying graph the letter a

Lemma 3.9.10. *Let $\mathcal{A} = (Q, A)$ be a complete strongly connected automaton. Suppose that there is a letter $a \in A$ such that all vertices of maximal a-height $L > 0$ have the same bud. Then $\mathcal{A}$ has a confluent pair.*

Proof. Let M be the set of all vertices of a-height L. Then $q \cdot a^L = q' \cdot a^L$ for all $q, q' \in M$ hence no pair of vertices from M is incompressible. Thus, any incompressible set in $\mathcal{A}$ has at most one common vertex with M. Take an incompressible set S of maximum size in $\mathcal{A}$ and choose any vertex $p \in S$. Since the automaton $\mathcal{A}$ is strongly connected, there is a path from p to a vertex in M. If $u \in A^*$ is the word that labels this path, then $S' = S \cdot u$ is an incompressible set of maximum size and it has exactly one common vertex with M (namely, $p \cdot u$). Then $S'' = S' \cdot a^{L-1}$ is an incompressible set of maximum size that has all its vertices except one (namely, $p \cdot ua^{L-1}$) in some a-cycles—the latter conclusion is ensured by our choice of L. If m is the least common multiple of the lengths of all simple a-cycles, then $r \cdot a^m = r$ for every r in every a-cycle but $(p \cdot ua^{L-1}) \cdot a = p \cdot ua^L \neq p \cdot ua^{L-1}$. We see that Lemma 3.9.9 applies (with S'' playing the role P and a^m playing the role of w). $\square$

Now we are ready to start the proof of Trahtman's road coloring theorem. Let us first explicitly formulate the result.

Theorem 3.9.11. *Every strongly connected primitive graph Γ with constant out-degree admits a synchronizing labeling.*

Proof. If Γ has just one vertex, there is nothing to prove. Thus, we assume that Γ has more than one vertex and prove that it admits a labeling with a confluent pair of vertices. The result will then follow from Proposition 3.9.8.

Fix an arbitrary labeling of Γ by letters from an alphabet A and take an arbitrary letter $a \in A$. We induct on the number N of vertices that do not lie on any a-cycle in the chosen labeling.

Suppose first that $N = 0$. This means that all vertices belong to a-cycles (i.e., the a-height of every vertex is 0).

We say that a vertex p of Γ is *ramified* if it is the tail for two edges with different heads.

If we suppose that no vertex in Γ is ramified, then there is just one a-cycle (since Γ is strongly connected) and all cycles in Γ have the same length. This contradicts the assumption that Γ is primitive[3].

Thus, let p be a vertex that is ramified. Then there exists a letter $b \in A$ such that the vertices $q = p \cdot a$ and $r = p \cdot b$ are not equal. We exchange the labels of the edges $p \xrightarrow{a} q$ and $p \xrightarrow{b} r$, see Figure 3.5. It is clear that in the new labeling there is only one vertex of maximal a-height, namely, the vertex q. Thus, Lemma 3.9.10 applies and the induction basis is verified.

Now suppose that $N > 0$. We denote by L the maximum a-height of the vertices in the chosen labeling. Observe that $N > 0$ implies $L > 0$.

Let p be a vertex of height L. Since Γ is strongly connected, there is an edge $p' \to p$ with $p' \neq p$, and by the choice of p, the label of this edge is some letter $b \neq a$. Let $t = p' \cdot a$. One has $t \neq p$. Let $r = p \cdot a^L$ be the bud of p and let C be the a-cycle on which r lies.

The following considerations split in several cases. In each case except one we can re-label Γ by swapping the labels of two edges so that the new labeling either satisfies the condition of Lemma 3.9.10 (all vertices of maximal a-height have the same bud) or has more vertices on the a-cycles (and the induction assumption applies). In the remaining case finding a confluent pair will be easy.

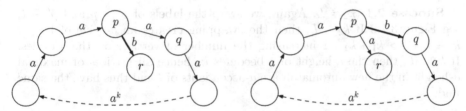

Figure 3.5 Relabeling in the basis of induction

Case 1: p' is not in C.

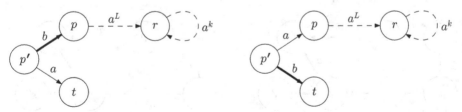

Figure 3.6 Relabeling in Case 1

---

[3] This is the only place in the whole proof where primitivity is used!

We swap the labels of $p' \overset{b}{\to} p$ and $p' \overset{a}{\to} t$, see Figure 3.6. If p' was on the a-path from p to r, then the swapping creates a new a-cycle increasing the number of vertices on the a-cycles. If p' was not on the a-path from p to r, then the a-height of p' becomes $L+1$ hence every vertex z of maximal a-height in the new automaton is an *aascendant* of p', that is $p' = z \cdot a^k$ for some $k \geq 0$, and thus has r as the bud.

Case 2: p' is in C. Let k_1 be the least integer such that $r \cdot a^{k_1} = p'$. The vertex $t = p' \cdot a$ is also in C. Let k_2 be the least integer such that $t \cdot a^{k_2} = r$. Then the length of C is $k_1 + k_2 + 1$.

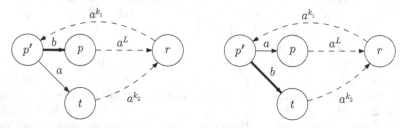

Figure 3.7 Relabeling in Subcase 2.1

Subcase 2.1: $k_2 \neq L$. Again, we swap the labels of $p' \overset{b}{\to} p$ and $p' \overset{a}{\to} t$, see Figure 3.7.3 If $k_2 < L$, then the swapping creates an a-cycle of length $k_1 + L + 1 > k_1 + k_2 + 1$ increasing the number of vertices on the a-cycles. If $k_2 > L$, then the a-height of t becomes k_2 hence all vertices of maximal a-height in the new automaton are a-ascendants of t and thus have the same bud.

Let s be the vertex of C such that $s \cdot a = r$.

Subcase 2.2: $k_2 = L$ and s is ramified. Since s is ramified, there is a letter c such that $s' = s \cdot c \neq r$.

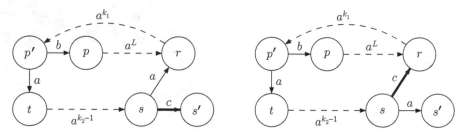

Figure 3.8 Relabeling in Subcase 2.2

We swap the labels of $s \overset{c}{\to} s'$ and $s \overset{a}{\to} r$, see Figure 3.7. If r still lies on an a-cycle, then the length of the a-cycle is at least $k_1 + k_2 + 2$ and the number of vertices on the a-cycles increases. Otherwise, the a-height of r becomes at least $k_1 + k_2 + 1 > L$ hence all vertices of maximal a-height in the new automaton are a-ascendants of r and have a common bud.

Let q be the vertex on the a-path from p to r such that $q \cdot a = r$.

Subcase 2.3: $k_2 = L$ **and** q **is ramified.** Since q is ramified, there is a letter c such that $q' = q \cdot c \neq r$.

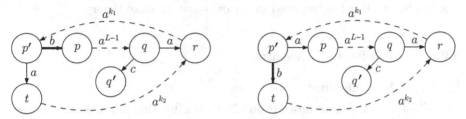

Figure 3.9 Relabeling reducing Subcase 2.3 to Subcase 2.2

If we swap the labels of $p' \overset{b}{\to} p$ and $p' \overset{a}{\to} t$, then we find ourselves in the conditions of Subcase 2.2 (with q and q' playing the roles of s and s' respectively), see Figure 3.9.

Subcase 2.4: $k_2 = L$ **and neither** s **nor** q **is ramified.**

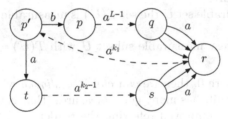

Figure 3.10 Subcase 2.4

In this subcase it is clear that q and s form a confluent pair whichever labeling of Γ is chosen, see Figure 3.10. This completes the proof. □

3.9.3 An Application of the Road Coloring Theorem to a Classification of Subshifts of Finite Type

We have mentioned that the classification of subshifts of finite type up to conjugacy is currently out of reach. We know that the problem reduces to edge subshifts of finite graphs (Theorem 1.6.23). But we do not know an algorithm that, given two finite graphs, decides whether the corresponding edge shifts are conjugate. While the problem seems to be purely syntactic like the word problem for semigroups, numerous attempts at solving it or proving that it is undecidable failed to produce a result. Thus several other, weaker, equivalences of subshifts have been studied. One of them was discussed in [6] and is related to the road coloring conjecture. Let us call two subshifts (X, T_X), (Y, T_Y) *almost conjugate* if there exist subshifts $X_0 \subseteq X$, $Y_0 \subseteq Y$ that are conjugate up to an absolutely negligible error. A more precise definition is given below. But first we need to review some measure theory.

3.9.3.1 Subshifts and Measures

Let (X, T) be a subshift.

Definition 3.9.12. A *good measure* on (X, T) is a function taking some subsets of X (called *measurable subsets*) to real numbers between 0 and 1. The set of measurable subsets and the measure can be defined simultaneously as follows.

(1) (*Probability.*) The whole set X is measurable and its measure is 1.
(2) The complement of every measurable subset of measure α is measurable, and its measure is $1 - \alpha$.
(3) The union of countably many disjoint measurable subsets U_i is measurable and its measure is the sum of measures of U_i.
(4) Every open ball $B(x, \epsilon) = \{y, \text{dist}(x, y) < \epsilon\}$ in X is measurable and has a positive measure.
(5) For every measurable set U the set $T(U)$ is measurable and has the same measure as U.
(6) (*Ergodicity.*) Every measurable subset U with $T(U) = U$ has measure 1 or 0.

If a set has measure 0, then we can call it *negligible* with respect to that measure. Every subshift has many good measures (that is not a trivial statement and we leave it unproved referring the reader to [203]). For example, consider the full subshift $(A^{\mathbb{Z}}, T)$. Assign probability $p_a > 0$ to every letter $a \in A$ so that $\sum p_a = 1$. For every word w let p_w be the product of probabilities of its letters. Then for every open ball $B = B(\omega, \frac{1}{2^k}) = \{\alpha \mid \text{dist}(\alpha, \omega) < \frac{1}{2^k}\}$ assign the measure

$$\mu(B) = p_{\omega(-k,k)}.$$

Using parts (1)–(5) of Definition 3.9.12 one can extend the function μ (by induction) to all measurable sets. The resulting measure on $A^{\mathbb{Z}}$ is good (see [26]). This measure is called a *Bernoulli measure* on $A^{\mathbb{Z}}$.

A subset that has measure 0 with respect to every good measure on (X, T) will be called *absolutely negligible*. The empty set is clearly always absolutely negligible. There are many other examples but we do not want to get too deep into the theory of measures on subshifts and we need only one kind of absolutely negligible sets that will be defined below in Section 3.9.3.2.

Given a measure μ on a subshift (X, T), we can *integrate* every *measurable* function $f: X \to \mathbb{R}$ (i.e., such a function that the preimage of every interval in $\mathbb{R}$ is measurable). Again we do not need the general definition of the integral $\int f d\mu$, just two properties, which are familiar to everybody who has studied calculus, and which can be found in any textbook on measure theory (see, for example, [26]).

Lemma 3.9.13. *(1) If X is a disjoint union of finitely many measurable subsets X_i, and f is constant on each X_i, then*

$$\int f d\mu = \sum_i \mu(X_i) f(X_i).$$

(2) If f is nonnegative on the whole X and positive on some measurable subset of positive measure, then $\int f d\mu > 0$.

The ergodicity of a good measure allows one to also apply the following famous ergodic theorem.

Theorem 3.9.14 (George Birkhoff, [268]). *Let μ be a good measure on a subshift (X, T), f be a measurable function $X \to \mathbb{R}$. Then for all $x \in X$ outside a subset of measure 0 we have*

$$\lim_{n \to \infty} \frac{1}{n} \sum_{i=0}^{n} f(T^i(x)) = \int f d\mu.$$

3.9.3.2 An Example of Absolutely Negligible Subsets in a Subshift

Lemma 3.9.15. *Let (X, T) be a subshift, w be a (finite) subword of a bi-infinite word from X. Let Y be the set of bi-infinite words ω one of whose infinite prefixes $\omega(-\infty, n), n \in \mathbb{Z}$ does not contain w. Then Y is an absolutely negligible subset of X.*

Proof. Indeed, suppose, by contradiction, that there exists a good measure μ on X such that $\mu(Y) > 0$. Consider the function $f: X \to \mathbb{R}$ that is equal to 1 on all bi-infinite words $\alpha \in X$ with $\alpha(1, |w|) \equiv w$ and 0 for all other $\alpha \in X$. By the assumption of the lemma, there exists $\omega \in X$ that contains w as a subword. Replacing ω by $T^s(\omega)$ if necessary, we can assume that $\omega(1, |w|) = w$. Therefore the function f is nonnegative on X and equal to 1 on the ball $\{\alpha \in X \mid \text{dist}(\alpha, \omega) < 2^{-|w|}\}$. By Lemma 3.9.13 then $\int f d\mu > 0$. On the other hand for every $\alpha \in Y$, and all sufficiently large i we have $f(T^i(\alpha)) = 0$ (by the definition of Y). Hence

$$\lim_{n \to \infty} \frac{1}{n} \sum_{i=0}^{n} f(T^i(x)) = 0,$$

which contradicts the Birkhoff ergodicity Theorem 3.9.14.   □

3.9.3.3 The AGW-Conjugacy of Subshifts

Now we can define *AGW-conjugacy* of subshifts (after Adler, Goodwyn and Weiss [6]).

Definition 3.9.16. We say that two subshifts (X, T_X), (Y, T_Y) are AGW-conjugate if there exists a third subshift (W, T_W) and two homomorphisms

$$\phi : (W, T_W) \to (X, T_X)$$

and

$$\psi : (W, T_W) \to (Y, T_Y)$$

such that

(1) Both ϕ and ψ are onto.
(2) Both ϕ and ψ are *finite-to-one*, that is the preimage of every point is finite.
(3) Each homomorphism is injective on a complement of an absolutely negligible shift-invariant subset.

Thus AGW-conjugate subshifts are conjugate "up to an absolutely negligible error". Clearly, conjugate subshifts are AGW-conjugate. Note that AGW-conjugation is an equivalence relation on subshifts (it is not obvious, for the proof see [6]).

Exercise 3.9.17 (Compare with Exercise 1.6.19). Show that any two AGW-conjugate subshifts have the same entropy.

3.9.3.4 Classification of Edge Subshifts of Primitive Graphs of Constant Out-Degree

The road coloring theorem leads to the following nice result.

Theorem 3.9.18 (Adler, Goodwyn, Weiss [6]). *Let Γ_1 and Γ_2 be finite primitive graphs of constant out-degrees k_1 and k_2. Then the edge subshifts of Γ_1 and Γ_2 are AGW-conjugate if and only if $k_1 = k_2$.*

Before we start the proof of that theorem, let us make some preliminary observations. Let Γ be a finite graph with out-degree of every vertex equal to k. Let $\mathcal{A} = (Q, A)$ be any complete automaton with underlying graph Γ. Then there exists a map ϕ from the edge subshift (X_Γ, T_Γ) of Γ to the subshift $(A^{\mathbb{Z}}, T)$. It maps every bi-infinite path in Γ to its label. Then ϕ is a homomorphism of subshifts.

The set S_n of all subwords of length n of bi-infinite words from X_Γ coincides with the set of paths of length n in Γ. Every such path is determined by its label and its initial vertex. Since $\mathcal{A}$ is complete, for every vertex x in Γ and every word u in the alphabet A there exists a path in $\mathcal{A}$ starting at x with label u (why?).

Therefore the complexity function $f(n)$ of X_Γ satisfies

$$k^n \le f(n) \le |Q|k^n.$$

Thus the entropy of (X_Γ, T_Γ) is equal to $\ln k$. Since every subshift is determined by the set of (finite) subwords of its bi-infinite words (Exercise 1.6.6), we conclude also that ϕ is surjective, so $\phi(X_\Gamma) = A^{\mathbb{Z}}$.

Now we can use the following lemma from symbolic dynamics that we present here without a proof.

Lemma 3.9.19 (See Corollary 8.1.20 from [203]). *Suppose that* Γ, Γ' *are strongly connected finite graphs, and the edge subshifts* (X_Γ, T_Γ), $(X_{\Gamma'}, T_{\Gamma'})$ *have the same entropy. Let* ϕ *be a surjective homomorphism from* (X_Γ, T_Γ) *to* $(X_{\Gamma'}, T_{\Gamma'})$. *Then* ϕ *is finite-to-one.*

By Lemma 3.9.19 our homomorphism ϕ is finite-to-one.

Proof of Theorem 3.9.18. By Exercise 3.9.17, if the edge-shifts of two graphs Γ_1, Γ_2 with constant out-degrees k_1, k_2 are AGW-conjugate, then $k_1 = k_2$.

Let us prove that the converse statement is true for primitive graphs. So suppose that $k_1 = k_2$. By the road coloring theorem (Theorem 3.9.11), there are labelings of Γ_1, Γ_2 that turn these graphs into complete synchronized automata $\mathcal{A}_1 = (Q_1, A)$ and $\mathcal{A}_2 = (Q_2, A)$ (the same alphabet because the out-degrees of Γ_1 and Γ_2 are the same). Then as before we have maps ϕ_i from the edge subshifts $(X_{\Gamma_i}, T_{\Gamma_i})$ onto $(A^{\mathbb{Z}}, T)$, $i = 1, 2$. By Lemma 3.9.19 these maps are finite-to-one.

Let w_1, w_2 be the reset words for $\mathcal{A}_1, \mathcal{A}_2$. Let Y_i, $i = 1, 2$, be the set of bi-infinite words ω from $A^{\mathbb{Z}}$ such that some prefix $\omega(-\infty, n)$ does not contain w_i. By Lemma 3.9.15 each Y_i is absolutely negligible. Let $\bar{Y}_i$ be the complement of Y_i. We claim that $\phi_i(\omega)$ in $\bar{Y}_i$ uniquely determines ω. Indeed, suppose that there are two bi-infinite paths ω, ω' such that $\phi_i(\omega) = \phi_i(\omega') \in \bar{Y}_i$. Since the bi-infinite words $\phi_i(\omega)$ and $\phi_i(\omega')$ have infinitely many occurrences of the reset word w_i to the left of the position 0, the infinite to the left paths $\omega(-\infty, 0)$, $\omega'(-\infty, 0)$ visit the same vertex infinitely many times. Since the automaton $\mathcal{A}_i$ is deterministic, and labels $\phi_i(\omega(-\infty, 0))$ and $\phi_i(\omega'(-\infty, 0))$ coincide, the paths $\omega(-\infty, 0)$ and $\omega'(-\infty, 0)$ coincide. But then the paths $\omega(0, \infty)$ and $\omega'(0, \infty)$ must coincide because these paths start at the same vertex and have the same labels. This implies that $\omega = \omega'$. Thus the map ϕ_i is one-to-one on $\phi_i^{-1}(\bar{Y}_i)$. Note that the complement of $\phi_i^{-1}(\bar{Y}_i)$ in X_{Γ_i} is absolutely negligible as well (prove that!).

Thus we found two surjective finite-to-one maps ϕ_i from X_{Γ_i} to $A^{\mathbb{Z}}$, which are one-to-one on complements of absolutely negligible sets $\phi_i^{-1}(\bar{Y}_i)$, $i = 1, 2$.

Now consider the *equalizer* of the maps ϕ_1, ϕ_2. Namely, consider the alphabet of pairs $E_1 \times E_2$ where E_i is the set of edges of the graph Γ_i (as we did in Exercise 2.5.25). Then there exists a natural subshift homomorphism π_i from $(E_1 \times E_2)^{\mathbb{Z}}$ to $E_i^{\mathbb{Z}}$ which maps every letter from $E_1 \times E_2$ to its ith coordinate.

Let $(W,T) \subseteq ((E_1 \times E_2)^Z, T)$ be a subshift consisting of all bi-infinite words ω such that $\pi_i(\omega) \in X_{\Gamma_i}$ and

$$\phi_1(\pi_1(\omega)) = \phi_2(\pi_2(\omega)).$$

The next exercise shows that (W,T) and the homomorphisms π_i from (W,T) to $(X_{\Gamma_i}, T_{\Gamma_i})$ give us the AGW-conjugacy of the edge subshifts.

Exercise 3.9.20. Show that each π_i is finite-to-one and is one-to-one on a complement of an absolutely negligible set.

□

3.10 Further Reading and Open Problems

3.10.1 Other Applications of Burnside Properties

There are quite a number of applications of Theorem 3.3.4. Here we present two, more - in Section 3.10.6.

3.10.1.1 The Restricted Burnside Problem for Semigroup Varieties

What is now called the *restricted Burnside problem* was first formulated by Magnus [216] (see Kostrikin [185]):

Problem 3.10.1. *Is there an upper bound for the size of a m-generated finite group of exponent n?*

It is easy to see that this is equivalent to the following problem (see the survey [177]).

Problem 3.10.2. *Is it true that locally finite groups in the variety* var$\{x^n = 1\}$ *form a variety?*

In this form, the question is interesting for any variety of universal algebras.

Problem 3.10.3. *Let $\mathcal{V}$ be a variety of universal algebras. Is it true that the locally finite algebras from $\mathcal{V}$ form a variety?*

Of course the answer depends on the variety: for the variety of commutative groups, for example, the answer is negative because there are arbitrary large finite cyclic groups.

Thus it is natural to ask the question in the following algorithmic form.

Problem 3.10.4. *Given a finite system of identities of universal algebras, can we decide if the locally finite algebras from this variety form a subvariety,* *i.e., for every $m \geq 1$ there is an upper bound for the size of m-generated finite algebras in the variety.*

The history of attempts to solve Problems 3.10.1, 3.10.2 can be found in [185]. These problems were first reduced to the case when n is a power of a prime (modulo the classification of finite simple groups) in [143] and then solved for all prime powers by Zel'manov [333, 334]. Since the classification of finite simple groups is now complete [15], we have positive answers to Problems 3.10.1, 3.10.2.

For arbitrary universal algebras, Problem 3.10.4 is *undecidable*:

Theorem 3.10.5. *There is no algorithm to decide, given a finite set of identities, whether all locally finite algebras satisfying these identities form a variety.*

Proof. Let $S = \mathrm{sg}\langle A \mid u_1 = v_1, \ldots, u_n = v_n \rangle$ be a finitely presented semigroup. Consider the class $\mathcal{C}$ of universal algebras with $|A|$ unary operations labeled by the letters from A. Then S corresponds to a finitely based subvariety $\mathcal{S}$ of $\mathcal{C}$. Finitely generated algebras from $\mathcal{S}$ are just sets on which the semigroup S acts with finitely many *orbits* i.e., subsets of the form $x \cdot S$. Therefore locally finite algebras in $\mathcal{S}$ form a variety if and only if there exist only finitely many finite acts of the semigroup S generated by one element. This is equivalent to the following property: S has only finitely many finite homomorphic images (prove it!). Thus if there was an algorithm solving Problem 3.10.3, there would be an algorithm that, given a finite semigroup presentation decided whether the semigroup has only finitely many finite homomorphic images. But the latter algorithm does not exist even in the case when S is a group which is proved by Bridson and Wilton [52].   □

Nevertheless for varieties of semigroups, Problem 3.10.4 is decidable. That follows from the next theorem and Theorem 3.3.4.

Theorem 3.10.6 (Sapir, [279]). *The following properties of a finitely based variety $\mathcal{V}$ of semigroups are equivalent (the last two properties are equivalent by Theorem 3.3.4):*

(1) *the locally finite semigroups from $\mathcal{V}$ form a variety;*
(2) *$\mathcal{V}$ contains only a finite number of finite semigroups with a fixed number of generators;*
(3) *there exists a recursive function $f(k)$ giving an upper bound for the order of the k-generated finite semigroups from $\mathcal{V}$;*
(4) *the nilsemigroups from $\mathcal{V}$ are locally finite.*
(5) *Z_{n+1} is not an isoterm for the identities of $\mathcal{V}$, where n is the number of variables in a finite basis of identities of $\mathcal{V}$.*

The following problem is still open.

Problem 3.10.7. *Is Problem 3.10.4 decidable for varieties of monoids?*

3.10.2 Free Semigroups in Periodic Varieties. The Brzozowski Problem

Every periodic variety of semigroups is contained in the variety given by one identity $x^m = x^{m+n}$. Thus the varieties $\mathcal{B}_{m,n} = \mathrm{var}\{x^m = x^{m+n}\}$ are of special interest. Relatively free semigroups in these varieties are among periodic semigroups admitting the easiest syntactic description. Quite surprisingly, there are some natural problems concerning these semigroups that have been open for more than 40 years in spite of a lot of attention.

The main problem about the relatively free semigroups $S_{m,n,k}$ with $k \geq 2$ generators in the variety $\mathcal{B}_{m,n}$ was formulated by Brzozowski in 1969 in the case of $m \geq 2, n \geq 1$ [58].

Problem 3.10.8. *Is the word problem decidable in $S_{m,n,k}$? More precisely, is it true that for every element of $S_{m,n,k}$ the language of all words (in the generating set of $S_{m,n,k}$) representing this element is regular?*

In the case $m = 1$ the word problem for $S_{m,n,k}$ reduces to the word problem in the free Burnside group $B_{n,k}$, which is discussed below in Section 5.2.3 (this was observed by Green and Rees in [116]). In particular, if $m = 1$, $n = 1, 2, 3, 4, 6$, then $S_{m,n,k}$ is finite, if $n \geq 665$ and odd, then the word problem is decidable, etc. Thus the case $m = 1$ is not really a semigroup theoretic problem, hence we start with $m = 2$.

The current status of Problem 3.10.8 is the following. It was proved for every $n \geq 1$, $m \geq 3$ and every k by Guba [128], but the case $m = 2$ is still open. Moreover, it is still open in the "simplest" case $m = 2, n = 1$. In that case, Plyushchenko [270] recently reduced the problem to the case of 2 generators (see do Lago and Simon [90] for a more complete history of the problem and attempts to solve it). In the case $m \geq 3$, do Lago [89] found a Church–Rosser presentation of $S_{m,n,k}$ with regular language of left-hand sides (it can also be deduced from Guba's proof although it is not explicitly mentioned in the proof, this is explained in [132]). This not only solved Problem 3.10.8 in that case, but also allowed do Lago to describe $\mathcal{L}$- and $\mathcal{R}$-classes of $S_{m,n,k}$.

3.10.3 The Finite Basis Problem for Semigroups

There are many papers devoted to the finite basis problems for semigroups (see detailed surveys [296, 325, 326]). One of the goals is to find smallest in some sense infinitely based varieties. For example, while the 6-element semigroup B_2^1 is not finitely based (Theorem 3.6.28), every 5-element semigroup (there are $1,915$ of those) is finitely based (Trahtman, [316], see also Lee [196]) and among 6-element semigroups (there are $28,634$ of those) exactly four are non-finitely based [197] (including A_2^1, B_2^1 of course).

One of the most interesting results about "small" non-finitely based varieties was obtained by Kaďourek [165]. He proved that the variety var$\{x^2y = xy\}$ contains infinitely based subvarieties. This variety is remarkable because (a) it does not contain nontrivial groups and (b) all nilpotent semigroups in it are semigroups with zero product (prove (a) and (b)!). All other examples of non-finitely based varieties of semigroups had either nontrivial groups or nilpotent semigroups of arbitrary large class or both.

Since inherently non-finitely based finite semigroups are completely described (Theorem 3.6.34), it is interesting to study identities of non-inherently non-finitely based finite semigroups.

Here are a few outstanding open problems.

Problem 3.10.9. *Is the set of all finite finitely based semigroups recursive?*

There are several classes of "comparatively easy" semigroups for which the question is also open and is worth studying.

For example, let L be a finite language closed under taking subwords and $S(L)$ be the finite semigroup obtained from L by the Dilworth construction, i.e., the semigroup $L \cup \{0\}$ where the product $u \cdot v$ is uv if $uv \in L$ or 0 otherwise. The semigroup $S(L)$ is finite and nilpotent, hence the variety generated by $S(L)$ is always finitely based (prove it!). On the other hand the monoid $S(L)^1$ may not be finitely based as was first proved by Perkins [266]. He proved that if L consists of all subwords of the words from $\{abtba, atbab, abab, aat\}$, then $S(L)^1$ is not finitely based. Since monoids $S(L)^1$ are so easy to describe (and all of them are not inherently non-finitely based by Theorem 3.6.34), it would be very interesting to describe finite languages L for which $S(L)^1$ is finitely based. It has been done when L consists of one word in a 2-letter alphabet and all its subwords (it is proved that the semigroup is finitely based if and only if the word is of the form $x^m y^n$ or $x^m y x^n$, $m, n \geq 0$), and in some more complicated cases [160, 283, 284]. While classes of languages L get more complicated the description gets more and more complicated also, which may indicate that perhaps there is no general algorithm to check whether $S(L)^1$ is finitely based.

Problem 3.10.10. *Is there an algorithm to decide, given a finite collection of words L, whether the monoid $S(L)^1$ (i.e., the monoid consisting of all words that do not contain subwords from L) is finitely based?*

Remark 3.10.11. Note a similarity between Problems 3.10.10 and 1.6.16. We do not know whether there is any deep connection between these problems.

Another class of finite semigroups for which Problem 3.10.9 has been studied is the class of finite completely simple semigroups $M(G, I, J, P)$. Mashevitsky [222] discovered that this problem is closely related to the finite basis problem for finite groups with an additional 0-ary operation, i.e., *pointed groups*, considered earlier by R. Bryant [57]. He proved that if G is a non-finitely based pointed finite group with 0-ary operation g, then the Rees–Sushkevich semigroup $M(G, L, R, P)$ where $L = R = \{1, 2\}$, $P = \begin{pmatrix} 1 & g \\ 1 & 1 \end{pmatrix}$ is not finitely based.

3.10.4 Inherently Non-finitely Based Finite Monoids

Although inherently non-finitely based finite semigroups have been completely described (Theorem 3.6.34) the similar question for finite monoids is still open.

Problem 3.10.12. *Is there an algorithm to decide if a finite monoid is inherently non-finitely based?*

3.10.5 Identities of Finite Inverse Semigroups

Identities of inverse semigroups including identities of finite inverse semigroups are discussed, in particular, in [269, Chapter VII]. It was discovered by E.I. Kleiman [178, 179] that the Brandt monoid B_2^1 plays a very important role. Namely every variety of inverse semigroups that does not contain B_2^1 is finitely based if and only if its groups form a finitely based variety of groups. Moreover, the variety generated by a finite inverse semigroup S contains B_2^1 if and only if B_2^1 is a divisor of S, i.e., a factor-semigroup of one of the subsemigroups of S. The variety of groups from var S is generated by the subgroups of S, hence it is always finitely based by Theorem 1.4.33. Thus if B_2^1 is not a divisor of S, then var S is finitely based. E.I. Kleiman proved [179] that the Brandt monoid B_2^1 is not finitely based. He also formulated the following

Problem 3.10.13. *Is it true that a finite inverse semigroup S is finitely based (as an inverse semigroup) if and only if var S does not contain B_2^1 ?*

That would follow if B_2^1 was inherently non-finitely based as an inverse semigroup. But we have seen (Theorem 3.8.5) that there are no inherently non-finitely based finite inverse semigroups at all. This makes Problem 3.10.13 quite difficult. Nevertheless, Kaďourek [166] made a breakthrough by proving that the answer to Problem 3.10.13 is positive provided every subgroup G of S is *solvable*, i.e., has a sequence of normal subgroups $\{1\} = G_n < G_{n-1} < \ldots < G_0 = G$ where all factor-groups G_i/G_{i+1} are commutative. For inverse semigroups S with non-solvable subgroups, Problem 3.10.13 is still open.

3.10.6 Growth Functions of Semigroups

3.10.6.1 The Set of All Growth Functions of Finitely Generated Semigroups

If $D = (X, T)$ is a subshift, and $S(D)$ is the corresponding semigroup, then the growth function $f_S(n)$ of $S(D)$ is related to the complexity function $f_D(n)$ by the formula

$$f_S(n) = \sum_{i=1}^{n} f_D(i)$$

(prove it!) By Exercise 1.6.10, the complexity function $f_D(n)$ is either 0 (if the subshift is empty), or bounded from above by a positive integer (if the subshift is finite), or satisfies $f_D(n) \geq n + 1$. This implies the following result.

Lemma 3.10.14. *The growth function $f_S(n)$ is either bounded from above by cn for some constant c or is bounded from below by the quadratic function* $2 + 3 + \ldots + (n + 1) = \frac{(n+1)(n+2)}{2} - 1$.

It turns out that this statement is true for every finitely generated semigroup. It was proved independently by several people, see, for example, [43, 184]. In fact it can be proved in the same way as the statement from Exercise 1.6.10.

In [318, 319], Trofimov proved that finitely generated semigroups can have very weird growth functions strictly between n^2 and $\exp n$.

Here is the main result of [318] (in particular, it gives many more examples of semigroups of intermediate growth).

Theorem 3.10.15. *Let f_1, f_2 be two increasing functions $\mathbb{N} \to \mathbb{N}$ such that $n^2 < f_1$ and $f_2 < 2^n$ (the partial order $<$ on functions was defined in Section 3.7.1), f_1 is not equivalent to n^2 and f_2 is not equivalent to 2^n. Then there exists a finitely generated semigroup S whose growth function $f_S(n)$ satisfies $f_S(n) < f_1(n)$ for infinitely many values of n and $f_S(n) > f_2(n)$ for infinitely many values of n.*

For more information about growth functions of finitely generated semigroups one can look in Shneerson [302], Ufnarovsky [320, 321] and Belov-Kanel, Borisenko and Latyshev [34] (where a strengthened version of Theorem 3.10.15 is proved, in particular).

Problem 3.10.16. *What is (up to the equivalence) the set of growth functions of finitely presented semigroups? Can these growth functions behave like in Theorem 3.10.15?*

3.10.6.2 Relatively Free Semigroups. Zimin Words and Growth

A surprising connection between Theorem 3.3.4 and growth functions in semi-groups was discovered by Shneerson. He started by exhibiting the first example of a relatively free semigroup of intermediate growth.

Theorem 3.10.17 (Shneerson, [300]). *Let S_k be the k-generated relatively free semigroup in the variety $xyzyx = yxzxy$ (this variety has appeared before in a paper by B. H. Neumann and T. Taylor [245] describing subsemigroups of nilpotent groups). Then the growth function of S_k is intermediate if $k > 2$.*

Note that Z_4 is not an isoterm for $xyzyx = yxzxy$, so all periodic semi-groups in this variety are locally finite. It turned out that this is not a random coincidence because of the following strong and general theorem.

Theorem 3.10.18 (Shneerson, [301]). *Let $\mathcal{V}$ be a finitely based non-periodic semigroup variety where all periodic semigroups are locally finite. Then the growth function of every finitely generated semigroup in $\mathcal{V}$ is strictly subexponential.*

Note that there are no associative rings and no known examples of groups, which satisfy nontrivial identities and have intermediate growth. Every ring satisfying a nontrivial identity has polynomial growth (Section 4.4.3). The Grigorchuk group of intermediate growth considered in Section 5.7.4 does not satisfy nontrivial identities. It follows from Abért's Theorem 5.4.1 (see [1]).

Non-periodic varieties where semigroups have polynomial growth were completely described in [303]. It turned out that these varieties must satisfy one additional identity besides $Z_n = W$. For every $\gamma > 0$ and $s > 0$ consider the following word $V_{\gamma,s}$ in the alphabet $\{x, y, u, w_1, w_2, \ldots, w_s\}$:

$$V_{\gamma,s} = x^\gamma u y^\gamma w_1 x^\gamma w_2 u y^\gamma w_3 \ldots x^\gamma u y^\gamma w_s x^\gamma u y^\gamma$$

(that is $V_{\gamma,s}$ is obtained by substituting $x^\gamma u y^\gamma$ for u in $u w_1 u w_2 \ldots w_s u$).

Theorem 3.10.19 (Shneerson, [303]). *The following conditions are equivalent for every finitely based non-periodic semigroup variety $\mathcal{V}$ given by identities in n variables.*

(1) *All finitely generated semigroups of $\mathcal{V}$ have polynomial growth.*
(2) *$\mathcal{V}$ does not contain finitely generated semigroups of intermediate growth.*
(3) *Z_{n+1} and $V_{\gamma,n}$ (for some $\gamma > 0$) are not isoterms for $\mathcal{V}$.*

The proof is very nontrivial. In particular, it strengthens and makes much more precise several parts of the proof of Theorem 3.3.4.

The following natural problem in this area still remains unsolved and seems difficult.

Problem 3.10.20. *Is it true that every finitely based non-periodic variety of semigroups containing a non-locally finite periodic semigroup contains a finitely generated semigroup of exponential growth?*

The answer to Problem 3.10.20 is positive in the case when both sides of every identity of V are avoidable. Indeed, in this case, using Exercise 2.5.25, we can find a bi-infinite uniformly recurrent word ω, in, say, k, letters avoiding both sides of every identity of V. Each subword of ω is then an isoterm for V. Let A be an alphabet with k letters, S_k be the relatively free semigroup of V with generating set A. Take any letter $a \in A$ and any letter $x \notin A$. For every subword w of ω we can produce exponentially many words by replacing some occurrences of a in w by x. Each of these words is again an isoterm for V (prove it!), so these words represent different elements in S_{k+1}, hence S_{k+1} has exponential growth. That argument does not work, unfortunately, if both sides of some identity of V are unavoidable (say, for the Mal'cev identity from Exercise 3.3.8).

3.10.7 The Road Coloring and the Černý Conjecture

The road coloring problem and related topics (including symbolic dynamics and coding) are discussed, in particular, in [7] and also in the books [176, 203].

A very natural question to ask is the following: *given a positive integer n, how long can reset words be for synchronizing automata with n vertices?* Černý [65] found a lower bound by constructing, for each $n > 1$, a synchronizing automaton C_n with n vertices and 2-letter alphabet whose shortest reset word has length $(n-1)^2$. The graph on Figure 3.1 is the graph C_4 in the sequence.

The automaton C_n can be described as follows. Its vertex set is

$$Q = \{0, 1, 2, \ldots, n-1\},$$

the alphabet is $\{a, b\}$, and the action of $\{a, b\}^+$ on Q is given by the following formulas:

$$i \cdot a = \begin{cases} i & \text{if } i > 0, \\ 1 & \text{if } i = 0; \end{cases} \quad i \cdot b = i + 1 \pmod{n}.$$

Exercise 3.10.21. Prove that the word $(ab^{n-1})^{n-2}a$ of length $(n-1)^2$ resets C_n.

Theorem 3.10.22 ([65, Lemma 1]). *Any reset word for C_n has length at least $(n-1)^2$.*

If we define the *Černý function* $\mathfrak{C}(n)$ as the maximum length of shortest reset words for synchronizing automata with n vertices, the above property of the series $\{C_n\}$, $n = 2, 3, \ldots c$, yields the inequality $\mathfrak{C}(n) \geq (n-1)^2$.

Conjecture 3.10.23 (The Černý conjecture). *$\mathfrak{C}(n) = (n-1)^2$ for all $n > 1$.*

The conjecture is proved by Kari [172] for complete automata where the in-degree and out-degree of every vertex coincide (in fact in this case the bound turned out to be $(n-1)(n-2) + 1 < (n-1)^2$). For more information, see the survey by Volkov [327].

Chapter 4
Rings

In this chapter, all rings are assumed to be associative. We start with a study of free associative algebras. This gives us the main syntactic tool to study rings. The first major result proved in this chapter is Shirshov's height theorem, which is used to prove a theorem of Kaplansky. Then we prove the Dubnov–Ivanov–Nagata–Higman theorem about associative algebras satisfying the identity $x^n = 0$. Unlike semigroups, associative rings satisfying this identity are nilpotent. But there exist finitely generated non-nilpotent associative nil-algebras, and we describe the classical example of such an algebra constructed by Golod.

Another result showing a fundamental difference between rings and semigroups is the Kruse–L'vov theorem (proved in this chapter) that every finite associative ring is finitely based. Finally we give an example of a non-finitely based variety of associative rings due to Belov-Kanel (which was a solution of a long-standing problem).

Theorems proved in this Chapter include

- Theorem 4.3.1 of Bergman about centralizers of elements in the free associative algebra.
- Theorem 4.4.6 of Kaplansky about finitely generated rings satisfying a nontrivial identity.
- Shirshov's height Theorem 4.4.7.
- Theorem 4.4.19 by Dubnov, Ivanov, Nagata and Higman about rings satisfying nil-identities.
- Theorem 4.4.20 by Golod solving the Kurosh problem about nil-rings.
- Baer's Theorem 4.5.3 about the structure of rings.
- The ring part of Theorem 1.4.33 by Kruse and L'vov.
- Theorem 4.5.17 by Belov-Kanel giving an example of a non-finitely based variety of rings.

© Springer International Publishing Switzerland 2014
M.V. Sapir, *Combinatorial Algebra: Syntax and Semantics*,
Springer Monographs in Mathematics, DOI 10.1007/978-3-319-08031-4_4

4.1 The Basic Notions

We briefly recall some basic notions about rings. A subgroup I of the additive group of a ring R is called a *left* (resp. *right*) *ideal* if for every $x \in I$, $y \in R$ we have $yx \in I$ (resp. $xy \in I$). If I is both a left ideal and a right ideal, we call I a *two-sided ideal* or just an *ideal*. By Exercise 1.4.19, ideals are precisely the kernels of ring homomorphisms. If $X \subseteq R$, the least ideal of R containing X is called the *ideal generated by* X.

Exercise 4.1.1. Prove that the ideal generated by a nonempty subset X of a ring R consists of all sums of the form

$$\sum_i x_i + \sum_j p_j x_j + \sum_k x_k q_k + \sum_\ell p_\ell x_\ell q_\ell,$$

where $x_i, x_j, x_k, x_\ell \in X \cup -X$ and $p_j, p_\ell, q_k, q_\ell \in R$.

Given a finite sequence of ideals $I_1, \ldots, I_k$ of a ring R, one can form their *sum*

$$I_1 + \ldots + I_k = \{x_1 + \ldots + x_k \mid x_j \in I_j, \ j = 1, \ldots, k\}$$

and their *product*

$$I_1 \cdots I_k = \{\sum_s x_{1,s} \cdots x_{k,s} \mid x_{j,s} \in I_j, \ j = 1, \ldots, k\}.$$

In particular, we can define the power I^k for every $k \geq 1$.

Exercise 4.1.2. Verify that the sum and the product of any finite sequence of ideals of a ring R are ideals of R.

The sum $I_1 + \ldots + I_k$ of the ideals $I_1, \ldots, I_k$ is said to be *direct* if for each $x \in I_1 + \ldots + I_k$, its presentation in the form $x = x_1 + \ldots + x_k$ where $x_j \in I_j$ for $j = 1, \ldots, k$ is unique.

Exercise 4.1.3. Prove that the sum $I_1 + \ldots + I_k$ is direct if and only if $I_j \cap (I_1 + \ldots + I_{j-1} + I_{j+1} + \ldots + I_k) = \{0\}$ for each $j = 1, \ldots, k$.

In particular, the sum of two ideals is direct if and only if their intersection is $\{0\}$.

For a ring R and a positive integer n, we let $M_n(R)$ denote the ring of all $n \times n$ matrices with entries in R. The operation in $M_n(R)$ are the usual matrix addition (entry-wise) and matrix multiplication ("row-by-column") familiar from Linear Algebra. The ring $M_n(R)$ has many nice properties. In particular, if R has an identity element, then the ideals of $M_n(R)$ can be easily described in terms of the ideals of R.

Exercise 4.1.4. Let R be a ring with identity element. Prove that for each ideal I of R, the set of all $n \times n$ matrices with entries in I is an ideal of $M_n(R)$, and if R has an identity element, then each ideal of $M_n(R)$ arises in this way. **Hint:** Let I be an ideal of $M_n(R)$, $I(R)$ be the ideal of R generated by all entries of all matrices of I. You need to show that $I = M_n(I(R))$. Let A be any matrix in I and r be the (i,j)-entry of A. Multiplying the matrix A by *matrix units* $e_{i,j}$, that is matrices with (i,j)-entry 1 and all other entries 0, on the left and on the right show that the matrix $re_{1,1}$ is in I. Deduce that the whole set $I(R)e_{1,1}$ is contained in I. Again multiplying by matrix units, deduce that for every i,j, $I(R)e_{i,j}$ is contained in I. Therefore every matrix from $M_n(I(R))$ is in I.

A ring R is called *nilpotent* if $R^k = \{0\}$ for some k i.e., its multiplicative semigroup is nilpotent. A ring R is *simple* if $R^2 \ne \{0\}$ and R has no ideals except R and $\{0\}$. Exercise 4.1.4 implies that if R is a simple ring with an identity element, then so is the matrix ring $M_n(R)$. In particular, since every field is a simple ring (prove it!), we see that the ring of all $n \times n$ matrices with entries in any given field is simple.

Let K be a field. Recall that a ring R is a *K-algebra* if each $\alpha \in K$ defines a unary operation $\alpha\cdot$ on R and these unary operation are consistent with the multiplication and the addition in both K and R in the following sense: (a) $\alpha(x+y) = \alpha x + \alpha y$, (b) $(\alpha + \beta)x = \alpha x + \beta x$, (c) $(\alpha\beta)x = \alpha(\beta x)$, (d)$1 \cdot u = u$, and (e) $x(\alpha y) = (\alpha x)y = \alpha(xy)$ for every $\alpha, \beta \in K$ and every $x, y \in R$ (see Section 1.4). When we speak about an ideal (one-sided or two-sided) I of a K-algebra R, we always assume that I is closed also under the unary operations $\alpha\cdot$ for each $\alpha \in K$.

4.2 Free Associative Algebras

Let K be a commutative ring with an identity element. The *absolutely free associative K-algebra over an alphabet X* is the set KX^+ of all (finite) linear combinations of words from X^+ with natural addition and distributive multiplication. Note that this is a subalgebra of $K\langle\!\langle X \rangle\!\rangle$ defined in Section 1.8.8 (in the case when K is a field).

Exercise 4.2.1. Prove that KX^+ is indeed a free algebra in the variety of all associative algebras over K. Prove that if X consists of one element, KX^+ is isomorphic to the (commutative) algebra of all polynomials over K with one variable.

We call linear combinations of words (non-commutative) *polynomials*. Accordingly words are sometimes called *monomials*. The *degree* of a polynomial is the biggest length of its monomial (with non-zero coefficient).

We call any map $X \to KX^+$ a *substitution*. As for every free universal algebra, every substitution can be extended to an endomorphism of KX^+ (see Section 1.4.2).

We say that polynomial p *does not avoid* polynomial q if $p = r\phi(q)s$ for some substitution ϕ and some polynomials r and s.

If we take any set of polynomials $W \subseteq KX^+$ and consider the set $KI(W)$ of all linear combinations of polynomials that do not avoid W, then $KI(W)$

is an ideal of KX^+ that is called the *verbal* ideal defined by W (it is also called the T-ideal defined by W). The following theorem is similar to the Theorem 3.2.1.

Theorem 4.2.2. *The quotient algebra KX^+/I is relatively free if and only if $I = KI(W)$ for some set W of polynomials from KX^+. An ideal I is verbal if and only if it is stable under all endomorphisms of KX^+.*

4.3 Commuting Elements in Free Associative Algebras over a Field

Free associative algebras are similar to free semigroups but more complicated. To illustrate the similarity and the difference let us consider the problem of describing commuting elements. Recall that by Theorem 1.2.9 two words in a free semigroup X^+ commute if and only if both are powers of the same word. Therefore the set of all words commuting with a given word u, i.e., the *centralizer* $C(u)$, is just a cyclic subsemigroup (prove it using Theorem 1.2.16!). On the other hand if u is a polynomial from the free algebra KX^+ then every linear combination of powers of u commutes with u. Therefore the centralizer of u contains a subalgebra isomorphic to the algebra of polynomials in one variable (denoted, as usual, $K[x]$), which is precisely the free associative algebra generated by one element, i.e., an analog of the cyclic semigroup of natural numbers $\mathbb{N}$ under addition. It turned out that the description of centralizers in the free associative algebra is very similar to Theorem 1.2.9.

Theorem 4.3.1 (Bergman, [42]). *Let K be a field. The centralizer of any non-zero element u in KX^+ is a subalgebra generated by one element $v \in KX^+$.*

The proof from [42] is not easy (see also a somewhat different but still complicated proof in Cohn [72]). We shall present only a small particular case of it — when v is a *homogeneous* polynomial, i.e., the degrees (i.e., lengths) of all monomials in v are the same. Below we are following the outline of the proof sent to us by George Bergman.

Lemma 4.3.2. *Let a, u, v, c be homogeneous polynomials such that, $a \neq 0$, $c \neq 0$, $au = vc$ and the degree of u is not smaller than the degree of c. Then there exists a homogeneous polynomial b such that $u = bc, v = ab$ (that is both au and vc become abc).*

Proof. Let X be the finite set of letters that appear in monomials of a, u, v, c. Denote the degrees of polynomials a, c, u by $d_a, d_c, d_c + d_b$, where $d_b \geq 0$. For every $d \geq 0$ let P_d be the vector space of homogeneous polynomials of degree d in X (if $d = 0$, then P_d is the field K). The dimension of P_d is $|X|^d$ (prove it!). Consider three vector spaces $A = P_{d_a}, B = P_{d_b}, C = P_{d_c}$. Consider bases $\mathcal{A}, \mathcal{B}, \mathcal{C}$ in A, B, C respectively, such that $a \in \mathcal{A}, c \in \mathcal{C}$ (recall that

$a, c \neq 0$ and every non-zero element of a vector space belongs to a basis by Exercise 1.4.13). Let $\mathcal{A} = \{a_1, \ldots, a_p\}$, $\mathcal{B} = \{b_1, \ldots, b_q\}$, $\mathcal{C} = \{c_1, \ldots, c_r\}$ where $a_1 = a, c_1 = c$. Then the set $\mathcal{ABC} = \{a_i b_j c_k, 1 \leq i \leq p, 1 \leq j \leq q, 1 \leq k \leq r\}$ is a basis of the vector space $P_{d_a + d_b + d_c}$.

Exercise 4.3.3. Prove the last statement. **Hint:** Show that the elements in $\mathcal{ABC}$ are linearly independent, and the dimension of the space $P_{d_a + d_b + d_c}$, i.e., $|X|^{d_a + d_b + d_c}$, coincides with the number of elements in $\mathcal{ABC}$, then apply Exercise 1.4.13.

Similarly, the sets $\mathcal{AB} = \{a_i b_j, 1 \leq i \leq p, 1 \leq j \leq q\}$ and $\mathcal{BC} = \{b_j c_k, 1 \leq j \leq q, 1 \leq k \leq r\}$ are bases of the spaces $P_{d_a + d_b}$ and $P_{d_b + d_c}$ respectively.

Note that $u \in P_{d_b + d_c}, v \in P_{d_a + d_b}$. Represent u and v as linear combinations of elements of the corresponding bases: $u = \sum \alpha_{j,k} b_j c_k, v = \sum \beta_{i,j} a_i b_j$. Then the equality $au = vb$ becomes

$$\sum \alpha_{j,k} a_1 b_j c_k = \sum \beta_{i,j} a_i b_j c_1$$

(recall that $a = a_1, c = c_1$).

Since elements of $\mathcal{ABC}$ are linearly independent, we have $\alpha_{j,k} = 0$ for every $k > 1$, $\beta_{i,j} = 0$ for every $i > 1$ and $\alpha_{j,1} = \beta_{1,j}$ for every j. Denote $b = \sum \alpha_{j,1} b_j = \sum \beta_{1,j} b_j$. Then $u = bc$ and $v = ab$ as required. □

Let us order all words (monomials) in the alphabet X using the ShortLex order. The biggest monomial of a polynomial p is called the *leading term* of p, and the coefficient from K of this monomial, the *leading coefficient*. A polynomial p is called *monic* if its leading coefficient is 1. Clearly if we divide any polynomial by its leading coefficient, we get a monic polynomial.

Exercise 4.3.4. Show that a product of two monic homogeneous polynomials is again a monic homogeneous polynomial.

Therefore the set of all monic homogeneous polynomials is a subsemigroup of the multiplicative semigroup of the free associative algebra KX^+.

Lemma 4.3.5. *The semigroup M of monic homogeneous polynomials from KX^+ is free with free generating set consisting of all its irreducible elements.*

Proof. Let us apply Theorem 1.8.6. The semigroup M is cancellative and generated by its indecomposable elements (prove it!). Let a, u, v, c be four monic homogeneous polynomials such that $au = vc$ in M. Let $d_1, d_3, d_3 + d_2$ be degrees of a, u, c. If $d_2 > 0$, then by Lemma 4.3.2 there exists a homogeneous polynomial b of degree d_2 such that $u = bc$. If $d_2 < 0$, then, again by Lemma 4.3.2, there exists a homogeneous polynomial b of degree d_2 such that $c = bu$. Finally if $d_2 = 0$, then applying Lemma 4.3.2, we obtain a homogeneous polynomial β of degree 0, i.e., a constant from K, such that $u = \beta c$, but since both u and c are monic, we get $\beta = 1$ (why?). Hence $c = u$. Thus all conditions of Theorem 1.8.6 hold and we can conclude that S is indeed a free semigroup freely generated by the set of indecomposable elements. □

Now let u be a homogeneous polynomial in KX^+, α be its leading coefficient. Then $u' = \frac{1}{\alpha}u$ is a monic polynomial from S. The set of all elements from S commuting with u' is a cyclic semigroup C generated by a monic polynomial v (by Lemma 1.2.9 applied to the free semigroup S). Let p be any polynomial of KX^+ commuting with u. Then p commutes with u' as well (check it!). There exists a unique representation of p as a sum of homogeneous polynomials $p_1 + p_2 + \ldots + p_m$ (group monomials of the same length in one homogeneous polynomial) called the *homogeneous components*. Let α_i be the leading coefficient of p_i.

Exercise 4.3.6. Prove that p commutes with u if and only if each monic homogeneous polynomial $\frac{1}{\alpha_i}p_i$ commutes with u'.

By Exercise 4.3.6 then p is a linear combination of powers of the homogeneous polynomial v. Therefore the centralizer $C(u)$ coincides with $K\langle v \rangle$, i.e., the subring of all linear combinations of powers of v.

This completes the proof of Theorem 4.3.1 in the case when u is homogeneous.

Remark 4.3.7. An analog of Theorem 4.3.1 is true for the algebra $\mathbb{Q}\langle\!\langle A \rangle\!\rangle$ of formal infinite linear combinations of words from A^+ defined in Section 1.8.8: the centralizer of any non-constant element $u \in \mathbb{Q}\langle\!\langle A \rangle\!\rangle$ is equal to the subring $\mathbb{Q}\langle\!\langle v \rangle\!\rangle$ for some $v \in \mathbb{Q}\langle\!\langle A \rangle\!\rangle$, see [72].

4.4 Burnside-Type Problems for Associative Algebras

4.4.1 Preliminaries

The Burnside type problem for associative algebras was formulated by Kurosh in the 30s:

Question   Suppose all 1-generated subalgebras of a finitely generated associative algebra A are *finite dimensional* (i.e. is spanned by a finite set of elements). Is A finite dimensional?

The answer is negative (Golod [112]) but not as easy as in the case of semigroups. For example, in the semigroup case we saw that there exists a 3-generated infinite semigroup satisfying the identity $x^2 = 0$. In the case of associative algebras over a field of characteristic $\neq 2$ this is impossible. Indeed, the following theorem holds.

Theorem 4.4.1. *Every algebra over a field of characteristic $\neq 2$ that satisfies the identity $x^2 = 0$ is nilpotent of class 3, that is every product of 3 elements in this algebra is 0. Every nilpotent finitely generated algebra is finite dimensional.*

Proof. Indeed, let A satisfy the identity $x^2 = 0$. Then for every x and y in A we have $(x + y)^2 = 0$ or $x^2 + y^2 + xy + yx = 0$. We know that $x^2 = y^2 = 0$, so $xy + yx = 0$. Let us multiply the last identity by x on the right: $xyx + yx^2 = 0$. Since $yx^2 = 0$, we have $xyx = 0$. Now take $x = z + t$ where z and t are arbitrary elements of A. We get $(z + t)y(z + t) = 0$. Expanding again yields: $zyz + tyt + zyt + tyz = 0$. Since we already know that $zyz = tyt = 0$, we have $zyt + tyz = 0$. Now the identity $xy + yx = 0$, which has been established above, implies $zyt = -yzt = -(yz)t = tyz$, so $tyz + tyz = 2tyz = 0$. Since the characteristic is > 2 or 0 we deduce $tyz = 0$. Recall that t, y, z were arbitrary elements of A. Thus A is nilpotent of class 3.

Let $A = \langle X \rangle$ be a finitely generated nilpotent of class k algebra. Then every element of A is a linear combination of products of no more than $k - 1$ elements from X. Since X is finite there are finitely many products of length $< k$ of elements of X. Thus A is spanned as a vector space by a finite set of elements, so A is finite dimensional. □

Just as an acorn "contains" a huge oak tree, the preceding proof "contains" all the other proofs of results related to Kurosh's question. We will see later that in Theorem 4.4.1 one can replace 2 by any other natural number n. This is the so-called Dubnov–Ivanov–Nagata–Higman theorem (Theorem 4.4.19).

Our next goal is to prove that the answer to Kurosh's question is positive if the algebra satisfies any nontrivial identity.[1] This result was obtained by Kaplansky [167] and Shirshov [297–299].

4.4.1.1 What Is a Nontrivial Ring Identity?

Usually we call an identity $u = v$ nontrivial if u is not identically equal to v in the absolutely free algebra of a given type. That definition is just fine for associative algebras over a field. But in the case of rings and, more generally, algebras over commutative rings with identity element, we need to be more careful. Indeed, suppose that u is a polynomial from the absolutely free algebra $\mathbb{Z}X^+$ (for simplicity we consider the case $K = \mathbb{Z}$, the general case is similar), such that the greatest common divisor d of the coefficients of u is not 1. Let p be a prime number dividing d. Then the identity $u = 0$ follows from the identity $px = 0$. But that identity holds in every associative (and not only associative) algebra over any field of characteristic p (why?). So there is no hope to deduce any substantially nontrivial algebraic property of a ring satisfying such an identity. Therefore Shirshov suggested to call an identity $u = 0$ of rings *nontrivial* if one of the coefficients of a monomial of highest

---

[1] Note that every ring identity $f = g$ is equivalent to the identity $f - g = 0$, so it is enough to consider only identities of the form $f = 0$

degree (length) of the identity is 1. Notice that this is not a restriction if K is a field: in that case we can always divide u by the non-zero coefficient of a monomial of highest degree.

4.4.1.2 What Does It Mean that Every 1-Generated Subalgebra of an Algebra A Is Finite Dimensional?

Lemma 4.4.2. *Let A be an associative K-algebra, $a \in A$. The subalgebra generated by a is finite dimensional if and only if $f(a) = 0$ where $f(x)$ is a polynomial with coefficients from the commutative ring K and with the leading coefficient 1.*

Proof. Indeed, if a is a root of such a polynomial f of degree n, then every power of a is a linear combination of powers a^m with $m < n$. Since the subalgebra generated by a is spanned by the powers of a, it is spanned by finitely many powers of a, and so it is finite dimensional.

Conversely, if $\langle a \rangle$ is finite dimensional, then it is spanned by finitely many powers of a, so there exists a natural number m such that a^m is a linear combination of smaller powers of a. Thus a satisfies the equality $a^m + \sum_{i<m} \alpha_i a^i = 0$ for some $\alpha_i \in K$. Thus a is a root of the polynomial $x^m + \sum_{i<m} \alpha_i x^i$. □

An algebra where every element is a root of a polynomial with the leading coefficient 1 is called *algebraic*.

4.4.1.3 The Standard Identity

Now let us turn to identities of associative algebras.

Let σ be a permutation of a finite set $\{1, \ldots, n\}$. Then σ is a product of *transpositions* (i, j), i.e., permutations that switch i and j and leave all other numbers fixed (why?). A permutation is called *even* (resp. *odd*) if it is a product of even (respectively, odd) number of transpositions. The following exercise is in every Abstract Algebra book.

Exercise 4.4.3. Prove that being even or odd does not depend on the way a permutation is represented as a product of transpositions.

Let us call an associative algebra over a commutative ring n-*dimensional* if it is spanned by n elements.

Lemma 4.4.4. *Every n-dimensional algebra over a commutative ring K satisfies the following standard identity $\mathcal{S}_{n+1} = 0$ of degree $n + 1$ where $\mathcal{S}_{n+1}$ is the following polynomial:*

$$\sum_{\sigma \in S_{n+1}} (-1)^\sigma x_{\sigma(1)} x_{\sigma(2)} \cdots x_{\sigma(n+1)} = 0. \tag{4.4.1}$$

Here S_{n+1} is the group of permutations of numbers $\{1,\ldots,n+1\}$, $(-1)^\sigma$ is the oddness of σ: it is 1 if σ is even and -1 if σ is odd.

In particular every 2-dimensional algebra satisfies the identity $xyz - xzy - yxz + yzx + zxy - zyx = 0$.

Proof. First of all notice that S_{n+1} is a *multilinear* polynomial, that is every letter occurs in every monomial in S_{n+1} exactly once. Therefore

$$S_{n+1}(x_1,\ldots,x_{i-1},\alpha x + \beta y,x_{i+1},\ldots,x_{n+1}) =$$

$$\alpha S_{n+1}(x_1,\ldots,x_{i-1},x,x_{i+1},\ldots,x_{n+1})+$$

$$\beta S_{n+1}(x_1,\ldots,x_{i-1},y,x_{i+1},\ldots,x_{n+1})$$

for arbitrary i.

Now let us take an n-dimensional algebra A spanned by a set X. Every element of A is a linear combination of elements from X. Therefore if we want to prove that S_{n+1} is identically equal to 0 in A, we have to substitute linear combinations of elements of X for x_i and prove that the result of this substitution is 0. The fact that our identity is multilinear allows us to take elements of X instead of these linear combinations (indeed, the sum of zeroes is zero).

Now let us take $t_1,\ldots,t_{n+1} \in X$. We have to prove that

$$S_{n+1}(t_1,\ldots,t_{n+1}) = 0.$$

Since $|X| = n$ at least two of the elements t_i are equal. Suppose that $t_i = t_j$, $i \neq j$. Let us divide all permutations from S_{n+1} into pairs. Two permutations belong to the same pair if one of them can be obtained from another one by switching i and j. One of the permutations in each pair is even and another one is odd, so the terms of S_{n+1} corresponding to these permutations have opposite signs. The "absolute values" of these terms are equal since $t_i = t_j$. Thus these terms cancel. Since every term belongs to one of these pairs, all terms will cancel, and the sum will be equal to 0. $\qquad\square$

In particular the algebra of all $m \times m$-matrices satisfies the identity $S_{n+1} = 0$ where $n = m^2$. Indeed this algebra is spanned by m^2 matrix units.

Remark 4.4.5. In fact by a theorem of Amitsur and Levitzki [13], for every $m \geq 1$ the algebra of all $m \times m$-matrices over a commutative ring K satisfies the identity $S_{2m} = 0$. Moreover if K is a field of characteristic 0, then $2m$ is the minimal degree of a nontrivial identity satisfied by the algebra.

Thus every finite dimensional algebra is finitely generated, algebraic, and satisfies a nontrivial identity.

The following theorem states that the converse statement also holds.

Theorem 4.4.6 (Kaplansky [167]). *Every finitely generated algebraic algebra over a commutative ring that satisfies a nontrivial identity is finite dimensional.*

This will follow from Shirshov's height theorem which is stated and proved in the next section.

4.4.2 Shirshov's Height Theorem

Theorem 4.4.7 (Shirshov, [297, 298]). *Let $A = \langle X \rangle$ be a finitely generated associative algebra over a commutative ring K and suppose that A satisfies a nontrivial identity of degree n. Then there exists a number H depending only on $|X|$ and n such that every element $a \in A$ can be represented as a linear combination of words of the form $v_1^{n_1} \cdot \ldots \cdot v_h^{n_h}$ where $h \leq H$ and each word v_i is of length less than n.*

Exercise 4.4.8. Show that Theorem 4.4.7 implies Theorem 4.4.6.

Let us prove Shirshov's theorem. First of all let us show that instead of arbitrary identities we can consider only multilinear identities (we have seen that multilinear identities are very convenient).

Lemma 4.4.9. *Every nontrivial identity of rings implies a multilinear identity of the same or smaller degree.*

Proof. The idea of the proof is similar to the one used in the proof of Theorem 4.4.1. There we proceeded from a nonlinear identity $x^2 = 0$ to a multilinear identity $xy + yx = 0$.

Let $f(x_1, \ldots, x_n) = 0$ be a non-multilinear identity. We can represent f as a sum of *normal components* f_i such that all terms (monomials) in f_i have the same content and different f_i have different contents. Then we can take a component f_i with a minimal (under the inclusion) content. Substitute 0 for variables, which are not in the content of f_i. This makes all other components 0 (we used a similar trick in the proof of Theorem 1.4.39). Thus if an algebra satisfies the identity $f = 0$, then it satisfies the identity $f_i = 0$. Subtracting f_i from f, we can proceed as before and prove that $f = 0$ implies $f_j = 0$ for every component f_j. One of these components must have a monomial of highest degree with coefficient 1, so one of the identities $f_j = 0$ is nontrivial.

Therefore we can assume that f is normal. For every variable x_i in f let x_i-*degree* of f be the largest number of occurrences of x_i in a highest degree monomial of f. For example, if $f = x_1^5 x_2^3 x_1 + x_1^4 x_2^4 x_1 + x_1^7 x_2 + x_2^6 x_1$, then the x_1-degree is 6 and the x_2-degree is 4.

Since f is normal, if every monomial of highest degree in f is multilinear, then the whole f is multilinear (prove it!). So suppose that for some i the x_i-degree d_i of f is at least 2. We can assume that d_i is the maximal possible. To simplify notation, let us assume that $i = 1$. Consider the polynomial $f' = f(x_1' + x_1'', x_2, \ldots, x_n) - f(x_1', x_2, \ldots, x_n) - f(x_1'', x_2, \ldots, x_n)$. This polynomial is identically 0 in every algebra that satisfies the identity $f = 0$, it has the same degree and the same coefficients of monomials of highest degree as f (prove it!), i.e., $f' = 0$ is also a nontrivial identity. Moreover the x_1'-degree and the x_1''-degree of f' are smaller than d_1. Thus the number of variables x

with x-degree d_1 in f' is smaller than the similar number for f. Continuing this process (which is called the process of *linearization*) we shall finally get a multilinear polynomial identity. □

Thus we can consider only multilinear identities. Let $f = 0$ be a nontrivial multilinear identity. Then every monomial of f is a product of variables in some order (each variable occurs exactly once in each monomial). Hence all monomials are permutations of the monomial $x_1 x_2 \cdots x_n$. Therefore every nontrivial multilinear identity has the following form:

$$\sum_{\sigma \in S_n} \alpha_\sigma x_{\sigma(1)} \cdots x_{\sigma(n)} = 0$$

where S_n is the group of all permutations of the set $\{1, \ldots, n\}$ and one of the coefficients α_σ is 1. This identity can be then transformed into the form

$$x_1 x_2 \cdots x_n = \sum_{\sigma \in S_n \setminus \{1\}} \beta_\sigma x_{\sigma(1)} \cdots x_{\sigma(n)} \qquad (4.4.2)$$

by moving a monomial with coefficient 1 to the left and renaming the variables. Now if we have a word u that is represented as $p v_1 \ldots v_n q$, then we can *apply* the identity (4.4.2) to it and obtain the polynomial $p(\sum_{\sigma \in S_n \setminus \{1\}} \beta_\sigma v_{\sigma(1)} \cdots x_{\sigma(n)}) q$. Notice that all monomials in this polynomial have the same length as u.

Thus if an algebra A over a field satisfies the identity (4.4.2), then every product $u_1 u_2 \cdots u_n$ of elements of A is a linear combination of permutations of this product. Theorem 4.4.7 claims that every element of A is a linear combination of words of a special kind.

Our strategy in proving that all words are linear combinations of "good" words modulo (4.4.2) would be the following: if u is not "good enough", then all the monomials in the polynomial $p(\sum_{\sigma \in S_n \setminus \{1\}} \beta_\sigma v_{\sigma(1)} \cdots x_{\sigma(n)}) q$ should be smaller than u. Thus we could use an induction if

(1) we had a partial order on the words in X^+, and
(2) every word in X^+ that is not equal in the algebra A to a linear combination of short products of powers of short words, is equal in A to a linear combination of smaller words of the same length.

Now we will make this idea work. Let $A = \langle X \rangle$ be a finitely generated algebra over a commutative ring K satisfying an identity (4.4.2). We will use the Lex partial order $\leq_\ell$ introduced in Section 1.2.3.

Definition 4.4.10. We call a word u n-*divisible* if one of its subwords can be represented as a product of n words $u_1 \ldots u_n$ and $u_1 >_\ell u_2 >_\ell \ldots >_\ell u_n$.

In this case the word $u_1 \ldots u_n$ is greater in the Lex order than any product

$$u_{\sigma(1)} \cdots u_{\sigma(n)}$$

where $\sigma \in S_n \setminus \{1\}$.

Therefore if a word is n-divisible and we can apply a multilinear identity of degree n, then this word is a linear combination of Lex smaller words of the same length. Thus it would be enough to show that if a word over X is not n-divisible then it is a product of a bounded number of powers of words of bounded lengths.

We present here a short and elegant proof of this fact due to Belov-Kanel [29].

Let us call words of lengths $\leq n$ *basic words*.

Lemma 4.4.11. *Suppose that a word v is represented in the form $v_0 t v_1 t \ldots v_n$ and the word t contains n subwords such that any two of these subwords are Lex comparable. Then v is n-divisible.*

Proof. Indeed let us represent $t = p_i t_i q_i$, $i = 1, \ldots, n$, where $t_1 >_\ell t_2 >_\ell \ldots >_\ell t_n$. Then

$$v = v_0 p_1 (t_1 q_1 v_1 p_2)(t_2 q_2 v_2 p_3) \ldots (t_n q_n v_n)$$

and

$$(t_1 q_1 v_1 p_2) >_\ell (t_2 q_2 v_2 p_3) >_\ell \ldots >_\ell t_n q_n v_n.$$

So v is n-divisible.   □

For every word v let $v_p(i)$, $i = 0, \ldots, |v| - 1$, be the prefix of v of length i, and $v_s(i)$ be the corresponding suffix, so that $v \equiv v_p(i) v_s(i)$.

Lemma 4.4.12. *Suppose that $|v| \geq n$. Then either $v_s(0), \ldots, v_s(n-1)$ are pairwise Lex comparable or $v \equiv ab^k c$, where c is a prefix of b, $|a| + |b| < n$, k is a natural number.*

Proof. Suppose that $v_s(i)$ is not comparable with $v_s(j)$ for some i and j, $i < j$. Then $v_s(j)$ is a prefix of $v_s(i) \equiv b v_s(j)$. Notice that $|b| + i$ does not exceed $n - 1$.

For every natural ℓ we have that $b^\ell v_s(j)$ is a prefix of $b^{\ell+1} v(j)$. Therefore $v_s(i)$ is a prefix of b^ℓ for some ℓ. Therefore $v_s(i) \equiv b^r c$ for some r, where c is a prefix of b. Hence

$$v \equiv v_p(i) v_s(i) \equiv v_p(i) b^r c.$$

The proof is complete since $|v_p(i)| + |b| = i + |b| < n$.   □

Lemmas 4.4.11 and 4.4.12 imply the following statement.

Lemma 4.4.13. *If $|v| \geq n^2(k+1)|X|^{n(k+1)}$ and v is not n-divisible then v contains the k-th power of a basic word.*

Proof. Represent v as a product of $n|X|^{n(k+1)}$ words of length $\geq n(k+1)$. Suppose that v does not contain k-th powers of basic words. Then by Lemma 4.4.12, each of the parts contains n pairwise comparable subwords. There are at most $|X|^{n(k+1)}$ different words in X^+ of length $n(k+1)$. Therefore v contains at least n equal disjoint subwords each of which contains n pairwise Lex comparable subwords. By Lemma 4.4.11, v is n-divisible.   □

From Lemma 4.4.13, one can deduce by induction that if an algebra $A = \langle X \rangle$ satisfies a multilinear identity of degree n then every sufficiently long word w over X is equal in this algebra to a linear combination of words containing big powers of basic words. Indeed otherwise, applying the multilinear identity as above, we represent the word as a linear combination of smaller words of the same length. Then we can apply the identity to each of these words, etc. Since there are only finite number of words of any given length, this process must eventually stop. This is not exactly what we need because we do not have a bound for the number of powers of a given base u in w. This bound is provided by the following lemma.

If w is very long, not n-divisible and cannot be expressed as a short product of powers of basic words, then w contains many subwords of the form $u^n v$ where v is different from u, $|u| = |v| \leq n$. There are fewer than $2|X|^{2n}$ such words, so if a word w is very long, it must contain n non-overlapping occurrences of the same word $u^n v$ of this form. The following lemma states that this is impossible.

Lemma 4.4.14. *Let w be a word containing n non-overlapping occurrences of a word $u^n v$ where $|v| = |u|$ and $v \neq u$. Then w contains an n-divisible subword.*

Proof. Indeed, by the condition of the lemma $w = pu^n vp_1 u^n vp_2 \cdots u^n vp_n$. Since v has the same length as u, we can compare u and v. Suppose $u >_\ell v$. Then consider the following subword of w:

$$(u^n vp_1 u)(u^{n-1}vp_2 u^2)(u^{n-2}vp_3 u^3)\cdots(vp_n)$$

It is easy to see that $u^i vp_{n-i+1} u^{n+1-i} >_\ell u^j vp_{n+1-j} u^{n+1-j}$ for every $i > j$. Therefore this word is n-divisible.

Suppose $u <_\ell v$. Then we can apply a similar argument to the following subword of w:

$$(vp_1 u^{n-1})(uvp_2 u^{n-2})(u^2 vp_3 u^{n-3})\cdots(u^n v).$$

□

Now we know that every word w that is not n-divisible, can be represented in the following form: $p_0 u_1^{k_1} p_1 u_2^{k_2} p_2 \ldots u_m^{k_m} p_m$ where $k_i \geq n$, the number m is bounded by $2|X|^{2n}$, the lengths of u_i do not exceed n, and the lengths of p_i are bounded by $n^2(n+1)|X|^{n(n+1)}$. This immediately implies Shirshov's height theorem.

Exercise 4.4.15. Under the assumptions of Theorem 4.4.7 prove that if an element a can be represented by a polynomial of degree s in X, then a is a linear combination of words in X of the form $v_1^{n_1} \cdot \ldots \cdot v_h^{n_h}$ (as in Theorem 4.4.7) of length at most s.

Exercise 4.4.16. Based on the proof of the height theorem, give an estimate for H in this theorem.

For better estimates see Belov-Kanel and Rowen [36]. For even better, subexponential, estimates see Belov-Kanel and Kharitonov [35] (that result answered a question of Zel'manov from [88]). Existence of polynomial upper bounds is an outstanding open problem. Finally note that a (non-constructive) proof of a stronger version of Theorem 4.4.7 was found by de Luca and Varricchio [83] . The proof employs uniformly recurrent words.

4.4.3 Growth Function of a Ring Satisfying a Nontrivial Identity

Shirshov hight theorem has the following important corollary

Theorem 4.4.17. *If a finitely generated associative algebra A over a commutative ring satisfies a nontrivial identity, then A has polynomial growth function.*

Proof. Indeed, suppose that A is generated by k-element set X, and let H, n be the numbers from Theorem 4.4.7. Then the number of words of length ℓ in X of the form $v_1^{n_1} \cdot \ldots \cdot v_h^{n_h}$ where $h \le H$ and each word v_i is of length less than n does not exceed $C\ell^H$ where $C = k^{nH}$ (prove it!). Hence by Exercise 4.4.15 the dimension of the subspace spanned by all elements that are represented by words in X of length at most ℓ does not exceed $C\ell^H$.   □

4.4.4 Inherently Non-finitely Based Varieties of Rings

Kaplansky's theorem 4.4.6 immediately implies

Theorem 4.4.18. *There are no locally finite inherently non-finitely based varieties of rings.*

Proof. Indeed, every locally finite variety of rings satisfies an identity $x^m = x^{m+n}$ for some $m, n \ge 1$ (since the free ring with one generator of that variety is finite). On the other hand, by Kaplansky's theorem, every ring satisfying the identity $x^m = x^{m+n}$, $m, n \ge 1$, is locally finite. Hence every locally finite variety of rings is contained in a finitely based locally finite variety $\mathrm{var}\{x^m = x^{m+n}\}$ for some m, n.   □

4.4.5 The Dubnov–Ivanov–Nagata–Higman Theorem

Theorem 4.4.19 (See [92, 152]). *Every (not necessarily finitely generated) algebra over a field of characteristic $> n$ that satisfies the identity $x^n = 0$ is nilpotent of class $2^n - 1$.*

Proof. We have proved this theorem in the case $n = 2$ (Theorem 4.4.1). Notice that the restriction on the characteristic is important. For example, one can consider the algebra $\mathbb{F}_p[X]$ of polynomials in infinitely many variables over the field $\mathbb{F}_p$ of integers modulo p (p is a prime number), and factorize this algebra by the ideal generated by all polynomials x^p where $x \in X$. The factor algebra satisfies the identity $z^p = 0$ because of Exercise 1.1.3.

But it is not nilpotent: the product $x_1 \ldots x_n$ is not equal to 0 for any n ($x_i \in X$).

We will need one nice observation concerning identities. Suppose an algebra A satisfies an identity $f(x_1, \ldots, x_n) = 0$. We can represent this identity in the form $f_0 + f_1 + \ldots + f_m = 0$ where every monomial from f_i contains exactly i occurrences of x_1 (note that f_i is not a normal component of f any more). Suppose our field has at least $m + 1$ different elements $\alpha_0, \ldots, \alpha_m$. Let us substitute $x_1 \to \alpha_i x_1$ for $i = 1, 2, \ldots, m$. Then we get:

$$\begin{cases} \alpha_0^0 f_0 + \alpha_0^1 f_1 + \ldots + \alpha_0^m f_m = 0 \\ \alpha_1^0 f_0 + \alpha_1^1 f_1 + \ldots + \alpha_1^m f_m = 0 \\ \cdots \cdots \\ \alpha_m^0 f_0 + \alpha_m^1 f_1 + \ldots + \alpha_m^m f_m = 0 \end{cases}$$

This is a system of linear equations with coefficients α_i^j and unknowns f_i. The matrix of that system is the *Vandermonde matrix* and its determinant is well-known: $\prod_{i<j}(\alpha_i - \alpha_j) \neq 0$. Therefore every f_i is identically equal to 0 on A. Hence the identity $f = 0$ implies all identities $f_i = 0$, $i = 0, \ldots, m$.

Our identity $x^n = 0$ implies the identity $(x_1 + x_2)^n = 0$. Let $f(x_1, x_2) = (x_1 + x_2)^n$. Then the identity

$$f_1(x_1, x_2) = \sum_{i=0}^{n-1} x_2^i x_1 x_2^{n-1-i} = 0$$

also follows from $x^n = 0$. Here we used the fact that the characteristic of the field is $p > n$, and so our field contains at least $n + 1$ elements.

Now consider the following sum:

$$\sum_{i,j=1}^{n-1} x^{n-1-i} z y^j x^i y^{n-1-j} =$$

$$\sum_{j=1}^{n-1} \left(\sum_{i=1}^{n-1} x^{n-1-i} (z y^j) x^i \right) y^{n-1-j} =$$

$$-\sum_{j=1}^{n-1} x^{n-1} z y^j y^{n-1-j} = -(n-1) x^{n-1} z y^{n-1}.$$

On the other hand

$$\sum_{i,j=1}^{n-1} x^{n-1-i} z y^j x^i y^{n-1-j} =$$

$$\sum_{i=1}^{n-1} x^{n-1-i} z (\sum_{j=1}^{n-1} y^j x^i y^{n-1-j}) =$$

$$-\sum_{i=1}^{n-1} (x^{n-1-i} z x^i) y^{n-1} = x^{n-1} z y^{n-1}.$$

Thus $n x^{n-1} z y^{n-1} = 0$. Since the characteristic is greater than n we have that $x^{n-1} z y^{n-1} = 0$.

Now let us consider an algebra A over the field K satisfying $x^n = 0$. Suppose (by induction) that an arbitrary associative algebra over K that satisfies the identity $x^{n-1} = 0$ is nilpotent of class $2^{n-1} - 1$. Consider the ideal I of A generated by all powers x^{n-1}. This ideal is a vector space spanned by all products of the form $p x^{n-1} q$ for some $p, q \in A$. Since we have the identity $x^{n-1} z y^{n-1} = 0$ we have that $IAI = \{0\}$ (the product of every three elements a, b, c where $a, c \in I$, $b \in A$ is 0). By the induction hypothesis we have that A/I is nilpotent of class $2^{n-1} - 1$. Therefore the product of every $2^{n-1} - 1$ elements of A belongs to I. Hence the product of every $(2^{n-1} - 1) + 1 + (2^{n-1} - 1) = 2^n - 1$ elements of A may be represented as a product of three elements a, b, c where $a, c \in I$, $b \in A$. We have proved that this product is 0, so A is nilpotent of class $2^n - 1$. This completes the proof of the Dubnov–Ivanov–Nagata–Higman theorem.   □

Notice that the estimate $2^n - 1$ is not optimal. Razmyslov [272] proved that the upper bound is n^2 while Kuzmin [190] proved that the lower bound is $\frac{n(n-1)}{2}$ (and conjectured that this is an upper bound as well). More precise estimates can be found in a recent paper by Lopatin [205] (where one also can find a much more detailed history of the attempts to prove Kuzmin's conjecture) and by Belov-Kanel and Kharitonov [35].

4.4.6 Golod Counterexamples to the Kurosh Problem

Kaplansky's Theorem 4.4.6 shows that every finitely generated algebra over a commutative ring satisfying the identity $x^n = 0$ is finite dimensional. It turns out, however, that if we allow the exponent n to depend on x, then the situation is quite different.

Recall that an algebra A is called a *nil-algebra* if for every $x \in A$ there exists a number n_x such that $x^{n_x} = 0$. Specializing to nil-algebras, the Kurosh problem asks if all finitely generated nil-algebras are nilpotent. In 1964 Golod [112], using a method of Golod and Shafarevich [114], constructed a counterexample. The method of Golod and Shafarevich came from a study of Galois groups of algebraic extensions of the field of rational numbers.

Theorem 4.4.20. *For an arbitrary finite or countable field K there exists a finitely generated infinite dimensional nil-algebra over K.*

Let K be a finite or countable field and X a finite set. Consider the free algebra $F = KX^+$.

Every algebra A is a quotient algebra of F over some ideal I. Thus we have to construct an ideal I of F such that

(1) F/I is a nil-algebra,
(2) F/I is not finite dimensional.

In order to achieve the first goal we have to have the following: for every element $p \in F$ there must be a number n_p such that $p^{n_p} \in I$. In order to achieve the second goal we want I to be as small as possible (the smaller I we take, the bigger F/I we get). Thus we may suppose that I is generated by all these p^{n_p}. Therefore our problem may be rewritten in the following form.

1. List all elements from F (recall that K is finite or countable, X^+ is also countable, so $F = KX^+$ is countable):

$$p_1, p_2, \dots.$$

For every $i = 1, 2, \dots$ choose a natural number n_i. Let I be the ideal generated by $p_i^{n_i}$. The question is how to make a choice of n_i in order to obtain an infinite dimensional quotient algebra F/I.

To solve this problem we first of all will make it a bit harder. Recall that every polynomial p in F is a sum of homogeneous components $h_1 + \dots + h_m$ where for every i all monomials in h_i have the same length i (which is the degree of h_i).

Let $|X| = d \geq 2$. Then there exist exactly d^n words over X of length n. Thus for every n the space F_n of homogeneous polynomials of degree n from F has dimension d^n (it is spanned by the words of length n in the alphabet X).

If an ideal I is generated by a set R of homogeneous polynomials, then it is spanned by homogeneous polynomials. Indeed by the definition I is spanned by $X^* R X^*$. Each polynomial in $X^* R X^*$ is homogeneous. Therefore in this case $I = (I \cap F_1) \oplus (I \cap F_2) \oplus \dots$. Thus I has a nice decomposition into a sum of finite dimensional subspaces (a subspace of a finite dimensional space is finite dimensional). This shows that ideals generated by homogeneous polynomials are easier to study than arbitrary ideals.

Unfortunately the ideal I generated by $p_i^{n_i}$ won't be generated by homogeneous polynomials. Hence we shall make a sacrifice. Let us generate I not by $p_i^{n_i}$ but by all homogeneous components of $p_i^{n_i}$. For example if $p_i = x + y^2$ and we choose $n_i = 2$ then $p_i^2 = x^2 + y^4 + xy^2 + yx^2$, the homogeneous components will be x^2 (degree 2), $xy^2 + y^2x$ (degree 3), y^4 (degree 4). We will put all these components into I. It is clear that the ideal generated by these components is bigger than the ideal generated by $p_i^{n_i}$, so the quotient algebra will be a nil-algebra also.

Let R be the subspace spanned by the homogeneous components of $p_i^{n_i}$, $i = 1, 2, \ldots, I$ be the ideal generated by R, X' be the subspace spanned by X.

Now let us introduce the key concept of the proof of Golod's theorem.

With every subspace S of F spanned by homogeneous polynomials we associate the following *Hilbert series*:

$$H_S = \sum_{i=1}^{\infty} h_n t^n$$

where h_n is the dimension of $S \cap F_n$ and t is the unknown (compare with the growth series from Section 1.5). So H_S is a formal series in one unknown. For example

$$H_{X'} = dt, \quad H_F = \sum_n d^n t^n = \frac{dt}{1 - dt}. \tag{4.4.3}$$

It turns out that the algebraic properties of F/I are closely related to the properties of the Hilbert series of R.

Since $I = (I \cap F_1) \oplus (I \cap F_2) \oplus \ldots$, F/I is isomorphic as a vector space to the sum of complements $I^c = (I \cap F_1)^c \oplus (I \cap F_2)^c \oplus \ldots$ where $(I \cap F_j)^c \oplus (I \cap F_j) = F_j$ (every proper subspace has many complements, we choose one of them). The algebra F/I is finite dimensional if and only if I^c is finite dimensional, which happens if and only if $h_i = \dim(I^c \cap F_i)$ is zero for all sufficiently large i. In other words, F/I is finite dimensional if and only if the Hilbert series H_{I^c} is a polynomial. Therefore by a clever choice of n_i we have to make H_{I^c} not a polynomial.

Notice that we can choose n_i as we want. We can, for example, suppose that R does not have elements of degree ≤ 9 (it is enough to take all $n_i \geq 10$). We also can choose n_i in such a way that all coefficients in H_R are either 0 or 1: just choose n_i big enough so that all homogeneous components of $p_i^{n_i}$ have greater degrees than homogeneous components of $p_j^{n_j}$ for $j < i$. It is possible because we choose $n_1, n_2 \ldots$ one by one.

Now we need some elementary properties of the Hilbert series. We say that $H_S \leq H_T$ if every coefficient of H_S does not exceed the corresponding coefficient of H_T (i.e., the series $H_T - H_S$ has nonnegative coefficients).

(GS$_1$) If $S = T + U$ (a sum of two subspaces) then $H_S \leq H_T + H_U$. If this is a direct sum, then $H_S = H_T + H_U$. This follows from the fact that the dimension of the sum of two subspaces does not exceed the sum of dimensions of these subspaces and the dimension of a direct sum is equal to the sum of dimensions of the summands.

(GS$_2$) If $S = TU$ (recall that TU consists of linear combinations of products tu where $t \in T$, $u \in U$), then $H_S \leq H_T H_U$. This follows from the fact that if V, V' are finite dimensional spaces of F with bases $v_1, \ldots, v_m$ and $v'_1, \ldots, v_n$, then $v_i v'_j, 1 \leq i \leq m, j \leq j \leq n$ spans VV'.

(GS$_3$) If H, H', H'' are formal power series with nonnegative coefficients and $H \leq H'$ then $HH'' \leq H'H''$.

Exercise 4.4.21. Prove properties (GS_1), (GS_2), (GS_3).

Let us use these properties. Since I is generated by R, we have (by Exercise 4.1.1) that

$$I = R + FR + RF + FRF.$$

Since $I \oplus I^c = F$ we have that

$$I = R + I^c R + IR + RI^c + RI + FRF.$$

Since $FR \subseteq I, RI^c \subseteq IF, RI \subseteq IF, IR \subseteq IF$ we have

$$I = R + I^c R + IF.$$

Now since I is an ideal, for every word $w \in X^+$ we have $Iw \subseteq Ix$ where x is the last letter of w. Therefore $IF = IX'$, so

$$I = R + I^c R + IX'. \tag{4.4.4}$$

The equality $F = I \oplus I^c$ and property (GS_1) imply $H_I = H_F - H_{I^c}$. Applying (GS_1) and (GS_2) to (4.4.4) we obtain:

$$H_F - H_{I^c} \le H_R + H_{I^c} H_R + (H_F - H_{I^c})H_{X'}.$$

Expand and move everything to the right:

$$0 \le H_{I^c} + H_{I^c} H_R - H_{I^c} H_{X'} + H_R - H_F + H_F H_{X'}.$$

Therefore

$$0 \le H_{I^c}(1 + H_R - H_{X'}) + H_R - H_F + H_F H_{X'}.$$

Recall that $H_{X'} = dt$, $H_F = \frac{dt}{1-dt}$ (see (4.4.3)). Therefore $H_F H_{X'} - H_F = -dt$. Hence we have:

$$0 \le H_{I^c}(1 + H_R - dt) + H_R - dt.$$

Add 1 to both sides of this equality:

$$1 \le (H_{I^c} + 1)(1 - dt + H_R).$$

Let $P = H_{I^c} + 1$. We can conclude that $1 \le P(1 - dt + H_R)$. The function $GS_R(t) = 1 - dt + H_R$ is called the *Golod–Shafarevich function* of R. Suppose that P is a polynomial. Then the function $P(t)$ is defined on the whole real line. Take a real number $t > 0$ small enough so that the series $GS_R(t)$ converges. Then $1 < P(t)GS_R(t)$ where $<$ is the usual order on real numbers (why?). Since all coefficients of P are not negative, $P(t) \ge 0$. Thus we would get a contradiction if we find a $t > 0$ such that $GS_R(t) < 0$.

Since all coefficients of H_R are 0 or 1 and the first 9 coefficients are 0, we have

$$GS_R(t) \le 1 - dt + \sum_{i=10}^{\infty} t^i = 1 - dt + \frac{t^{10}}{1-t}.$$

Now take $t = \frac{1.5}{d}$. Then $GS_R(t)$ converges (why?) and

$$1 - dt + \frac{t^{10}}{1-t} = -.5 + \frac{1.5^{10}}{d^{10} - 1.5d^9} < 0$$

since $d \ge 2$.

This completes the proof of Theorem 4.4.20.

Exercise 4.4.22. Prove that the algebra F/I has infinite Gelfand–Kirillov dimension. **Hint:** Prove that the radius of convergence of H_{I^c} cannot exceed $\frac{1.5}{d}$, which implies that $\dim(I_n^c)$ grow exponentially.

Golod's construction was essentially the only known construction of counterexamples to Kurosh's problem untill 2007 when a new (syntactic!) method was developed by Lenagan and Smoktunowicz in [198]. In fact they managed to construct a finitely generated nil-algebra over any countable field given by homogeneous polynomial relations such that the algebra has polynomial growth, i.e., its Gelfand–Kirillov dimension is finite.

4.4.7 Zimin Words and the Baer Radical

Here we present some classic results about the so-called Baer radical (or lower nil-radical) of an associative ring. Roughly speaking, the idea of a radical in the structure theory of rings is the following. One assigns to every ring R its ideal $\mathfrak{r}(R)$ such that $\mathfrak{r}(R/\mathfrak{r}(R)) = \{0\}$. So if the elements of $\mathfrak{r}(R)$ are "bad" in some sense, then taking the quotient over $\mathfrak{r}(R)$ allows one to get rid of all bad elements and to pass to a ring that is "good" (for instance, the quotient may have a nice decomposition as a direct sum of ideals, see Theorem 4.5.3 below). The Baer radical is one of several successful implementations of this idea (others are the Köthe radical, the Levitzki radical, the Jacobson radical, etc.). Restricted to finite rings (and even to finitely generated associative rings satisfying a nontrivial identity), all these standard radicals coincide (see Theorem 4.6.2), so considering only one of them is well sufficient for our purposes. We will need the Baer radical in the next section. Also it is interesting for us because of the role of Zimin words. Actually, it is the description of the Baer radical where Zimin words seem to appear for the very first time.

An ideal P of a ring R is called *prime* if for any two ideals I and J of R, the inclusion $IJ \subseteq P$ implies that either $I \subseteq P$ or $J \subseteq P$. As the terminology suggests, this notion is inspired by the notion of a prime number.

Exercise 4.4.23. Prove that an ideal $I \subset \mathbb{Z}$ is prime if and only if either $I = \{0\}$ or $I = p\mathbb{Z}$ where p is a prime number.

The intersection of all prime ideals of a ring R is called the *Baer radical* of R. In the literature, this radical is often referred to as the *lower nil-radical*. The above definition is due to McCoy [224]; the fact that the radical defined by McCoy coincides with the one defined by Baer [16] was proved by Levitzki [201]. We denote the Baer radical of R by $B(R)$.

Our goal is to characterize the elements of $B(R)$. First, we characterize prime ideals in terms of elements.

Lemma 4.4.24. *An ideal P of a ring R is prime if and only if for every $a, b \notin P$ there exists $x \in R$ such that $axb \notin P$.*

Proof. Assume that P is prime and let $axb \in P$ for all $x \in R$, that is $aRb \subseteq P$. Then $(RaR)(RbR) \subseteq P$ hence either $RaR \subseteq P$ or $RbR \subseteq P$. Suppose that $RaR \subseteq P$ and let $I = a + aR + Ra + RaR$. Then I is the smallest ideal containing a and $I^3 \subseteq P$. This implies that $I \subseteq P$ hence $a \in P$. Similarly, if $RbR \subseteq P$, then $b \in P$.

Conversely, assume that P is an ideal satisfying the condition of the lemma. If I and J are ideals not contained in P, take elements $a \in I \setminus P$ and $b \in J \setminus P$. Then $axb \notin P$ for some $x \in R$ while $axb \in IJ$. Hence $IJ \nsubseteq P$ and the ideal P is prime.   □

Suppose that P is a prime ideal of a ring R and $a_1 \notin P$. By Lemma 4.4.24 there exists $x_2 \in R$ such that $a_1 x_2 a_1 \notin P$. If we denote $a_1 x_2 a_1$ by a_2, then by the same argument applied to a_2 there exists $x_3 \in R$ such that $a_2 x_3 a_2 \notin P$. We can iterate the process, finding an infinite sequence $x_2, x_3, \ldots, x_n, \ldots$ such that no element in the sequence $a_1, a_2, \ldots, a_n, \ldots$ defined by $a_n = a_{n-1} x_n a_{n-1}$ belongs to P. Observe that the sequence $a_1, a_2, \ldots, a_n, \ldots$ consists of values of Zimin words Z_i: indeed,

$$a_1 = Z_1(a_1), \ a_2 = Z_2(a_1, x_2), \ldots, a_n = Z_n(a_1, x_2, \ldots, x_n), \ldots c \qquad (4.4.5)$$

(where $Z_i(u_1, \ldots, u_i)$ is the result of substitution $x_1 \mapsto u_1, \ldots, x_i \mapsto u_i$ into Z_n). This observation suggests the following definition: for every sequence

$$a_1, x_2, x_3, \ldots, x_n, \ldots$$

of elements of a ring R, we say that the sequence (4.4.5) is a *Z-sequence starting with* a_1. Z-sequences were introduced by Levitzki [201] (under the name "m-sequences"). We say that a Z-sequence $a_1, a_2, \ldots, a_n, \ldots$ *vanishes* if $a_k = 0$ for some k—then of course $a_\ell = 0$ for all $\ell \geq k$. We are ready to present Levitzki's characterization of the elements contained in the Baer radical.

Proposition 4.4.25. *Let R be a ring. An element $a \in R$ belongs to the Baer radical $B(R)$ if and only if every Z-sequence starting with a vanishes.*

Proof. If $a \notin B(R)$, then there is a prime ideal P of R such that $a \notin P$. The argument preceding the definition of a Z-sequence shows that then there exists a Z-sequence starting with a all of whose elements do not belong to P. Clearly, this sequence does not vanish.

Conversely, let $a_1(= a), a_2, \ldots, a_n, \ldots$ be a Z-sequence starting with a that does not vanish. Consider the set of all ideals of R that contain no element of the Z-sequence. The set is nonempty (since it contains $\{0\}$) and is easily seen to satisfy the conditions of Zorn's Lemma 1.1.7. Let P be a maximal ideal in this set; we will show that P is prime. If I and J are ideals not contained in P, then both $I + P$ and $J + P$ contain some elements of the Z-sequence $a_1, a_2, \ldots, a_0 n, \ldots c$. Clearly, if a_k belongs to $I + P$, then so does a_ℓ for all $\ell \geq k$. Therefore we may assume that $I + P$ and $J + P$ contain a_k for some k. Since $a_{k+1} = a_k x_{k+1} a_k$ for some x_{k+1}, we have $a_{k+1} \in IJ + P$. On the other hand, $a_{k+1} \notin P$ by the choice of P hence $IJ \nsubseteq P$. Thus, P is prime and $a \notin B(R)$ since $a \notin P$.   □

With the above characterization in hand, we can easily deduce three basic properties of the Baer radical.

Corollary 4.4.26. *Let R be a ring.*

(1) $B(R)$ *is a nil-ring.*
(2) $B(R)$ *contains every nilpotent right or left ideal.*
(3) $B(R/B(R)) = \{0\}$; *in particular, the ring $R/B(R)$ has no non-zero nilpotent right (left) ideals.*

Proof. (1) Take an element $a \in B(R)$ and consider the Z-sequence $a = Z_1(a)$, $a^3 = Z_2(a, a), \ldots, a^{2^n - 1} = Z_n(a, a, \ldots, a), \ldots c$. Since the sequence vanishes, some power of a is equal to 0.

(2) If I is a right ideal of R (for left ideals the proof is similar) and a Z-sequence $a_1, a_2, \ldots, a_n, \ldots$ starts with $a_1 \in I$, then $a_k \in I^{2^{k-1}}$ for all $k = 1, 2, \ldots c$. Therefore if I is a nilpotent right ideal, then every Z-sequence starting with an arbitrary element of I vanishes hence $I \subseteq B(R)$.

(3) Take an arbitrary $\bar{a} \in B(R/B(R))$ and let $a \in R$ be a preimage of $\bar{a}$. If $a_1(= a), a_2, \ldots, a_n, \ldots$ is a Z-sequence starting with a, then the sequence of images $\bar{a}_1, \bar{a}_2, \ldots, \bar{a}_n, \ldots$ is a Z-sequence in $R/B(R)$ starting with $\bar{a}$ so it vanishes. This means that $a_k \in B(R)$ for some k. It is clear that $a_k, a_{k+1}, \ldots, a_n, \ldots$ is a Z-sequence starting with a_k so it vanishes. Therefore the sequence $a_1(= a), a_2, \ldots, a_n, \ldots$ vanishes and $a \in B(R)$, that is, $\bar{a} = 0$. The "in particular" statement follows now from Part (2).

□

4.5 The Finite Basis Problem

Here we discuss the finite basis problem for identities of associative rings. It is worth comparing this discussion with a similar discussion for the group case in the next chapter. One can observe that in a sense associative rings tend to be "more finitely based" than groups: even though methods applied to the study of the finite basis problem in rings and groups are quite similar, as a rule, "positive" results are easier to prove in the ring case while "negative" examples are easier for groups.

4.5.1 Basic Facts About Finite Associative Rings

An associative ring R is called a *division ring* if $R \setminus \{0\}$ is a group under multiplication (called the *multiplicative group* of R).

Lemma 4.5.1 (Wedderburn [214]). *Every finite division ring R is a field (i.e., it is commutative) and $R \setminus \{0\}$ is a cyclic group (under multiplication).*

There are very many short proofs of this important result including [11, 151] and the proof published by the future unabomber Ted Kaczynski [164], so we do not include the proof here.

Lemma 4.5.1 can be used to show that for every n, the ring of all $n \times n$-matrices over a finite field is generated by two elements. A concrete 2-element generating set is presented in the following exercise where $e_{i,j}$ stands for the usual matrix unit.

Exercise 4.5.2. If α is a generator of the multiplicative group $F \setminus \{0\}$, then the ring of all $n \times n$-matrices over F is generated by the matrices αe_{12} and $e_{12} + e_{23} + \ldots + e_{n-1\,n} + e_{n\,1}$. **Hint:** Prove that every $n \times n$-matrix unit $e_{i,j}$ belongs to the ring generated by these two matrices.

The following theorem contains facts which clarify the structure of finite rings.

Theorem 4.5.3. *The Baer radical $B(R)$ of every finite ring R is a nilpotent ideal. If $R \neq B(R)$, then the quotient $R/B(R)$ decomposes as a direct sum of ideals $R/B(R) = S_1 + \ldots + S_m$ where each S_i is isomorphic to the ring of all $n_i \times n_i$-matrices over a finite field F_i. In particular, $R/B(R)$ has no non-zero nilpotent quotients.*

Proof. By Corollary 4.4.26, Part (1), $B(R)$ is a nil-ideal. Hence, $B(R)$ is nilpotent by Lemma 3.1.16.

Let I be a minimal right ideal of $\overline{R} = R/B(R)$. By Part (3) of Corollary 4.4.26, I is not nilpotent. Then by Lemmas 3.1.14 and 3.1.16, I contains a non-zero idempotent e. Then the right ideal generated by e, i.e., $e\overline{R}$, is a non-zero right ideal of $\overline{R}$ contained in I, hence $I = e\overline{R}$, that is I is generated as a right ideal by an idempotent. Recall that there is a natural partial order on the set of idempotents: $e \leq f$ if $ef = fe = e$.

Exercise 4.5.4. Prove that if I is a minimal right ideal of $\overline{R}$, then the idempotent e is minimal in the multiplicative semigroup $\overline{R}$.

The same statement of course is true for minimal left ideals. Let us prove the converse statement of Exercise 4.5.4. Suppose that e is a minimal idempotent in the multiplicative semigroup $\overline{R}$ but $e\overline{R}$ is not a minimal right ideal. Then there exists an idempotent $f \neq 0$ such that $f\overline{R} \subsetneq e\overline{R}$, and $f\overline{R}$ is a minimal right ideal of $\overline{R}$. Then $ex = f$ for some x or $ef = f$. Hence $fefe = fe$. If $fe = 0$, then $f^2 = efef = 0$, a contradiction. So fe is a non-zero idempotent in $f\overline{R}$. But $efe = fe, fee = fe$, so $fe \leq e$. Hence $fe = e$, and $f\overline{R} = e\overline{R}$, a contradiction. Thus minimal idempotents in $\overline{R}$ are precisely the idempotents generating minimal right (and, of course, minimal left) ideals.

A set of idempotents $e_1, \ldots, e_n$ is called *orthogonal* if $e_i e_j = e_j e_i = 0$ whenever $i \neq j$. Note that the sum of an orthogonal set of idempotents is an idempotent (prove it!).

We claim that an arbitrary non-zero right ideal J of $\overline{R}$ is generated (as a right ideal) by an idempotent that is the sum of an orthogonal set of minimal idempotents. If J is minimal, then J is generated by a minimal idempotent as shown above. Suppose that J is not minimal. Inducting on $|J|$, we may assume that the claim holds for all non-zero right ideals that are strictly contained in J. Now take a minimal non-zero right ideal $I_0 \subsetneq J$ and let e_0 be an idempotent such that $I_0 = e_0\overline{R}$. Consider the set $I = \{x - e_0 x, x \in J\}$ which is clearly a right ideal in $\overline{R}$. Note that $e_0 I = \{0\}$ and $J = I + I_0$. If $I = \{0\}$, then $x = e_0 x$ for all $x \in J$, that is, $J = I_0$, a contradiction. Thus, $I \neq 0$; on the other hand, $I \subsetneq J$ since $e_0 \notin I$. Therefore, the induction assumption applies to I. Let f be an idempotent that generates I as a right ideal and can be represented as the sum of an orthogonal set of minimal idempotents $f_1, \ldots, f_n$. It is easy to see that each $g_i = f_i - f_i e_0$, $i = 1, \ldots, n$ is an idempotent generating the same right ideal as f_i since $g_i f_i = f_i$, hence g_i generates a minimal right ideal. Also the set g_i, $i = 1, \ldots, n$ is clearly orthogonal, $e_0 g_i = g_i e_0 = 0$, and $g = g_1 + \ldots + g_n = f - f e_0$ generates I as a right ideal of $\overline{R}$. Therefore $e = e_0 + g = e_0 + g_1 + \ldots + g_n$ is the sum of an orthogonal set of minimal idempotents, and e generates $J = I_0 + I$ as a right ideal since $ee_0 = e_0$ and $eg = g$. This proves our claim.

Now suppose that S is a non-zero two-sided ideal in $\overline{R}$. Since S is a right ideal, $S = e\overline{R}$ for some idempotent e as in the previous paragraph. Then $s = es$ for all $s \in S$. We claim that $s = se$ as well. Indeed, the set $L = \{s - se \mid s \in S\}$ is a left ideal in $\overline{R}$ and $L^2 \subseteq LS = LeS = \{0\}$. Since $\overline{R}$ has no non-zero nilpotent left ideals by Corollary 4.4.26, Part (3), we conclude that $L = \{0\}$ and $s = se$ for all $s \in S$. Thus, e is a two-sided identity element of S. Moreover, for each $x \in \overline{R}$, we have $ex, xe \in S$ hence $ex = exe$ and $xe = exe$. This means that e commutes with an arbitrary $x \in \overline{R}$, that is, e is a *central* idempotent of $\overline{R}$.

For $S = \overline{R}$ the above argument shows that $\overline{R}$ has a two-sided identity element which we denote by 1. This immediately implies the "in particular" statement of the theorem (because nilpotent rings cannot have identity elements).

Now let M be a minimal non-zero two-sided ideal of $\overline{R}$. The following exercise is similar to Exercise 3.6.6.

Exercise 4.5.5. Prove that if I is a minimal non-zero two-sided ideal of a ring, then either $I^2 = \{0\}$ or I is a simple ring.

Since $\overline{R}$ has no no-zero nilpotent ideals, Exercise 4.5.5 implies that M is a simple ring. Further, $M = e\overline{R} = \overline{R}e$ for some idempotent $e = e_1 + \ldots + e_n$ where $e_1, \ldots, e_n$ is an orthogonal set of minimal idempotents. Since e_i is a minimal idempotent, $G_i = e_i\overline{R}e_i$ is a group with zero element adjoined. It is also a subring of $\overline{R}$. By Lemma 4.5.1 G_i is a finite field. Consider the set $N = \bigcup_{i,j=1}^{n} e_i\overline{R}e_j$. It is easy to see that N is a subsemigroup of the multiplicative semigroup $\overline{R}$. Suppose that for some i, j, we have $e_i\overline{R}e_j = \{0\}$. Then $\overline{R}e_i\overline{R}\,\overline{R}e_j\overline{R} = \{0\}$. But M as a simple ring is generated as a two-sided ideal by any of its non-zero elements, in particular, by e_i and by e_j. Thus, if $x, y \in M$, we can write x as a sum of elements from $\overline{R}e_i\overline{R}$ and write y as a sum of elements from $\overline{R}e_j\overline{R}$ hence multiplying x by y yields 0. Hence $M^2 = \{0\}$, a contradiction. Thus $e_i\overline{R}e_j \neq 0$. Let $x = e_ipe_j$ be a non-zero element of N. Since e_i is a minimal idempotent, we have $xt = e_i$ for some t. For every $\ell = 1, \ldots, n$ there exists $y \in \overline{R}$ such that $e_iye_\ell \neq 0$. Since e_ℓ is a minimal idempotent, then $ze_iye_\ell = e_\ell$ for some z. Therefore the ideal generated by every non-zero element of N is N. Therefore N is a completely 0-simple semigroup by Theorem 3.6.16. Clearly an element $e_ixe_j \in N$ is an idempotent if and only if it is equal to e_i. Since the product of every two different idempotents of N is 0, we conclude, using Theorem 3.6.16 and Exercise 3.6.19, that N is the Brandt semigroup over the group $G = e_1\overline{R}e_1 \smallsetminus \{0\}$. We can represent elements of N as triples (i, g, j), $i, j = 1, \ldots, n, g \in G$, as in Theorem 3.6.16.

Every element of M is a sum of triples $(i, g_{i,j}, j)$. With every sum t like that, we associate the $n \times n$-matrix $m_t = (g_{i,j})$ where $g_{k,\ell} = 0$ if the summand $(k, g_{k,\ell}, \ell)$ is missing. It is easy to see that the map $\phi: t \mapsto m_t$ is a well-defined, that is if $t = t'$ in R then $m_t = m_{t'}$. Indeed, if

$$t = \sum_{i,j}(i, g_{i,j}, j) = \sum_{i,j}(i, g'_{i,j}, j) = t',$$

then multiply both sums from the left by $(i, 1, i)$ and from the right by $(j, 1, j)$. We get $(i, g_{i,j}, j) = (i, g'_{i,j}, j)$ for every i, j, hence $m_t = m_{t'}$. Also clearly ϕ is a surjective homomorphism. Since M is a simple ring by Exercise 4.5.5, ϕ must be injective, and M is isomorphic to a matrix ring over the field $G \cup \{0\}$. Since $\overline{R} = M + (1 - e)M$ (a direct sum), we conclude by induction on $|\overline{R}|$ that $\overline{R}$ is a direct sum of rings of matrices over finite fields. $\square$

4.5.2 Positive Result. Identities of Finite Rings

In this subsection we prove the Kruse–L'vov parts of Theorems 1.4.33, 1.4.36:
every finite associative ring generates a finitely based and even Cross variety.
The proof presented here follows the original proof by L'vov with a neat
simplification suggested by Latyshev [192].

Remark 4.5.6. Note that for rings the condition (c) in the definition of a
Cross variety (see Section 1.4.7) can be substituted by a formally weaker
condition

(c') there exists a positive integer N such that every critical ring in $\mathcal{V}$ can
be generated by N elements.

Indeed, in a locally finite variety of rings there are only finitely many
N-generated rings (prove it!).

Thus, to prove Theorem 1.4.36 for finite rings, it is sufficient to construct,
given a finite ring A, a finite set Σ of identities holding in A such that the
variety var Σ is locally finite and satisfies (c'). Then var Σ is a Cross variety
and so is var A by Proposition 1.4.35.

4.5.2.1 The Number of Generators of Critical Rings

The following observation by Latyshev [192] serves as the main tool for
bounding the number of generators for critical rings.

Lemma 4.5.7. *Let R be a finite ring and $R_1, \ldots, R_m$ be proper subrings
of R such that each element $r \in R$ can be represented as a sum $r = \sum_{i=1}^{m} r_i$
with $r_i \in R_i$ (so $R = R_1 + \ldots + R_m$). Suppose that for each nonempty subset
$\Lambda \subseteq \{1, \ldots, m\}$ either the subrings R_j with $j \in \Lambda$ do not generate R or
$R_{i_1} R_{i_2} \cdots R_{i_p} = \{0\}$ whenever $\{i_1, \ldots, i_p\} = \Lambda$. Then the ring R is not critical.*

Proof. By the definition of critical algebras, we need to show that R is in
the variety generated by its proper subrings. Thus we must prove that if an
identity $f(x_1, \ldots, x_n) = 0$ holds in every proper subring of R, then it holds
in R. Let $a_1, \ldots, a_n$ be elements of R. Since every element of R is a sum of
elements from R_i, each a_i can be represented as $\sum_{j=1}^{m} a_{i,j}$ where $a_{i,j} \in R_j$.
Our goal is to show that $f(a_1, \ldots, a_n) = 0$. Note that $f(a_1, \ldots, a_n)$ is a sum
of words in $a_{i,j}$ and we can write

$$f(a_1, \ldots, a_n) = \sum_{\emptyset \neq \Lambda \subseteq \{1, \ldots, m\}} f_\Lambda,$$

where each f_Λ is a sum of words involving precisely the letters $a_{i,j}$ with $j \in \Lambda$
(i.e., f_Λ is a normal component of $f(\sum a_{1,j}, \sum a_{2,j}, \ldots, \sum a_{n,j})$).

It is enough to prove that each f_Λ is equal to 0 in R. We induct on $|\Lambda|$. If
$|\Lambda| = 1$, $\Lambda = \{k\}$, then $f_\Lambda = f(a_{1,k}, \ldots, a_{n,k})$. Hence $f_\Lambda = 0$ since R_k satisfies

the identity $f(x_1, \ldots, x_n) = 0$ being a proper subring of R. Now suppose that $|\Lambda| > 1$ and let R' be the subring generated by R_j, $j \in \Lambda$. If $R' = R$, then by the assumption of the lemma, $R_{i_1} R_{i_2} \cdots R_{i_p} = \{0\}$ whenever $\{i_1, \ldots, i_p\} = \Lambda$. Thus every summand in f_Λ is 0 and so is f_Λ. If R' is a proper subring, for each $i = 1, \ldots, n$ let a_i' be obtained from a_i by deleting the summands whose second indices are not in Λ. Then $a_i' \in R'$ and $f(a_1', \ldots, a_n') = 0$ (since R' satisfies the identity $f(x_1, \ldots, x_n) = 0$). However, it is easy to see that

$$f(a_1', \ldots, a_n') = f_\Lambda + \sum_{\varnothing \neq \Theta \subsetneq \Lambda} f_\Theta$$

(check it!). We can apply the induction assumption to each f_Θ, where $\varnothing \neq \Theta \subsetneq \Lambda$ since $|\Theta| < |\Lambda|$. So $\sum_{\varnothing \neq \Theta \subsetneq \Lambda} f_\Theta = 0$. Hence $f_\Lambda = 0$. $\square$

As a first application of Lemma 4.5.7, we have the following reduction.

Proposition 4.5.8. *Let R be a critical ring and $I \subsetneq R$ be a nilpotent of class k ideal in R. If the quotient ring R/I can be generated by s elements, then the ring R can be generated by $k + s - 1$ elements.*

Proof. We may assume that $k > 1$ as otherwise the result is obvious. Take a generating set $\bar{a}_1, \ldots, \bar{a}_s$ of R/I and fix a preimage $a_i \in R$ for each $\bar{a}_i$, $i = 1, \ldots, s$. If R_0 is the subring in R generated by $a_1, \ldots, a_s$, then for each element $r \in R$ there is an element $q \in R_0$ such that $r - q \in I$. Clearly, the set $(I \setminus I^2) \cup R_0$ generates R. Let $\{b_1, \ldots, b_m\}$ be the smallest subset of $I \setminus I^2$ such that $\{b_1, \ldots, b_m\}$ together with R_0 generates R. If $m < k$, the desired conclusion holds so we may assume that $m \geq k$. Let $R_j = b_j + R_0 b_j + b_j R_0 + R_0 b_j R_0 + I^2$ for each $j = 1, \ldots, m$. Then $R_0, R_1, \ldots, R_m$ are proper subrings of R and each element $r \in R$ can be represented as $r = \sum_{j=0}^{m} r_j$ with $r_j \in R_j$. By minimality of $\{b_1, \ldots, b_m\}$, for any $\ldots$ subset $\Lambda \subsetneq \{0, 1, \ldots, m\}$ the rings R_j with $j \in \Lambda$ do not generate R, and the products of all the rings $R_0, R_1, \ldots, R_m$ (in any fixed order) are contained in I^m. The assumption that $m \geq k$ implies that $I^m = \{0\}$ but then Lemma 4.5.7 applies showing that the ring R is not critical, a contradiction. $\square$

In a similar way one can bound the number of generators of a nilpotent critical ring in terms of its nilpotency class.

Proposition 4.5.9. *Every nilpotent critical ring of class k can be generated by $k - 1$ elements.*

Proof. We shall use the following simple property of nilpotent rings:

Exercise 4.5.10. Prove that every nilpotent ring R is generated by the set $R \setminus R^2$.

Now let R be a nilpotent critical ring of class k and let $\{b_1, \ldots, b_m\}$ be the smallest subset of $R \setminus R^2$ that generates R.

Suppose that $m \geq k$. Let $R_j = \{nb_j \mid n \in \mathbb{Z}\} + R^2$ for each $j = 1, \ldots, m$. Then $R_1, \ldots, R_m$ are proper subrings of R and each element $r \in R$ can be represented as $r = \sum_{j=0}^{m} r_j$ with $r_j \in R_j$. By minimality of $\{b_1, \ldots, b_m\}$, for any ... subset $\Lambda \subsetneqq \{1, \ldots, m\}$ the rings R_j with $j \in \Lambda$ do not generate R. The product of the rings $R_1, \ldots, R_m$ in any order is contained in R^m (by the definition of R^m). Since $m \geq k$, we have $R^m = \{0\}$, so every product of $R_1, \ldots, R_m$ is $\{0\}$. Lemma 4.5.7 implies that R is not critical, a contradiction. Hence $m < k$. $\qquad\qquad\qquad\qquad\qquad\qquad\qquad\qquad\qquad\qquad\qquad\qquad\qquad\quad$ $\square$

Proposition 4.5.11. *Let R be a non-nilpotent critical ring whose Baer radical $B(R)$ is nilpotent of class k. Then the number of summands in the decomposition $R/B(R) = S_1 + \ldots + S_m$ from Theorem 4.5.3 does not exceed $2k-1$.*

Proof. Suppose that $m \geq 2k$. Let R_i be the preimage of S_i in R, $i = 1, \ldots, m$, and $R_0 = B(R)$. Then $R_0, R_1, \ldots, R_m$ are proper subrings of R and each element $r \in R$ can be represented as $r = \sum_{j=0}^{m} r_j$ with $r_j \in R_j$ (why?) For any ... subset $\Lambda \subsetneqq \{0, 1, \ldots, m\}$ the rings R_j with $j \in \Lambda$ do not generate R because their images do not generate $R/B(R)$. Now consider a product of all the rings $R_0, R_1, \ldots, R_m$ (in any fixed order). Since $R_i R_j \subseteq R_0$ for all $i \neq j$, $i, j = 1, \ldots, m$, and $m \geq 2k$, the product is contained in $R_0^k = (B(R))^k = \{0\}$. Lemma 4.5.7 then implies that the ring R is not critical, a contradiction. $\quad\square$

Now we can collect all the above information in order to bound the number of generators of an arbitrary critical ring in terms of the nilpotency class of its Baer radical.

Proposition 4.5.12. *A critical ring whose Baer radical is nilpotent of class k can be generated by at most $5k - 3$ elements.*

Proof. Let R be a critical ring whose Baer radical $B(R)$ is nilpotent of class k. If $R = B(R)$, the number of generators of R does not exceed $k-1 < 5k-3$ by Proposition 4.5.9. If $R \neq B(R)$, Proposition 4.5.11 and Exercise 4.5.2 imply that the quotient $R/B(R)$ can be generated by at most $4k - 2$ elements, and the result follows from Proposition 4.5.8. $\qquad\qquad\qquad\qquad\qquad\qquad\quad$ $\square$

4.5.2.2 Using Identities to Bound the Nilpotency Class of Nilpotent Rings

Recall that we aim to construct, given a finite associative ring A, a finite set Σ of identities satisfied by A such that the variety var Σ is locally finite and, for every critical ring in var Σ, the number of generators can be bounded by some constant depending only on Σ. Proposition 4.5.12 implies that the second property is granted as soon as the nilpotency class of every nilpotent ring is bounded by some constant, and this can be easily expressed in the language of identities:

Lemma 4.5.13. *Let V be a ring variety, k a positive integer. The following are equivalent:*

(1) *Every nilpotent ring in V is nilpotent of class $\leq k$,*
(2) *Every k-generated nilpotent ring in V is nilpotent of class k,*
(3) *V satisfies an identity of the form*

$$x_1 x_2 \cdots x_k = f(x_1, x_2, \ldots, x_k), \tag{4.5.1}$$

where the length of each monomial of the polynomial $f(x_1, x_2, \ldots, x_k)$ is at least $k + 1$.

Proof. $(1) \to (2)$ is obvious.

$(2) \to (3)$. Let F_k be the relatively free ring of V with free generators $x_1, x_2, \ldots, x_k$. The quotient ring F_k/F_k^{k+1} is k-generated and nilpotent of class k. In particular, the product of the images of the free generators in F_k/F_k^{k+1} is equal to 0. Therefore the product $x_1 x_2 \cdots x_k$ of the generators themselves belongs to F_k^{k+1}. Elements of F_k^{k+1} are representable as polynomials in $x_1, x_2, \ldots, x_k$ whose monomials have length at least $k + 1$. Thus, $x_1 x_2 \cdots x_k = f(x_1, x_2, \ldots, x_k)$ where $f(x_1, x_2, \ldots, x_k)$ is such a polynomial. By Exercise 1.4.26 $x_1 x_2 \cdots x_k = f(x_1, x_2, \ldots, x_k)$ is an identity in V.

$(3) \to (1)$. In any ring R we have a decreasing sequence of ideals

$$R \supseteq R^2 \supseteq R^3 \supseteq \ldots c.$$

If R satisfies an identity of the form (4.5.1), then $R^k \subseteq R^\ell$ where $\ell > k$ is the minimum length of monomials of the polynomial $f(x_1, x_2, \ldots, x_k)$. Hence the above sequence stabilizes at R^k, that is, $R^k = R^{k+1} = \ldots c$. If R is nilpotent, this implies that $R^k = \{0\}$.   □

Now we observe that every variety generated by a finite ring has the above property for a suitable k.

Lemma 4.5.14. *Let A be a finite ring and let k be the maximum nilpotency class of nilpotent subrings of A. Then the nilpotency class of every nilpotent ring $R \in \operatorname{var} A$ does not exceed k.*

Proof. By Lemma 4.5.13 we may assume that R is k-generated. Let F_k be the relatively free ring of rank k in $\operatorname{var} A$. There is a surjective homomorphism $\varphi : F_k \to R$. Let I be the kernel of φ. If $B = B(F_k)$ is the Baer radical of F_k, then $\varphi(B)$ is an ideal in R and $R/\varphi(B) \cong F_k/(B + I)$ is a quotient of the ring F_k/B. The ring F_k is finite (since $\operatorname{var} A$ is locally finite) and the quotient of a finite ring over its Baer radical has no non-zero nilpotent quotients by Proposition 4.5.3. On the other hand, the ring $R/\varphi(B)$ is nilpotent. We conclude that $R/\varphi(B) = \{0\}$, that is $R = \varphi(B)$. Since F_k embeds into a Cartesian product of several copies of the ring A, the ideal B embeds into a Cartesian product of nilpotent subrings of A hence the nilpotency class of B does not exceed k. Therefore the nilpotency class of $R = \varphi(B)$ does not exceed k as well. $\square$

4.5.2.3 The Conclusion of the Proof of Kruse–L'vov Theorem

Now let us conclude the proof of the ring part of Theorem 1.4.33.

Proof. Let A be a finite ring. By Lemmas 4.5.13 and 4.5.14, A satisfies an identity of the form (4.5.1) for some k. By Exercise 1.8.10 applied to the additive group A, the ring A satisfies an identity $nx = 0$ for $n = |A|$. Let $\mathcal{V}$ be the variety defined by (4.5.1) and $nx = 0$; we claim that $\mathcal{V}$ is a Cross variety. Indeed, $\mathcal{V}$ is finitely based and Lemma 4.5.13 and Proposition 4.5.12 imply that for every critical ring in $\mathcal{V}$, the number of generators is bounded by $5k-3$. It remains to verify that $\mathcal{V}$ is locally finite. We take a finitely generated ring $R \in \mathcal{V}$ and prove that it is finite by induction on n.

The claim is obvious if $n = 1$. If $n > 1$, decompose it as $n = mp$ for some prime number p. Then the ring R is an extension of the ideal pR by the quotient R/pR. The ring R/pR satisfies $px = 0$ and can be considered as an algebra over the p-element field $\mathbb{F}_p$ of integers modulo p. Identifying all variables in (4.5.1), we get that R/pR satisfies the identity $x^k = x^{k+1}g(x)$ for some polynomial $g(x)$ hence R/pR is algebraic. By Theorem 4.4.6 (Kaplansky's theorem) R/pR is a finite dimensional algebra over $\mathbb{F}_p$ and hence finite. We may now use the following general statement.

Lemma 4.5.15 (Lewin [202]). *If I is an ideal of a finitely generated ring R such that the quotient R/I is finite, then I is finitely generated as a ring.*

Proof. Let $r = |R/I|$. Choose a finite set $u_1, \ldots, u_\ell \in R$ which includes a finite set of generators of R and representatives of all cosets of I (the set will contain at least one representative from each coset). Note that by Exercise 1.8.10, $ru_i \in I$ for every $i = 1, \ldots, \ell$. For all $i, j = 1, \ldots, \ell$ represent $u_i u_j$ as $u_i u_j = u_t + s_{ij}$ for some $t \in \{1, \ldots, \ell\}$ and $s_{ij} \in I$. Thus all elements

$$s_{ij}, \ u_m s_{ij}, \ s_{ij} u_m, \ u_m s_{ij} u_n \quad (i, j, m, n = 1, \ldots, \ell) \qquad (4.5.2)$$

belong to I.

Exercise 4.5.16. Prove that I is generated as a ring by the elements (4.5.2) together with elements ru_i, $i = 1, \ldots, \ell$, and those integer combinations $\sum_1^\ell k_i u_i$ with $0 \le k_i < r$ which belong to I. **Hint:** Consider any element t from I. Then t is a linear combination with integer coefficients of products of $u_i's$. Each of the products can be represented as an integer multiple of one of the u_j plus a product of elements from (4.5.2). Finally any linear combination of $u_1, \ldots, u_\ell$ with integer coefficients is a sum of several elements of the form ru_i and a linear combination $\sum_1^\ell k_i u_i$ with $0 \le k_i < r$.

$\square$

Lemma 4.5.15 implies that pR is a finitely generated ring. Since pR satisfies (4.5.1) and $mx = 0$, $m = n/p < n$, it is finite by the induction assumption. Thus, the ring R is finite too.

We have verified that $\mathcal{V}$ is a Cross variety. By Proposition 1.4.35, the variety var A is also a Cross variety. $\qquad\qquad\qquad\qquad\qquad\qquad\qquad$ □

4.5.3 Negative Result

The existence of a non-finitely based variety of associative rings was a long-standing open problem. In 1950 Specht [306] suggested this problem for the special case of associative algebras over a field of characteristic 0 while the general case have been studied since the 1960s. Kemer [173], see also [174], proved that every variety of associative algebras over a field of characteristic 0 is finitely based thus solving Specht's problem. This made the general case of the finite basis problem for associative rings even more intriguing. First examples of non-finitely based varieties of associative rings were found by three people almost simultaneously: Belov-Kanel [30, 31], Grishin [124, 125] and Shchigolev [291, 292]. The example presented here is a simplified version (due to Gupta and Krasilnikov [140]) of Belov-Kanel's example from [30]. Note that it uses ideas employed earlier by Vaughan-Lee, Bryant and Yu. Kleiman in the case of varieties of groups (see Sections 5.3.1, 5.3.2).

Recall that (x, y) denotes the ring commutator $xy - yx$. Consider the following sequence of ring polynomials:

$$w_n = (x, y^2)x_1^2 \cdots x_n^2 (x, y^2)^3, \quad n = 0, 1, 2, \ldots c. \qquad (4.5.3)$$

Theorem 4.5.17. *The variety $\mathcal{V}$ of associative rings defined by the system of identities $\{w_n = 0 \mid n = 0, 1, 2, \ldots\}$ is not finitely based.*

Proof. If the variety $\mathcal{V}$ were finitely based, it would be defined by a finite subsystem of the system $\{w_n = 0 \mid n = 0, 1, 2, \ldots\}$ (prove that!). In order to show that this is impossible, we construct a sequence of finite associative algebras B_n over $\mathbb{F}_2$, $n = 1, 2, \ldots c$, such that for each n, the algebra B_n satisfies the identities $w_k = 0$ for all $k < n$ but does not satisfy the identity $w_{n+1} = 0$.

We fix an integer $n \geq 1$. The algebra B_n is constructed in two steps. In the first step, we construct B_n assuming the existence of an associative algebra R over $\mathbb{F}_2$ with 1 satisfying the following two properties:

(R$_1$) the identities $(x, y^2) = ((x, y), z) = 0$ hold in R.[2]

---

[2] Recall that an associative ring with the derived operation (x, y) is a Lie ring. The identity $((x, y), z) = 0$ then means that the Lie ring is *nilpotent* of class 2.

(R$_2$) for every $n \geq 1$, there exist $s_1, \ldots, s_{n+1} \in R$ such that the product
$s_1^2 \cdots s_{n+1}^2$ does not lie in the subspace M_n spanned by the set of all products
of at most n squares in R;

In the second step, we construct an algebra R that satisfies (R$_1$) and (R$_2$).
This step is postponed till Section 5.3.2.

Thus let us fix an algebra R with 1 satisfying properties (R$_1$) and (R$_2$), and
consider the subalgebra T of the algebra of 5×5-matrices over R consisting
of the following matrices

$$\begin{pmatrix} 0 & * & * & * & * \\ 0 & * & * & * & * \\ 0 & 0 & 0 & * & * \\ 0 & 0 & 0 & * & * \\ 0 & 0 & 0 & 0 & 0 \end{pmatrix}$$

where $*$ indicates an arbitrary element of R.

Exercise 4.5.18. Show that T is indeed a subalgebra in the algebra of
5×5-matrices over R.

The structure of matrices in T ensures that $e_{15}T = Te_{15} = \{0\}$ hence the
set C_n of matrices

$$\begin{pmatrix} 0 & 0 & 0 & 0 & m \\ 0 & 0 & 0 & 0 & 0 \\ 0 & 0 & 0 & 0 & 0 \\ 0 & 0 & 0 & 0 & 0 \\ 0 & 0 & 0 & 0 & 0 \end{pmatrix}$$

where $m \in M_n$ forms an ideal in T. It is convenient to think of the quotient
of T over the ideal C_n as of the ring of 5×5-matrices of the form

$$\begin{pmatrix} 0 & * & * & * & \diamond \\ 0 & * & * & * & * \\ 0 & 0 & 0 & * & * \\ 0 & 0 & 0 & * & * \\ 0 & 0 & 0 & 0 & 0 \end{pmatrix},$$

where $* \in R$, and $\diamond \in R/M_n$, the quotient vector space of the vector space R
over the subspace M_n.

We define B_n to be the subalgebra of T/C_n generated by the matrix

$$d = \begin{pmatrix} 0 & 1 & 0 & 0 & 0 \\ 0 & 0 & 1 & 0 & 0 \\ 0 & 0 & 0 & 1 & 0 \\ 0 & 0 & 0 & 0 & 1 \\ 0 & 0 & 0 & 0 & 0 \end{pmatrix}$$

and all the matrices

$$r = \begin{pmatrix} 0 & 0 & 0 & 0 & 0 \\ 0 & r & 0 & 0 & 0 \\ 0 & 0 & 0 & 0 & 0 \\ 0 & 0 & 0 & r & 0 \\ 0 & 0 & 0 & 0 & 0 \end{pmatrix}, \tag{4.5.4}$$

where r runs over R. Recall that our goal is to show that B_n satisfies the identities $w_k = 0$ for all $k < n$ but does not satisfy the identity $w_{n+1} = 0$.

Clearly, the set of all elements (4.5.4) is a subalgebra $\mathbf{R}$ isomorphic to R. By the property (R_1) the algebra $\mathbf{R}$ satisfies the identity $(x, y^2) = 0$. Now let I be the ideal of B_n generated by d. We have $B_n = \mathbf{R} + I$ hence for every $b_1, b_2 \in B_n$ we have $(b_1, b_2^2) \in I$ (why?).

Every element of I can be represented as $\alpha d + \mathbf{v}_1 d + d\mathbf{u}_1 + \mathbf{v}_2 d\mathbf{u}_2 + t$, where $\alpha \in \mathbb{F}$, $\mathbf{u}_i, \mathbf{v}_i$ are of the form (4.5.4) for some $u_i, v_i \in R$ and $t \in I^2$. It is easy to calculate that $\mathbf{v}_2 d\mathbf{u}_2 = 0$ and that t is of the form

$$t = \begin{pmatrix} 0 & 0 & * & * & \diamond \\ 0 & 0 & 0 & * & * \\ 0 & 0 & 0 & 0 & * \\ 0 & 0 & 0 & 0 & 0 \\ 0 & 0 & 0 & 0 & 0 \end{pmatrix},$$

hence every element of I is of the form

$$\begin{pmatrix} 0 & u & * & * & \diamond \\ 0 & 0 & v & * & * \\ 0 & 0 & 0 & u & * \\ 0 & 0 & 0 & 0 & v \\ 0 & 0 & 0 & 0 & 0 \end{pmatrix} \tag{4.5.5}$$

for some $u, v \in R$. In particular, for every $b_1, b_2 \in B_n$, the element (b_1, b_2^2) is of the form (4.5.5). It is easy to calculate that the cube of a matrix of the form (4.5.5) is of the form

$$\begin{pmatrix} 0 & 0 & 0 & uvu & \diamond \\ 0 & 0 & 0 & 0 & vuv \\ 0 & 0 & 0 & 0 & 0 \\ 0 & 0 & 0 & 0 & 0 \\ 0 & 0 & 0 & 0 & 0 \end{pmatrix}.$$

Now let k be a positive integer. Take arbitrary $c_i \in B_n$ $(i = 1, 2, \ldots, k)$ and represent it as $c_i = r_i + d_i$, where $r_i \in \mathbf{R}$ and $d_i \in I$. Then the product

$$(b_1, b_2^2) c_1^2 c_2^2 \cdots c_k^2 (b_1, b_2^2)^3$$

is of the form

$$
\begin{pmatrix}
0 & 0 & 0 & 0 & f+M \\
0 & 0 & 0 & 0 & 0 \\
0 & 0 & 0 & 0 & 0 \\
0 & 0 & 0 & 0 & 0 \\
0 & 0 & 0 & 0 & 0
\end{pmatrix}, \tag{4.5.6}
$$

where $f = ur_1^2 r_2^2 \cdots r_k^2 vuv$ (check it!). By the property (R_1) $(u, r_i^2) = 0$, so u commutes with r_i^2 and we can rewrite f as the product of $k+1$ squares: $f = r_1^2 r_2^2 \cdots r_k^2 (uv)^2$. Recall that M_n is the linear span of the set of all products of at most n squares in R hence $f \in M_n$ whenever $k < n$. Thus, we have

$$
(b_1, b_2^2) c_1^2 c_2^2 \cdots c_k^2 (b_1, b_2^2)^3 = 0
$$

for all $b_1, b_2, c_1, \ldots, c_k \in B_n$. This means that the algebra B_n satisfies the identities $w_k = 0$ for all $k < n$.

Now take $s_1, \ldots, s_{n+1} \in R$ with $s_1^2 \cdots s_{n+1}^2 \notin M$ (such $s_1, \ldots, s_{n+1}$ exist by property (R_2)) and consider the element

$$
h = (d, \mathbf{1}^2) \mathbf{s}_1^2 \ldots \mathbf{s}_{n+1}^2 (d, \mathbf{1}^2)^3,
$$

where $\mathbf{1}$, $\mathbf{s}_i$ are the matrices of the form (4.5.4) corresponding to 1 and to s_i respectively. Then $(d, \mathbf{1}^2) = d$ (check it! **Hint:** use the fact that our algebras are over a field of characteristic 2). Therefore h is of the form (4.5.6) with $f = s_1^2 \cdots s_{n+1}^2$. By the choice of the elements $s_1, \ldots, s_{n+1}$, we have $f \notin M$ hence $h \neq 0$ and B_n does not satisfy the identity $w_{n+1} = 0$. This completes the first step of the proof.   □

As we have mentioned, the construction of algebra R is postponed till Section 5.3.2.

4.6 Further Reading

In this chapter, we only touched the surface of the large area related to the Burnside properties and identities of rings. Fortunately, there are numerous surveys and books devoted to these topics, both classical and very recent: Jacobson [161], Rowen [274], Ufnarovskii [320, 321], Belov-Kanel and Rowen [36], Belov-Kanel, Borisenko and Latyshev [34], Krause and Lenagan [187], Zel'manov [336], etc.

4.6.1 The Kurosh Problem

During the last 10–15 years we have learned quite a bit about the Kurosh problem and nil-rings in general. In particular, thanks to Agata Smoktunow-icz [305], we now know that there exists a nil-ring which is simple, and a nil-ring R such that its *ring of polynomials* $R[x] = \{a_0 + a_1 x + \ldots + a_n x^n \mid a_i \in R\}$ (with natural addition and multiplication) is not nil.

Still there are some outstanding simple to formulate open problems in this area. For example, the following problem has been open for more than 80 years.

Problem 4.6.1 (Köthe's problem [183]). *Let A, B be left ideals of a ring R. Suppose A and B are nil-rings. Is it true that the sum $A + B$ is also a nil-subring of R?*

4.6.2 Identities of Rings

While arbitrary rings can be arbitrarily "bad", rings satisfying nontrivial identities are nice [36, 274]. For example, the various radicals of any such ring are quite manageable. The most important radical of a ring is the Jacobson radical [161]: the intersection of all maximal left ideals. A more syntactic definition of a Jacobson radical is this: it is the maximal right ideal consisting of elements x such that for some y, $xy + x + y = 0$.[3] Note that Jacobson's radical of a ring always contains the Baer radical (prove it! **Hint:** Use the fact that $x(-x + x^2 - x^3 + \ldots) = -x^2 + x^3 - \ldots$ and if x is a nil-element, then the sums are finite).

Theorem 4.6.2 (Braun, [50]). *The Jacobson radical of every finitely generated ring satisfying a nontrivial identity is nilpotent, hence coincides with the Baer radical (and all the other radicals mentioned above).*

The original proof of Theorem 4.6.2 involved a lot of nontrivial ring theory. A more syntactic (but still nontrivial) proof was found by Belov-Kanel [32]. Thus for finitely generated rings satisfying nontrivial identities, the Jacobson radical (= Baer radical) can be defined as the largest nilpotent ideal.

The current status of the finite basis problem for (associative) rings is well presented in the introduction of Belov-Kanel's paper [33]. That paper also contains the strongest, as of today, results in the area (see also a more detailed presentation of these results including an interesting connection between

---

[3] If the ring contains 1, then this condition is equivalent to invertibility of $1 + x$.

identities of rings and the theory of quivers in [37–40]). Still some natural problems about identities of rings are unsolved. For example,

Problem 4.6.3. *Are identities of every finitely generated ring finitely based?*

Amazingly (especially compared with the situation in semigroups or groups), one would expect a positive answer to this question.

Chapter 5
Groups

In this chapter we introduce a new tool – van Kampen diagrams (although these are sibling of diagrams introduced in Section 1.7.4). We start with defining van Kampen diagrams and explaining basic methods of using them: the bands, the Swiss cheese method and the small cancelation theory. Then we explain Golod's solution of the unbounded Burnside problem and a road map of Olshanskii's proof of the Novikov–Adian theorem solving the bounded Burnside problem: there exists an infinite finitely generated group satisfying the identity $x^n = 1$ for all odd $n \geq 665$. The theorem was one of the main achievements in group theory of the 20th century. Its initial proof occupied more than 300 pages and was extremely complicated. Olshanskii managed to find a much simpler proof (for much bigger n). Our road map explains the main ideas and "points of interest" of Olshanskii's proof.

Then we discuss the free groups. We show how to use inverse automata to describe subgroups of free groups and to prove the main properties of these subgroups (due to Schreier, Howson and Hall).

From free groups, i.e., groups consisting of words, we move to groups consisting of 2-dimensional words, diagram groups. One of the most famous diagram group is the R. Thompson group, and we prove some basic properties of that group in this chapter.

Next we present a (negative) solution of the finite basis problem for groups due to Bryant and Yu. Kleinman. Again, this is not the first solution of the problem (the first was by Olshanskii [256]), but it is one of the easiest.

One of the most active parts of group theory deals with growth functions of groups. Here we present a proof of the Bass–Guivarc'h theorem giving precise estimates of the growth functions of nilpotent groups. We also present Grigorchuk's solution of Milnor's problem about groups of intermediate growth.

Another hot topic in group theory is amenability, that is the last topic of the chapter. It is related to growth and to the Burnside problem. In particular, we discuss Adian's solution of the von Neumann-Day problem.

© Springer International Publishing Switzerland 2014   197
M.V. Sapir, *Combinatorial Algebra: Syntax and Semantics*,
Springer Monographs in Mathematics, DOI 10.1007/978-3-319-08031-4_5

Theorems proved in this Chapter include

- Theorem 5.1.17 of Greendlinger about groups with presentations satisfying a small cancelation condition.
- Theorem 5.2.2 of Burnside and Theorem 5.2.3 of Sanov about groups of exponents 3 and 4.
- Theorem 5.2.1 of Golod solving the unbounded Burnside problem.
- Theorem 5.3.1 of Bryant and Yu. Kleiman giving an example of a non-finitely based variety of groups.
- Theorem 5.4.1 of Abért giving a general condition implying that a group does not satisfy any nontrivial identity.
- Theorem 5.7.28 of Bass and Guivarc'h describing growth functions of nilpotent groups.
- Theorem 5.7.31 of Grigorchuk giving an example of an infinite periodic finitely generated group of intermediate growth.
- Theorem 5.8.3 of Hausdorff, Banach and Tarski about doubling a sphere.
- Ph. Hall's marriage Lemma 5.8.25.
- Theorem 5.8.29 of Følner about amenable groups.
- Theorem 5.8.40 of Jónsson and Dekker about groups with Tarski number 4.
- Theorem 5.8.45 of Adian about non-amenability of groups given by certain Dehn presentations.

5.1 Van Kampen Diagrams

5.1.1 Group Presentations

Let X be a set. If R is a set of group words in X (i.e., words in the alphabet $X \cup X^{-1}$), then we can consider the normal subgroup N generated by R in the free group F_X, i.e., the smallest normal subgroup containing R. In this case we shall say that the group $G = F_X/N$ is *given by a presentation* $\mathrm{gp}\langle X \mid R \rangle$ (compare with Section 1.8.5).

Exercise 5.1.1. Prove that N is generated as a subgroup by all the *conjugates* of elements of R, that is elements of the form $r^z = z^{-1}rz$, $z \in F_X$, $r \in R$.

Note that if $r \equiv uv \in R$, then $vu = 1$ in G because $vu = r^u$. The word vu is a cyclic shift of r. Hence we can and will assume that each cyclic shift of every $r \in R$ is reduced, that is each $r \in R$ is *cyclically reduced*.

Exercise 5.1.2. Prove that a group word W in the alphabet X belongs to the normal subgroup N if and only if it can be represented in the free group in the form

$$u_1 r_1 u_2 \ldots u_m r_m u_{m+1} \tag{5.1.1}$$

where

- $u_1 u_2 \ldots u_{m+1} = 1$ in the free group,
- each r_i is a cyclic shift of a word from $R^{\pm 1}$.

5.1.2 Van Kampen Diagrams: The Definition

Let G be a group given by a presentation $\mathcal{P} = \mathrm{gp}\langle X \mid R \rangle$. Then G is a semigroup given by the presentation

$$\mathcal{P}' = \mathrm{sg}\langle X \cup X^{-1} \cup \{1\} \mid 1 \cdot 1 = 1, x1 = x = x1, xx^{-1} = 1, x^{-1}x = 1, r = 1, x \in X, r \in R \rangle$$

where X^{-1} is a copy of X, and there is an involution $^{-1}: X \leftrightarrow X^{-1}$ (see Section 1.8.5). Note that the 0-*relations* $1 \cdot 1 = 1, x \cdot 1 = x$, etc. will play an important role later, see Section 5.1.4.6. If a freely reduced word W in $X \cup X^{-1}$ is equal to 1 in G, then there exists a $(W, 1)$-diagram over $\mathcal{P}'$. Adding inverse edges labeled by letters from $X \cup X^{-1}$, we turn it into a labeled graph in the sense of Serre. Let us collapse every edge e that is labeled by 1, that is, identify $\iota(e)$ and $\tau(e)$, and remove e, e^{-1} from the set of edges. If $\iota(e) = \tau(e)$ in the diagram, then there could be a ... subgraph bounded by the edge e. In this case we remove e together with that subgraph. Also if there exist two edges e, f with $\iota(e) = \iota(f), \tau(e) = \tau(f)$ and having the same label $x \in X$, then we can identify these edges and remove the whole subgraph bounded by the path ef^{-1}. Note that these transformations do not destroy planarity of the graph and do not change the boundary path labeled by W. The result is a *van Kampen* diagram Δ over $\mathcal{P}$. Clearly, since Δ is planar, if we remove Δ from the plane, the plane decomposes into a number of bounded components called *cells* and one infinite component. Then Δ has the property that the boundary of each cell is labeled by a word of R, and the boundary of the infinite component (called the *boundary* of Δ, and denoted $\partial(\Delta)$), is the word W. In this case we shall call Δ a van Kampen diagram over the presentation $\mathcal{P}$ for the word W. A van Kampen diagram is called *minimal* if it has the smallest number of cells among all van Kampen diagrams over $\mathcal{P}$ with the same boundary label.

Figure 5.1 shows how a typical van Kampen diagram may look like (without labels of edges). One can see that the cells may be of different sizes and shapes, a cell can touch itself, etc. It is important also that a diagram itself may not be an embedded disc: several pieces as on Figure 5.1 can be connected by paths to form a tree of discs.

Thus we have shown how to turn a word that is equal to 1 in a group into a diagram. This is the first half of the van Kampen lemma.

Lemma 5.1.3. *If a freely reduced group word W over the alphabet X is equal to 1 in G, then there exists a van Kampen diagram over the presentation of G with boundary label W.*

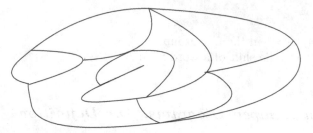

Figure 5.1 A van Kampen diagram without labels

This is a half of the so-called *van Kampen lemma* [51, 210, 260]. The converse statement (that the boundary label of every van Kampen diagram is equal to 1 in G) constitutes the second half. In order to prove it, we have to, given a van Kampen diagram, produce an equality of the form (5.1.1) and apply Exercise 5.1.2. The proof is by cutting a van Kampen diagram along the edges. In a sense it shows how to turn a diagram into a word. This is the second half of the van Kampen lemma.

Proposition 5.1.4. *Let Δ be a van Kampen diagram over a presentation $gp\langle X \mid R \rangle$ where $X = X^{-1}$, R is closed under cyclic shifts and inverses. Let W be the boundary label of Δ. Then W is equal in the free group to a word of the form $u_1 r_1 u_2 r_2 \ldots u_m r_m u_{m+1}$ where:*

(1) Each r_i is a word from $R^{\pm 1}$;
(2) $u_1 u_2 \ldots u_{m+1} = 1$ in the free group;
(3) $\sum_{i=1}^{m+1} |u_i| \leq 4|E|$ where E is the set of positive edges of Δ.

In particular, W is in the normal subgroup N and is equal to 1 in G.

Remark 5.1.5. Part (3) of Proposition 5.1.4 shows that the way we cut the diagram is close to the most economical.

Proof. If Δ has an internal edge (i.e., an edge which belongs to the boundaries of two cells) then it has an internal edge f one of whose vertices belongs to the boundary (why?). Let us cut Δ along f leaving the second vertex of f untouched. We can repeat this operation until we get a diagram Δ_1 which does not have internal edges. The boundary label of Δ_1 is equal to W in the free group because each cut inserts a subword xx^{-1}. The number of edges of Δ_1 which do not belong to contours of cells (let us call them *edges of type 1*) is the same as the number of such edges in Δ, and the number of edges which belong to contours of cells in Δ_1 (*edges of type 2*) is at most twice the number of such edges of Δ (we cut each edge from a contour of a cell at most once, after the cut we get two external edges instead of one internal edge).

Suppose that for a cell Π in Δ_1 there are at least two edges each of which has a common vertex with Π but does not belong to the contour of Π. Take any point O on the boundary $\partial(\Pi)$ of the cell Π which belongs to one of the edges not on $\partial(\Pi)$. Let p be the boundary path of Δ_1 starting at O

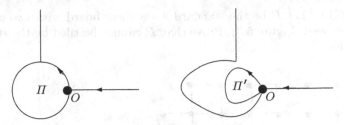

Figure 5.2 Cutting Π' inside Π

and let q be the boundary path of Π starting at O. Consider the path $qq^{-1}p$. The subpath $q^{-1}p$ bounds a subdiagram of Δ_1 containing all cells but Π. Replace the path q in $qq^{-1}p$ by a loop q' with the same label starting at O and lying inside the cell Π. Let the region inside q' be a new cell Π'. Then the path $q'q^{-1}p$ bounds a diagram whose boundary label is equal to W in the free group $F(X)$. Notice that Π' has exactly one edge having a common vertex with Π' and not belonging to the contour of Π' (see Figure 5.2). Thus this operation reduces the number of cells Π such that more than one edge of the diagram has a common vertex with Π but does not belong to the contour of Π.

After a number of such transformations we shall have a diagram Δ_2 which has the form of an apple tree T with cells hanging like apples (each has exactly one common vertex with the tree).

The number of edges of type 1 in Δ_2 cannot be bigger than the number of all edges in Δ_1, so it cannot be more than two times bigger than the total number of edges in Δ.

The boundary label of Δ_2 is equal to W in the free group, and it has the form

$$u_1 r_{i_1} u_2 r_{i_2} \ldots u_m r_{i_m} u_{m+1}$$

where m is the number of cells in Δ, $u_1 u_2 \ldots u_{m+1}$ is the boundary label of a tree consisting of edges of type 1 in Δ_2 (traced counterclockwise), so $u_1 u_2 \ldots u_{m+1} = 1$ in the free group. The sum of lengths of u_i is at most four times the number of edges in Δ because the word $u_1 u_2 \ldots u_{m+1}$ is written on the tree T, and when we trace the tree counterclockwise, we pass through each edge twice.   □

5.1.3 Van Kampen Diagrams and Tilings.
An Elementary School Problem
and Its Non-elementary Solution

As an easy application of van Kampen diagrams, consider the following elementary problem.

Example 5.1.6. Let P be the standard 8×8 chess board with two opposite squares removed (Figure 5.3). Prove that P cannot be tiled by the standard 2×1 dominos.

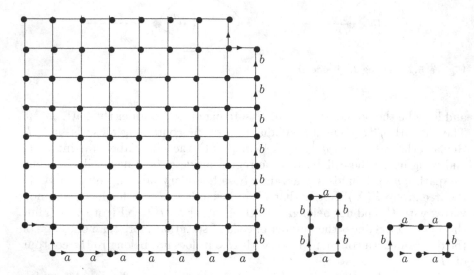

Figure 5.3 The chess board with two squares removed and two dominos

Proof. The elementary solution of this problem is well known. Color squares of P in black and white in the usual (chessboard) way. Then the number of squares of one color (black or white) in P differs from the number of squares of the other color by 2. Since each domino covers exactly one white and exactly one black square, P cannot be tiled by dominos.

Here is a solution which, although less elementary, can be applied to many regions of the plane square grid for which the elementary proof above does not work; it also applies to regions of non-square (say, hexagonal) lattices on the plane. This solution first appeared in a paper by W. Thurston [315]. The ideas of that paper have many applications in several areas of mathematics from combinatorics to probability to mathematical physics.

Let us label horizontal edges pointed rightward in P by the letter a and vertical edges pointed upward by the latter b. Then the (counterclockwise) boundary of P has label $W = a^7 b^7 a^{-1} b a^{-7} b^{-7} a b^{-1}$. Every domino can be placed either vertically or horizontally. In the first case its boundary label is $ab^2 a^{-1} b^{-2}$, and in the second case its boundary label is $a^2 b a^{-2} b^{-1}$. Now consider the group G with the presentation $\mathrm{gp}\langle a, b \mid ab^2 a^{-1} b^{-2} = 1, a^2 b a^{-2} b^{-1} = 1 \rangle$. Suppose that P can be tiled by the dominos. Then every such tiling turns P into a van Kampen diagram over the presentation of G.

Hence, by the van Kampen lemma (more precisely by Proposition 5.1.4), the word W would be equal to 1 in G. But consider the 6-element symmetric group S_3 and two permutations

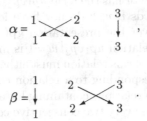

in it (as in Exercise 1.8.43). Both relations of G hold if we replace a by α and b by β because $\alpha^2 = \beta^2 = 1$. Hence the substitution $a \mapsto \alpha$, $b \mapsto \beta$ extends to a homomorphism $G \to S_3$. Note that $W(\alpha, \beta) = (\alpha\beta)^4 = \alpha\beta$ is the permutation

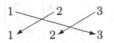

which is not trivial. Hence W is not equal to 1 in G, a contradiction.   □

This example shows a similarity between the word problem in groups and the tiling problem. Indeed, by the van Kampen lemma, a word W is equal to 1 in $G = \mathrm{gp}\langle X \mid R \rangle$ if and only if we can tile a disc with boundary labeled by W by pieces whose boundaries are labeled by words from R. Nevertheless the word problem differs from the tiling problem as in Example 5.1.6 because, as we have seen, when we draw a van Kampen diagram, we do not fix the shapes or sizes of the tiles, only the labels of the boundaries of them, so one can view a general van Kampen diagram as a tiling of a disc by tiles made of soft rubber, while the traditional tiling problems such as Example 5.1.6 are about tiles made of hard plastic.

5.1.4 The Three Main Methods of Dealing with van Kampen Diagrams: Bands, Swiss Cheese, and Small Cancelation

The number of papers dealing with van Kampen diagrams is very large. Still one can distill a few methods that are commonly used in many of these papers. Let $G = \mathrm{gp}\langle X \mid R \rangle$ be a group presentation. We shall always assume that R is closed under taking cyclic shifts and inverses (indeed, if a word is equal to 1 in a group, then so are its cyclic shifts and the inverse).

5.1.4.1 The Band Method. HNN Extensions

Suppose that there exists a subset $P \subset X$ so that every relation $r \in R$ either does not have letters from P or has the form $up_1vp_2^{-1}w$ where p_1, p_2 are letters from P of X, and the words u, v, w do not have letters from P. Let Δ be a van Kampen diagram over the presentation $\mathrm{gp}\langle X \mid R \rangle$. Suppose that a cell corresponding to the relation $up_1vp_2^{-1}w = 1$ is in Δ. Then the edges e, f labeled by the letters p_1, p_2 of this relation must either belong to the boundary $\partial\Delta$ or to another cell corresponding to a relation containing letters from P. Thus cells corresponding to relations containing letters from P form P-*bands*, i.e., sequences of cells where every two consecutive cells share an edge labeled by a letter from P, see Figure 5.4.

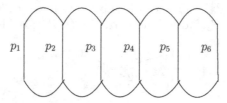

Figure 5.4 A P-band with 5 cells, $p_i \in P$

Bands are partially ordered by inclusion. The minimal bands are just edges, they do not have any cells. If we do not say otherwise, all bands we will be dealing with will be maximal. Every P-band has the *start edge* and the *end edge* labeled by letters from P. The boundary of the subdiagram which is the union of cells from the band is subdivided by the start and end edges into two parts, the *sides* of the band. These are labeled by words not containing letters from P.

For example, consider the presentation $\mathcal{P}_{ab} = \mathrm{gp}\langle a, b \mid a^{-1}b^{-1}ab = 1 \rangle$ of the commutative group $\mathbb{Z} \times \mathbb{Z}$. Then we can talk about a-bands and b-bands, i.e., both sets $\{a, a^{-1}\}$ and $\{b, b^{-1}\}$ can play the role of the set P above. If Δ is a van Kampen diagram over that presentation, then every cell in Δ is an intersection of an a-band and a b-band. The next two exercises give two basic properties of bands.

Exercise 5.1.7. Prove that if Δ is a minimal van Kampen diagram over $\mathcal{P}_{ab}$, then no a-band and no b-band forms an annulus, see Figure 5.5.

Exercise 5.1.8. Prove that if Δ is a minimal van Kampen diagram over $\mathcal{P}_{ab}$, then no a-band intersects a b-band twice, see Figure 5.6.

Exercise 5.1.9. Use Exercises 5.1.7, 5.1.8 to show that every minimal van Kampen diagram over $\mathcal{P}_{ab}$ with boundary of length n has as most $\frac{n^2}{16}$ cells.

One immediate application of bands is the following classical theorem first proved by Higman, B.H. Neumann and H. Neumann [154]. Recall that for every element t of a group G, the conjugacy map $a \mapsto a^t$, $a \in G$, is an

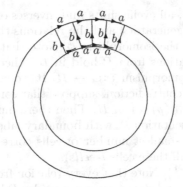

Figure 5.5 A b-annulus

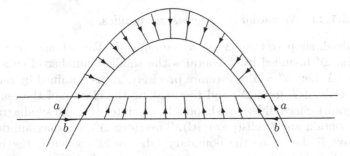

Figure 5.6 A b-band intersects an a-band twice

automorphism. Hence if A and B are two subgroups of G, G is a subgroup of H and $t^{-1}At = B$ for some $t \in H$, then A and B are isomorphic (such subgroups are called *conjugate*). Theorem 5.1.10 shows that the converse is also true: every two isomorphic subgroups in a group G are conjugate in some bigger group $H > G$.

Theorem 5.1.10. *Let G be a group and A, B be subgroups of G. Suppose that A and B are isomorphic and $\phi: A \to B$ is an isomorphism. Then there exists a group $H > G$ and $t \in H$ such that $t^{-1}At = B$. Moreover $t^{-1}at = \phi(a)$ for every $a \in A$.*

Proof. Let us consider the following *triangular* presentation $\mathcal{P}$ of G. The set of generators X consists of all elements of G, and the set R of relations consists of all relations of the form $ab = c$ which are true in G where $a, b, c \in G$. Thus $\mathcal{P}$ is just the multiplication table of G. Note that if G is infinite, then $\mathcal{P}$ is infinite.

Now consider the following presentation $\mathcal{P}'$ of a group H:

$$\mathcal{P}' = \text{gp}\langle X \cup \{t\} \mid \mathcal{P} \cup \{t^{-1}at = \phi(a), a \in A\}\rangle$$

(we also add, by default, all cyclic shifts and inverses of the relations). Thus we add a new "formal" generator t and the relations that we want: the conjugation by t "induces" the isomorphism ϕ. Note that by definition X is a subset of H and all relations from P hold in H. Therefore the map $a \mapsto a$ from G to H is a homomorphism $\psi \colon G \to H$. It would be enough to show that ψ is injective.[1] By contradiction, suppose that some element $p \in G = X$ is "killed" by ψ, that is $\psi(p) = 1$ in H. Then there must be a van Kampen diagram Δ over the presentation $\mathcal{P}'$ with boundary label p. We can of course assume that Δ has the smallest number of cells corresponding to relations containing t (we shall call these cells t-cells).

Let P be the set $\{t, t^{-1}\}$. Note that every relation from $\mathcal{P}'$ that contains t has the form $utvt^{-1}w$ or $ut^{-1}vtw$ so we can consider P-bands (which we shall call t-bands).

Lemma 5.1.11. *No t-band in Δ forms an annulus.*

Proof. Indeed, suppose that Δ has a t-annulus $\mathcal{T}$. We can assume that the subdiagram Δ' bounded by $\mathcal{T}$ contains the smallest number of cells for all t-annuli in Δ. Let $\Delta'' = \Delta' \smallsetminus \mathcal{T}$ (more precisely, Δ'' is obtained by removing all closed cells of $\mathcal{T}$ from Δ', and then taking the closure of the remaining open diagram). Since different t-bands do not intersect, the subdiagram Δ'' does not contain any t-cells(prove it!). Therefore Δ'' is a diagram over the presentation $\mathcal{P}$. Let u be the boundary label of Δ'' which is the internal boundary of $\mathcal{T}$. Let v be the external boundary label of $\mathcal{T}$. Note that every letter from the word u occurs in a relation containing t. Therefore either all letters from u are in A or all letters from u are in B depending on the direction of the t-edges of $\mathcal{T}$ (prove it!). In the first case, v is the word $\phi(u)$, and in the second case v is the word $\phi^{-1}(u)$. By Proposition 5.1.4, then $u = 1$ in G. Since ϕ is an isomorphism, then $v = 1$ in G. Hence there exists a van Kampen diagram Ψ over $\mathcal{P}$ with boundary label v – by Lemma 5.1.3. Let us replace the subdiagram Δ' in Δ by the diagram Ψ (it is possible because Δ' and Ψ have the same boundary labels!). The diagram Δ''' we get has the same boundary label as Δ and fewer t-cells, a contradiction with the minimality assumption about Δ.   □

Now we can easily finish the proof of the theorem. Since no t-bands of Δ form an annulus, each maximal t-band must start and end on the boundary of Δ. Hence $\partial(\Delta)$ must contain an edge (in fact at least two edges) labeled by $t^{\pm 1}$. But edges in $\partial(\Delta)$ are labeled by letters from X, a contradiction.   □

The group H constructed in the proof of Theorem 5.1.10 is called the *HNN extension of the group G with associated subgroups A and B and free letter t*, denoted $\mathrm{HNN}_\phi(G)$ or simply $G *_\phi$ for Higman, B.H. Neumann and

---

[1] To avoid confusion, note that the left a in the notation $a \mapsto a$ is from G, and the right a is from H; in general a, a' that are not equal in G may be equal in H, so the map is not automatically injective.

H. Neumann. For example, the famous *Baumslag–Solitar group* $BS_{m,n} =$ $\text{gp}\langle a, b \mid b^{-1}a^m b = a^n \rangle$ is an HNN extension of the infinite cyclic group $\text{gp}\langle a \rangle$, the associated subgroups are $\langle a^m \rangle$ and $\langle a^n \rangle$ (both are infinite cyclic), and $\phi \colon a^{mk} \mapsto a^{nk}$.

5.1.4.2 The Swiss Cheese Method. Amalgamated Products

Suppose that the set of relations R in a presentation $\mathcal{P}$ has a subset R', and Δ is a diagram over $\mathcal{P}$. Consider the union Δ' of all cells corresponding to the relations of R'. It is a planar labeled graph which is a union of several connected components which are diagrams with holes. The connected components look like pieces of Swiss cheese, hence the name of the method. The key idea of dealing with Swiss cheese subdiagrams is to show that the holes do not actually exist, and connected components are in fact van Kampen subdiagrams. Let us demonstrate this idea on another classical result about the amalgamated products.

It is very common to have two subgroups G, G' of a bigger group T so that $A = G \cap G'$ is nontrivial. Thus G and G' form an *amalgam*. Conversely, suppose that two groups G and G' have a common subgroup A and $G \cap G' = A$. We need to find a group T which contains subgroups G_1 and G_1' isomorphic to G and G' respectively, so that $G_1' \cap G_1$ is isomorphic to A. The next theorem shows that it is always possible.

Theorem 5.1.12. *Let G, G' be two groups and $A < G, A' < G'$ be two subgroups. Suppose that A is isomorphic to A' and $\phi \colon A \to A'$ is an isomorphism. Then there exists a group T with two subgroups G_1, G_1', and two isomorphisms $\psi_1 \colon G_1 \to G$, $\psi_2 \colon G_1' \to G'$, such that $\psi_1^{-1}(G_1 \cap G_1') = A$, $\psi_2^{-1}(G_1 \cap G_1') = A'$ and $\psi_2 \phi(a) = \psi_1(a)$ for every $a \in A$.*

Proof. Let $\mathcal{P} = \text{gp}\langle X \mid R \rangle$ and $\mathcal{P}' = \text{gp}\langle X' \mid R' \rangle$ be the triangular presentations of G and G' (see the proof of Theorem 5.1.10) where X and X' are disjoint. Consider the following presentation

$$\mathcal{T} = \text{gp}\langle X \cup X' \mid R \cup R' \cup \{a = \phi(a) \mid a \in A\} \rangle.$$

Let T be the group given by the presentation $\mathcal{T}$. Clearly the maps $x \mapsto x$ ($x \in X$) and $x' \mapsto x'$ ($x' \in X'$) are homomorphisms $\psi \colon G \to T$, $\psi' \colon G' \to T$. It is enough to show that ψ and ψ' are injective (why?).

Suppose that, say, for some element $u \in G$, $\psi(u) = 1$ in T. Then there exists a van Kampen diagram Δ over $\mathcal{T}$ with boundary label u. We can assume that Δ has the smallest possible number of cells corresponding to the relations $a = \phi(a)$, $a \in A$ (*extra cells* for short). Let Δ' be one of the connected components of the Swiss cheese subdiagram of Δ formed by all the $\mathcal{P}'$-cells in Δ. We shall show that Δ' does not have holes. In fact it is easier to prove the following stronger statement

Lemma 5.1.13. *Let Ψ be a diagram over $\mathcal{T}$ whose boundary label is a word over X (or a word over X', respectively). Suppose that Ψ has the smallest number of extra cells among all diagrams over $\mathcal{T}$ with the same boundary label. Then the Swiss cheese subdiagram of Ψ formed by all $\mathcal{P}'$-cells (resp. $\mathcal{P}$-cells) does not have holes (that means what we thought was Swiss cheese is in fact cheddar or provolone).*

Proof. Induction on the number of cells in Ψ that correspond to the relations $a = \phi(a)$, $a \in A$ (*extra cells* for short). Suppose that the label of $\partial(\Psi)$ is a group word in X, but one of the connected components Ψ' of the Swiss cheese subdiagram formed by $\mathcal{P}'$-cells has a ... hole Ω (the other case is similar). That hole Ω is a subdiagram of Ψ whose boundary is a group word in X' (prove it!). It has fewer extra cells than Ψ, so we can assume that the Swiss cheese subdiagram Ω' of Ω which is formed by all $\mathcal{P}$-cells from Ω does not have holes, hence it is a disjoint union of van Kampen diagrams. Suppose that Ω' is not empty. Then the boundary of Ω' consists of edges with labels from X. The only cells from $\Omega \smallsetminus \Omega'$ that can contain these edges are the extra cells. Therefore the boundary label v' of every connected component Ω_0' of Ω' is a group word over A. By Proposition 5.1.4 $v' = 1$ in G. Let Ω_0'' be the diagram Ω_0' together with all the extra cells which have common edges with $\partial(\Omega_0')$. The boundary label of Ω_0'' is $\phi(v')$. Since $v' = 1$ in G, $\phi(v') = 1$ in G'. Hence by Lemma 5.1.3 we can replace Ω_0'' by a van Kampen diagram over $\mathcal{P}'$ with the same boundary label. This operation reduces the number of extra cells, a contradiction. Hence we can assume that Ω' is empty, so Ω does not have $\mathcal{P}$-cells. Thus it must consist of extra cells. But every extra cell has an edge labeled by a letter from X, and two extra cells corresponding to relations $a = \phi(a)$ and $a' = \phi(a')$ which share an edge must correspond to the same relation, and so their union can be replaced by one edge labeled by a or $\phi(a)$ (see Figure 5.7) which would reduce the number of extra cells. Thus Ω is empty, a contradiction.   □

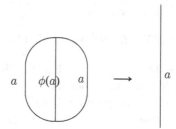

Figure 5.7 Cancelation of cells

Now the proof of Theorem 5.1.12 is easy and in fact is already contained in the proof of Lemma 5.1.13. Since Δ' does not have holes, it is a disjoint union of van Kampen diagrams over $\mathcal{P}'$. Let Δ_0' be a connected component of Δ'. The union Δ_0'' of Δ_0' and all extra cells that share an edge with Δ_0' is a van Kampen diagram with boundary label v over X. Since $\phi^{-1}(v)$ is equal to 1 in G' (why?), $v = 1$ in G, so by Lemma 5.1.3, there exists a van Kampen diagram Ψ over $\mathcal{P}$ with boundary label v. Replacing Δ_0'' by Ψ, we reduce the number of extra cells, a contradiction.   □

The group T constructed in the proof of Theorem 5.1.12 is called the *free product of G and G' with amalgamated subgroup* $A = A'$, denoted $G *_{A=A'} G'$. In the important particular case when A is the trivial subgroup, T is called the *free product of G and G'*, denoted $G * G'$. By induction, one can define free products and free products with amalgamation of several groups. For example, every free group is a free product of several infinite cyclic groups. Another classical example of free product with amalgamation is the group $SL(2, \mathbb{Z})$ (see Exercise 1.8.34). This group is the free product of the cyclic group of order 4 and the cyclic group of order 6 with amalgamated subgroup of order 2 (for more on this group see [48, 289]).[2]

Exercise 5.1.14. Let groups G_1, G_2 be given by Church–Rosser semigroup presentations $\text{sg}\langle X_1 \mid \mathcal{R}_1 \rangle$ and $\text{sg}\langle X_2 \mid \mathcal{R}_2 \rangle$ where X_1, X_2 are disjoint. Show that $\text{sg}\langle X_1 \cup X_2 \mid \mathcal{R}_1 \cup \mathcal{R}_2 \rangle$ is a Church–Rosser presentation of the free product $G_1 * G_2$ and that the canonical words are of the form $a_1 b_1 a_2 \ldots a_n b_n$ where a_i, b_i are canonical words in G_1, G_2 respectively, a_1, b_n may be empty, other a_i, b_i are not empty.

For more on HNN extensions and free products with amalgamation see [210, 289].

5.1.4.3 Auxiliary Planar Graphs. The Dehn–Greendlinger Algorithm

The third idea used in dealing with van Kampen diagrams is the following: construct a planar graph $\Gamma(\Delta)$ associated with a van Kampen diagram Δ, and then use the fact that the *Euler characteristic* of a tesselated polygon on a plane (i.e., the number of vertices minus the number of edges plus the number of cells) is 1 to deduce properties of $\Gamma(\Delta)$ and Δ.

For example, suppose that Δ is a van Kampen diagram over some presentation $\mathcal{P}$ and Δ contains several subdiagrams $\Delta_1, \ldots, \Delta_n$ such that if $i \neq j$, then Δ_i and Δ_j share only boundary vertices and edges (but not cells). In this case, for every $i \neq j$, $\partial(\Delta_i) \cap \partial(\Delta_j)$ is empty or a disjoint union of arcs $l_{i,j}^1, \ldots, l_{i,j}^{k(i,j)}$. The idea is then to consider the following auxiliary graph $\Gamma(\Delta)$: put a vertex v_i inside each subdiagram Δ_i and connect v_i and v_j

---

[2] The notation $G *_\phi$ for HNN extensions and $G *_{A=A'} G'$ for free product with amalgamation indicates a similarity between HNN extensions and free products with amalgams. Indeed (see [210]) every HNN extension is a subgroup of some natural free product with amalgamation, and every free product with amalgamation is inside some natural HNN extension.

by $k(i,j)$ edges, one edge per arc $l_{i,j}^s$. Clearly this graph is planar. One of the simplest use of the Euler characteristic is the following statement published by Heawood in 1890.

Theorem 5.1.15 (Heawood, [148]). *Every finite planar graph without multiple edges and loops (i.e., edges whose tail and head coincide) either is empty or has a vertex of degree at most 5.*

Proof. Let Γ be a finite planar graph without parallel edges and loops. Since the Euler characteristic of the plane is 1, the Euler formula gives $V - E + F = k$ where V is the number of vertices, E is the number of edges and F is the number of faces, i.e., finite connected regions obtained after removing Γ from the plane, $k \geq 1$ is the number of connected components of Γ. Suppose that every vertex of Γ has degree at least 6. Then $E \geq 3V$ (each vertex belongs to at least 6 edges, each edge belongs to 2 vertices). We also have $F \leq \frac{2}{3}E$ because every edge is on the boundary of at most 2 faces and every face has boundary consisting of at least 3 edges (since Γ has no parallel edges and no loops). Therefore, if the graph is not empty, $1 = V - E + F \leq \frac{E}{3} - E + \frac{2}{3}E = 0$, a contradiction. $\square$

Let us demonstrate how this idea applies. Consider the classical small cancelation condition. Let $\mathcal{P} = \mathrm{gp}\langle X \mid r = 1, r \in R \rangle$ be a group presentation. As usual we assume that R is closed under taking cyclic shifts and inverses. We shall also assume that every word in R is cyclically reduced (if a cyclic shift of a word is 1 in the group, then the word is 1 in the group also). A *piece* is the maximal common prefix of any two distinct words from R. An easy geometric picture illustrating this notion is the following. The label of a maximal subpath l in $\partial(\pi) \cap \partial(\pi')$ where π, π' are cells in a van Kampen diagram over $\mathcal{P}$ is a piece which is contained in some cyclic shifts of the labels of $\partial(\pi)$, $\partial(\pi')$ or their inverses, see Figure 5.8.

We say that $\mathcal{P}$ satisfies the *small cancellation condition* $C'(\lambda)$ for some number λ between 0 and 1 if for every piece u that is a subword of $r \in R$ we have $|u| < \lambda|r|$.

Exercise 5.1.16. The presentation

$$\mathrm{gp}\langle a, b \mid a^{-1}b^{-1}ab = 1 \rangle$$

(for brevity, we omit the cyclic shifts and inverses) satisfies the condition $C'(\lambda)$ for every $\lambda > \frac{1}{4}$) because the only pieces are a, a^{-1}, b, b^{-1}. The presentation

$$\mathrm{gp}\langle a_1, b_1, a_2, b_2 \mid [a_1, b_1][a_2, b_2] = 1 \rangle$$

satisfies $C'(\lambda)$ for every $\lambda > \frac{1}{8}$). Let γ be the substitution defined in Section 2.5.3.4. Then the presentation

$$\mathrm{gp}\langle a_{11}, \ldots, a_{rr} \mid \gamma(a_{11}) = 1, \ldots, \gamma(a_{rr}) = 1 \rangle$$

satisfies the condition $C'(\lambda)$ for every $\lambda > 1/r$).

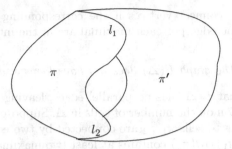

Figure 5.8 The label of each l_i is a piece

The general wisdom about the small cancellation condition is "the smaller the λ the better". It was proved by Goldberg [260] that every group has a presentation satisfying $C'(\lambda)$ for any $\lambda > \frac{1}{5}$. So only the conditions $C'(\lambda)$ with $\lambda \le \frac{1}{5}$ are meaningful.

The key fact about small cancellation presentation was proved by Greendlinger [118].

Theorem 5.1.17 (See Lyndon, Schupp [210]). *Let* $\mathcal{P} = \langle X \mid R \rangle$ *be a group presentation satisfying* $C'(\lambda)$, $\lambda \le 1/6$, *then every minimal van Kampen diagram* Δ *over* $\mathcal{P}$ *either has no cells or has a cell* π *where* $\partial(\pi) = uv$ *with* $|u| > |v|$, *and* u *is a subpath of* $\partial\Delta$ *(this cell is called a* Greendlinger *cell).* *It sticks out of the diagram as on Figure 5.9.*

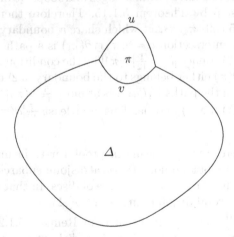

Figure 5.9 A Greendlinger cell: $|v| < |u|$

Proof. We shall give a sketch of the proof in the case $\lambda \le \frac{1}{12}$ which is a bit easier. Consider any minimal van Kampen diagram Δ over $\mathcal{P}$. Consider the auxiliary graph $\Gamma(\Delta)$ corresponding to all cells in Δ (i.e., as above, put a vertex in each cell, connect vertices if the corresponding cells share parts of the boundary, one edge per each maximal arc in the intersection of their boundaries).

Lemma 5.1.18. *The graph $\Gamma(\Delta)$ does not have loops and parallel edges.*

Proof. We prove that $\Gamma(\Delta)$ has no parallel edges, leaving the other part as an exercise. Induction on the number of cells in Δ. Suppose that two vertices v_1, v_2 corresponding to cells π_1, π_2 are connected by two edges. That means the intersection $\partial(\pi_1) \cap \partial(\pi_2)$ contains at least two maximal arcs l_1, l_2 as on Figure 5.8.

Then there exists a subdiagram Δ' of Δ with boundary of the form $\partial(\Delta_1) = p_1 p_2^{-1}$ where the path p_1 is part of $\partial(\pi_1)$ and path p_2 is a part of $\partial(\pi_2)$ which contains cells (here we use the fact that all words in R are cyclically reduced!). Since the diagram Δ' contains fewer cells than Δ, we can assume that Lemma 5.1.18, and hence Theorem 5.1.17, applies to Δ'. Therefore there exists a Greendlinger cell π in Δ', i.e., there exists a path $l \subseteq \partial(\pi) \cap \partial(\Delta')$ such that $|l| > \frac{1}{2}|\partial(\pi)|$. Then $l = l_1 l_2$ where $l_1 = \partial(\pi) \cap p_1$, $l_2 = \partial(\pi) \cap p_2$. Then one of the paths l_1 or l_2 has length greater than $\frac{1}{4}|\partial(\pi)|$. That means there exists a piece which is a subword of the label of $\partial(\pi)$ and is longer than $\frac{1}{4}|\partial(\pi)|$, a contradiction with $C'(\frac{1}{12})$. This contradiction proves the claim.

Exercise 5.1.19. Prove the remaining part of the lemma: that $\Gamma(\Delta)$ does not have loops.

Now let us show how to *deduce Theorem 5.1.17 from Lemma 5.1.18*. Since $\Gamma(\Delta)$ is planar, does not have parallel edges and loops, it must have a vertex v of degree at most 5 by Theorem 5.1.15. Therefore there exists a cell π which has at most 5 cells $\pi_1, \pi_2, \ldots$ which share a boundary edge with π. By Lemma 5.1.18, each intersection $l_i = \partial(\pi) \cap \partial(\pi_i)$ is a path. The label of this path is a piece (why?). Hence $|l_i| \le \frac{1}{12}|\partial(\pi)|$ by the condition $C'(\frac{1}{12})$. Note also that every edge of $\partial(\pi)$ either belongs to the boundary of Δ or to one of the l_i. The sum of lengths of the paths $l_1, l_2, \ldots$ is at most $\frac{5}{12}|\partial(\pi)|$. The complement $l = \partial(\pi) \smallsetminus \cup l_i$ is $\partial(\Delta) \cap \partial(\pi)$ and has length at least $\frac{7}{12}|\partial(\pi)| > \frac{1}{2}|\partial(\pi)|$.

$\square$

Remark 5.1.20. There are two gaps in this proof. First, the intersection $\partial(\Delta) \cap \partial(\pi)$ may not be an arc but a union of several disjoint subarcs of $\partial(\pi)$. Second, the diagram Δ may not be a disc but a tree of discs. In that case $\partial(\Delta) \cap \partial(\pi)$ may be a union of several disjoint subarcs of $\partial(\Delta)$.

Exercise 5.1.21. Fill the gaps mentioned in Remark 5.1.20. **Hint:** To fill the first gap, assume that $\partial(\Delta) \cap \partial(\pi)$ is a disjoint union of several arcs, and consider a (smaller) subdiagram bounded by $\partial(\pi)$ and $\partial(\Delta)$. To fill the second gap, prove that if a planar graph has no parallel edges or loops, and has at least 2 vertices, then it contains at least 2 vertices of degree ≤ 5 (this is not much stronger than Theorem 5.1.15). Alternatively, one can use 0-cells and contiguity subdiagrams (see Section 5.1.4.6 below).

Theorem 5.1.17 provides the following algorithm for solving the word problem in a group G given by a finite (or even recursive) presentation $\mathrm{gp}\langle X \mid R \rangle$ satisfying $C'(\lambda)$, $\lambda < \frac{1}{6}$. Consider a word w. If w is not trivial (i.e., equal to 1 in the free group) and does not contain more than a half of a word from R, then $w \neq 1$. If w contains a subword u such that $uv \in R$ for some v and $|v| < |u|$, then replace u by v in w. The new word w' is shorter than w. Moreover $w = 1$ in G if and only if $w' = 1$ in G, and we can proceed by induction on the length of $|w|$. This algorithm was first discovered by Dehn to solve the word problem in the *fundamental groups of the orientable surface S_g of genus $g > 1$*, or *surface group* for short i.e., the group $\pi_1(S_g) = \mathrm{gp}\langle a_1, \ldots, a_g, b_1, \ldots, b_g \mid [a_1, b_1] \cdots [a_g, b_g] = 1 \rangle$. That group satisfies $C'(\lambda)$ for every $\lambda > \frac{1}{4g}$ (prove it!).

Thus a presentation of a group G satisfying the conclusion of Theorem 5.1.17 is called a *Dehn presentation*. Groups admitting a Dehn presentations have been studied extensively (probably the first paper where these groups were introduced in full generality and studied was the paper by Adian [4]). Groups having finite Dehn presentations are known under the name of *Gromov hyperbolic groups* after Gromov discovered deep connection between these groups, topology and geometry in his seminal paper [127].

The surface groups play extremely important role in the modern group theory. The next exercise clarify their structure.

Exercise 5.1.22. Show that $\pi_1(S_g)$ can be represented as a free product of two free groups with cyclic amalgamated subgroup if $g \geq 2$ (say, one free group is generated by a_1 and b_1, another is generated by the rest of generating set), and as an HNN extension of a free group with cyclic associated subgroups if $g \geq 1$ (any generators can be considered as a free letter).

The next exercise shows that $\pi_1(S_g)$ has nice finite Church–Rosser presentation.

Exercise 5.1.23 (Hermiller, [150]). Let us denote a_i^{-1} by $\bar{a}_i$, b_i^{-1} by $\bar{b}_i$ (so that the defining relator of $\pi_1(S_g)$ has the form $\bar{a}_1\bar{b}_1a_1b_1\bar{a}_2\bar{b}_2a_2b_2 \ldots \bar{a}_g\bar{b}_ga_gb_g = 1$). Let us also denote $P = \bar{a}_2\bar{b}_2a_2b_2 \ldots \bar{a}_g\bar{b}_ga_gb_g$, $Q = \bar{b}_g\bar{a}_gb_ga_g \ldots \bar{b}_2\bar{a}_2b_2a_2$. Consider the monoid presentation with generators

$$a_1, \ldots, a_g, b_1, \ldots, b_g, \bar{a}_1, \ldots, \bar{a}_g, \bar{b}_1, \ldots, \bar{b}_g$$

and relations

$$a_i\bar{a}_i = 1, \ \bar{a}_ia_i = 1, \ b_i\bar{b}_i = 1, \ \bar{b}_ib_i = 1 \text{ for all } i,$$
$$a_1b_1 = b_1a_1Q,$$
$$\bar{a}_1\bar{b}_1 = Q\bar{b}_1\bar{a}_1,$$
$$\bar{a}_1b_1 = b_1P\bar{a}_1,$$
$$a_1Q\bar{b}_1 = \bar{b}_1a_1.$$

Show that this is a finite Church–Rosser presentation of the surface group $\pi_1(S_g)$. **Hint:** The most difficult part of this exercise is to prove that the rewriting system is terminating. For this you can use the recursive path ordering on the set of words (see Section 1.2.3).

5.1.4.4 Small Cancellation and Conjugacy. Annular (Schupp) Diagrams

Let $G = \mathrm{gp}\langle X \mid R \rangle$ be a group presentation. Suppose that two words u, v over X are conjugate in G, that is there exists a word w such that $wuw^{-1} = v$ in G. Then there exists a van Kampen diagram with $\partial(\Delta) = p_1 p_2 p_3^{-1} p_4^{-1}$ where the labels of p_1, p_2, p_3, p_4 are w, u, w, v. Thus we can identify the arcs p_1 and p_3 making the diagram into an annulus Δ' with two boundary components: one labeled by u and another labeled by v. The paths $p_1 = p_3$ form a cut of the diagram Δ'. The only difference between Δ' and van Kampen diagrams is that Δ' is a tesselated annulus, not a tesselated disc. Such annular diagrams are sometimes called Schupp diagrams (Schupp introduced them in [288]). Note that if there exists an annular diagram Δ' with boundary labels u, v, and p is any path without self-intersections, connecting the two boundary components, so that the word u on a boundary component of Δ starting (and ending) at p_- and the word v starts and ends at p_+, we can cut Δ' along p and obtain a van Kampen diagram Δ with boundary label $wuw^{-1}v^{-1}$. So $wuw^{-1} = v$ in G. See Figure 5.10. Thus we obtain

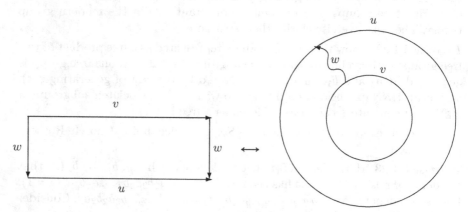

Figure 5.10 Turning a conjugacy van Kampen diagram into an annular diagram and back

Lemma 5.1.24. *The words u, v are conjugate in G if and only if there exists an annular diagram over the presentation of G with boundary labels u and v.*

Note that another way of getting annular diagrams is by removing subdiagrams from van Kampen diagrams. One can deal with annular diagrams in the same manner as with van Kampen diagrams. In particular, the auxiliary graphs are defined similarly. Only one needs to be aware that a cell can now touch itself forming an annulus surrounding the hole in the diagram, and also two cells can touch twice, again, forming an annulus surrounding the hole. So the auxiliary graph may have loops (at most one per vertex) and

parallel edges: at most two per vertex. Since the graph is still planar, the
Euler characteristic method still works, only one needs smaller λ in $C'(\lambda)$.
In particular, the following analog of Theorem 5.1.17 still holds.

Theorem 5.1.25. *Let $\mathcal{P} = \langle X \mid R \rangle$ be a group presentation satisfying $C'(\lambda)$.
Then every minimal annular diagram Δ over $\mathcal{P}$ either has no cells or has a
cell π which shares more than $(1-7\lambda)$ of its boundary edges with the boundary
components of Δ, see Figure 5.11.*

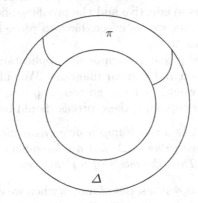

Figure 5.11 A Greendlinger cell π of an annular diagram Δ

This implies, in particular, the following interesting result.

Theorem 5.1.26. *Let $\mathcal{P} = \langle X \mid R \rangle$ be a group presentation satisfying $C'(\lambda)$,
$\lambda < \frac{1}{14}$. If two words u, v are conjugate modulo $\mathcal{P}$, then there exists a cyclic
shift u' of u, a cyclic shift v' of v, and a word w of length at most $\frac{7}{2}\lambda(|u|+|v|)$
such that $w^{-1}u'w = v'$ modulo $\mathcal{P}$.*

Proof. Indeed, let Δ be a minimal annular diagram over $\mathcal{P}$ with boundary
components p and q labeled by u and v. Applying Theorem 5.1.25, we find
a Greendlinger cell π which sticks out of Δ. We only consider the case when
$\partial(\pi)$ intersects both p and q leaving the other cases as an exercise. Thus
$\partial(\pi) = t_1 s_1 t_2 s_2^{-1}$ where $s_1 \subseteq p$, $s_2 \subseteq q$,

$$|p| + |q| \geq |s_1| + |s_2| > (1 - 7\lambda)|\partial(\pi)|.$$

Hence $|t_1| + |t_2| \leq 7\lambda|\partial(\pi)| \leq \frac{7\lambda}{1-7\lambda}(|p|+|q|)$. Thus at least one of the paths t_1 or
t_2 has length at most $\frac{7\lambda}{2(1-7\lambda)}(|u|+|v|)$. The label of that path is the desired
word w.   □

5.1.4.5 Diagrams and Elementary Topology[3]

A van Kampen diagram can be viewed as a CW complex on the plane. One can notice that, when dealing with van Kampen diagrams we have been using some topological properties of the plane. For example, we use the Euler characteristic in the proof of Theorem 5.1.15. We also repeatedly used Jordan's theorem: every simple closed loop on the plane separates the plane into two path connected components (find places above where Jordan's theorem is used!).[4] In fact the possibility to use topology when dealing with diagrams is what makes diagrams so effective and the proofs so short. If we work with words instead of diagrams, we, in effect, have to prove syntactic analogs of these topological statements!

The next two lemmas give other important applications of the topology of the plane to van Kampen and annular diagrams. We only sketch the proofs, a more detailed proof would make us go too deep into topology (and for a reader who knows basic topology, these proofs should be trivial).

Lemma 5.1.27. *Let Δ be a van Kampen or annular diagram over a presentation $\mathcal{P}$. Let p, p' be two paths in Δ that are homotopic in Δ (here we view Δ as a CW-complex). Then the labels of p, p' are equal modulo $\mathcal{P}$.*

Proof. Indeed, a homotopy is a series of moves when we either replace a path ee^{-1} by an empty path or insert a subpath ee^{-1} (e is an edge), or substitute a subpath which is a part of the boundary of a cell π by the complement of this subpath in $\partial(\pi)$. The effect of these moves on the label is: we either insert a word that is equal to 1 in the free group, or we insert a relation from $\mathcal{P}$. Both operations do not change the word modulo $\mathcal{P}$.   □

Lemma 5.1.28. *Let Δ be an annular diagram over a presentation $\mathcal{P}$ with boundary paths p, q whose labels are u and v. Suppose that Δ has two cuts c, c' with $c_- = c'_- \in p_-, c_+ = c'_+ \in q$ and labels w and w'. Then for some integer n, we have $w = u^n w'$ modulo $\mathcal{P}$.*

Proof. Indeed, as we know from basic topology, the fundamental group of an annulus is cyclic. Hence c is homotopic (in Δ) to c' times a power of p. It remains to use Lemma 5.1.27.   □

---

[3] Warning: Some knowledge of elementary topology is required to read this subsection.

[4] The full Jordan's theorem is a relatively complicated statement (although see [101]). Here we are using it only in the case when the curve is a polygon (all edges of a van Kampen diagram can be drawn as broken lines on the plane), which is relatively easy.

5.1.4.6 Small Cancellation, 0-Cells and the Baby Version of Contiguity Subdiagrams

The goal of this section is first to calm down potential critics who can say that cutting off subdiagrams and replacing them with subdiagrams with the same boundary labels is not that easy topologically because, as we have mentioned, a cell in a van Kampen diagram over a presentation $\mathrm{gp}\langle X \mid R \rangle$ is not necessarily an embedded disc (see Figure 5.1): the boundary of a cell can have multiple edges and vertices. Moreover the diagram itself may be a tree of discs connected by arcs. The second, more important, goal is to introduce a "baby version" (or, more precisely, rank 0 version) of the notion of contiguity subdiagram which will play a crucial role in the proof of the Novikov–Adian theorem below.

To avoid topological difficulties, Olshanskii introduced (see [260]) the notions of 0-edges and 0-cells. A 0-*edge* is an edge labeled by 1 (the identity element of the group). Let us allow 0-edges in a van Kampen diagram. That means, we can replace a relation $a_1 a_2 \cdots a_k = 1$ by the equivalent relation $a_1 1 a_2 1 \ldots a_n 1 = 1$. Letters from X will be called non-0-letters. A 0-*relation* is a relation of one of two forms $1^k = 1$ and $1^k a 1^l a^{-1} 1^m = 1$, $k, l, m \geq 0$. A cell corresponding to a 0-relation is called a 0-*cell*, see Figure 5.12

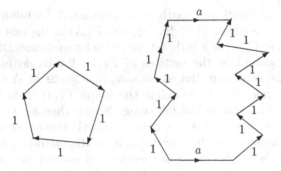

Figure 5.12 0-cells

A 0-*cell* is a cell corresponding to a 0-relation. Clearly adding 0-relations to $\mathcal{P}$ does not change the group. 0-cells allow us to assume that every cell and the diagram itself is an embedded disc. For example, Figures 5.13, 5.14 shows what to do when a cell touches itself and when a diagram consists of two disc diagrams connected by an arc.

As noted in [260, Chapter 4], using 0-cells, one can transform every van Kampen diagram Δ into a diagram Δ' over the "extended presentation" (including the 0-relations) with boundary label freely equal to the label of $\partial(\Delta)$ and such that

- No two non-0-edges $e \neq f^{\pm 1}$ share a vertex.
- No two different non-0-cells share a boundary vertex.

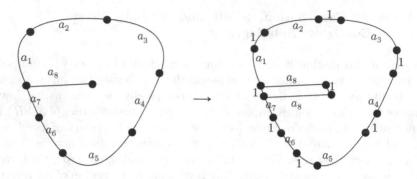

Figure 5.13 Turning a cell into an embedded disc

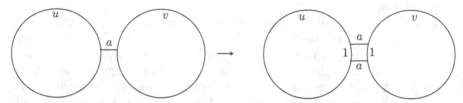

Figure 5.14 Turning a diagram into an embedded disc

Computing the length of a path in a diagram Δ containing 0-cells, we usually ignore the 0-edges. The "dual" graph $\Gamma(\Delta)$ in the case of small cancelation condition $C'(\lambda)$ is a little bit more tricky to define. Of course, only the non-0-cells should be the vertices of $\Gamma(\Delta)$: 0-cells do not satisfy the small cancelation condition. But as we saw, non-0-cells in Δ may not share parts of the boundary, so if we define the graph $\Gamma(\Delta)$ as before, it would have no edges. The cure is the following. Notice that for every generator $a \in X$ all 0-relations containing an a contain exactly two occurrences of a (see Figure 5.12). Therefore 0-cells containing a on the boundary form a-bands! The sides of the a-bands consist of 0-edges. An a-band can start (end) on the boundary of a non-0-cell or on $\partial(\Delta)$. Let us call an a-band connecting a non-0-cell π (or an arc p on $\partial(\Delta)$) with another non-0-cell π' (or another arc p' from $\partial(\Delta)$) a *bond*. Note that if we remove all 0-cells and 0-edges from Δ, then the start and end edges of a bond will coincide, so bonds just tell us which edges π or p shares with π' or p'. Now it is not difficult to define when two non-0-cells share an arc. If $\mathcal{T}_1$ and $\mathcal{T}_2$ are two not necessarily distinct bonds between non-0-cells π_1, π_2 (for arcs p_1, p_2 the definition is the same), then consider the subdiagram Δ' bounded by $\partial(\pi_1), \partial(\pi_2)$ and sides of the bands $\mathcal{T}_1, \mathcal{T}_2$, and containing $\mathcal{T}_1, \mathcal{T}_2$ but not containing π_1, π_2. The boundary of Δ' is subdivided into four arcs: $\partial(\Delta') = p_1 p_2 p_3^{-1} p_4^{-1}$ where p_1, p_3 are the sides of $\mathcal{T}_1, \mathcal{T}_2$, p_2 is a part of $\partial(\pi_1)$, p_4 is a part of $\partial(\pi_2)$, see Figure 5.15. Suppose that Δ' consists of 0-cells. Then we call Δ' a *contiguity subdiagram* between π_1 and π_2, p_1, p_3 are *sides* of Δ', p_2, p_4 are called the *contiguity arcs* of Δ'. There is a partial order on contiguity subdiagrams induced by inclusion. As in the case of bands, we shall consider only the maximal contiguity subdiagrams. Note that every bond is a contiguity subdiagram as well but not necessarily a maximal one.

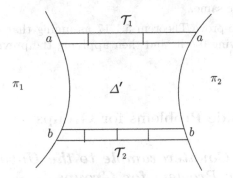

Figure 5.15 A contiguity subdiagram

Contiguity subdiagrams between a cell and a boundary arc of Δ or between two boundary arcs are defined in the same way. Note that p_1 and p_3 consist of 0-edges. Moreover the labels of arcs p_2 and p_3 are equal in the free group since Δ' consists of 0-cells. So these labels are pieces. The notions of a cell sharing an arc of its boundary with the boundary of the diagram and a cell sticking out is changed accordingly. The picture of a cell sticking out is on Figure 5.16 where Δ' is a contiguity subdiagram.

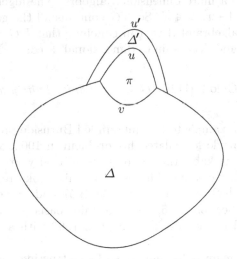

Figure 5.16 A cell π sticking out: $|v| < |u|$, $u = u'$ in the free group

Now we can define the "true" auxiliary graph $\Gamma(\Delta)$: put a vertex inside every non-0-cell, connect two vertices by an edge per contiguity subdiagram between the corresponding cells.

It is easy to see that the graph $\Gamma(\Delta)$ is the same as before insertion of 0-edges and 0-cells, so the formulation and the proof of Theorem 5.1.17 remain virtually the same.

Exercise 5.1.29. Re-prove Theorem 5.1.17 assuming that the diagram contains 0-cells (removing 0-cells and then applying the previous proof constitutes cheating!).

5.2 The Burnside Problems for Groups

5.2.1 Golod's Counterexample to the Unbounded Burnside Problem for Groups

Let K be a countable field of characteristic $p > 0$. Consider the algebra $A = F/I$ constructed in Section 4.4.6. Let B be the algebra A with an identity element adjoined. Consider the semigroup G generated by all elements $1 + x$ where $x \in X$. Every element of this semigroup has the form $1 + a$ where $a \in A$. Since A is a nil-algebra $a^{p^n} = 0$ in A for some n which depends on a. Since the characteristic is equal to p we have that $(1 + a)^{p^n} = 1 + a^{p^n} = 1$ (see Exercise 1.4.18). Therefore G is a group (every element $1 + a$ has an inverse $(1 + a)^{p^n - 1}$). By definition G is finitely generated. If G were finite, then KG would be a finite dimensional algebra. The algebra KG contains 1 and all elements $1 + x$, $x \in X$. So KG contains all the generators x of A. Since KG is a subalgebra of A we can conclude that $KG = A$. But A is not finite dimensional and KG is finite dimensional, a contradiction. Therefore we have

Theorem 5.2.1 (Golod, [112]). *G is an infinite finitely generated periodic group.*

This is a counterexample to the unbounded Burnside problem for groups.

Recall that Burnside formulated his problem in 1902, and Golod solved it in 1964. It is remarkable that a solution of a 60 years old problem can be that simple (other examples discussed in this book are the Hanna Neumann conjecture which was formulated in 1957 and proved with a 2-page proof in 2011, see Section 5.9.5, and the finite basis problems for varieties of rings formulated in the 50s and solved in 1999 with a 5-page proof, see Section 4.5.3).

Now there exist many other methods of constructing counterexamples to the unbounded Burnside problem for groups (Aleshin, Grigorchuk, and others using groups of automatic transformation (see, for example, [170, 230]),

Sushchansky using the so-called iterated wreath products [309], Olshanskii, Osin and Sapir (see [263]) using the so-called lacunary hyperbolic groups, Olshanskii and Osin [262] using the so-called large groups, etc. We shall return to this in Section 5.7.4).

5.2.2 The Bounded Burnside Problem. Positive Results

There are very few values of n such that every group satisfying the identity $x^n = 1$ is known to be locally finite, namely $n = 1, 2, 3, 4, 6$. This fact is obvious for $n = 1$, an easy exercise for $n = 2$, a medium difficulty exercise for $n = 3$ and nontrivial for $n = 4, 6$. We present here the proofs for $n = 3$ and $n = 4$ (since the variety $\text{var}\{x^4 = 1\}$ will appear again later). These cases as well as case $n = 6$ can be found, for example, in the classical book by M. Hall [142, Section 18.4] or in the book by Vaughan-Lee [324].

Theorem 5.2.2 (Burnside, [142]). *Every group satisfying the identity $x^3 = 1$ is locally finite.*

Proof. Let $G = \text{gp}\langle x_1, \ldots, x_k \rangle$ be a finitely generated group satisfying the identity $x^3 = 1$. Induction on k. The case $k = 1$ is obvious. Assume that all $k - 1$-generated groups with the identity $x^3 = 1$ are finite. The van Kampen diagram on Figure 5.17 (from [260]) shows that the identity $x^3 = 1$ implies the identity $[y, y^x] = 1$. Therefore every normal subgroup of G generated

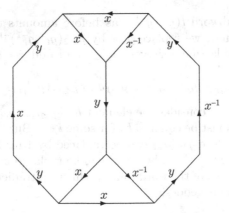

Figure 5.17 The identity $x^3 = 1$ implies the identity $[y, y^x] = 1$

(as a normal subgroup) by one element is commutative (prove it!). Let N be the normal subgroup generated by x_k. Since G/N is $(k - 1)$-generated, it is finite by the induction assumption. Every commutative group satisfying the identity $x^3 = 1$ is locally finite (Exercise 3.3.2). Hence by Corollary 3.4.2 G is finite.   □

Theorem 5.2.3 (Sanov, [276]). *Every group satisfying the identity $x^4 = 1$ is locally finite.*

Proof. We need to show that every finitely generated group satisfying the identity $x^4 = 1$ is finite. If the number of generators k of the group is 1, the statement is obvious. Suppose that we have proved the result for some k. Let G be a group generated by $\{x_1, \ldots, x_k, y\}$ and satisfying the identity $x^4 = 1$. Suppose that G is infinite. Let $H = \langle x_1, \ldots, x_k \rangle$ be the subgroup of G generated by $x_1, \ldots, x_k$. Then H is finite by the inductive assumption. Let M be the subgroup of G generated by H and y^2. Whether M is finite or infinite, we have two groups $G_1 < G_2$ both satisfying the identity $x^4 = 1$, G_1 is finite, G_2 is infinite and generated by G_1 and t such that $t^2 \in G_1$. If M is finite, then $G_2 = G$, $G_1 = M$, $t = y$. If M is infinite, then $G_2 = M$, $G_1 = H$, $t = y^2$.

We need to show that this situation is impossible. Since G_2 is infinite and is generated by a finite set $G_1 \cup \{t\}$, $t^2 \in G_1$, there exists an arbitrary long geodesic word of the form

$$g_1 t g_2 t \ldots g_s t g_{s+1} \qquad (5.2.1)$$

for G_2 where $g_i \in G_1$. The relation $(tg)^4 = 1$ implies $tgt = g^{-1} t g' t g^{-1}$ where $g' = t^2 g^{-1} t^2 \in G_1$ (check it!). Therefore for every $i = 2, \ldots, s-1$ we can rewrite the word (5.2.1) as follows:

$$g_1 t g_2 \ldots t (g_{i-1} g_i^{-1}) t g_i' t (g_i^{-1} g_{i+1}) \ldots t g_{s+1}.$$

Note that the length of that word in the alphabet $G_1 \cup \{t\}$ is not bigger than the length of (5.2.1).

Rewriting the subword $t(g_{i-1} g_i^{-1}) t$ as before amounts to changing some g_r again, in particular, we replace g_{i-2} by $g_{i-2}(g_{i-1} g_i^{-1})^{-1} = g_{i-2} g_i g_{i-1}^{-1}$. By induction we conclude that for every $i > 2$ and odd $j < i$, we can replace g_{i-i} by

$$h_{i,j} = g_{i-j} g_{i-j+2} \ldots g_{i-1} g_i^{-1} g_{i-2}^{-1} \ldots g_{i-j+1}^{-1}.$$

For every $i = 1, 2, \ldots$ consider the element $h_i = g_2 g_4 \ldots g_{2i} g_{2i-1}^{-1} g_{2i-3}^{-1} \ldots g_3^{-1}$. If $s > 2|G_1|$, then h_i must be equal to h_j for some $i < j$. But then $h_{2i,2j-1} = 1$ in G_1 (check it!). Therefore $g_{2i-2j+1}$ can be replaced by 1 in (5.2.1) without increasing the length. Then the subword t^2 can be replaced by an element from G_1, reducing the length of the word (5.2.1) which contradicts the assumption that the word (5.2.1) is geodesic.   □

5.2.3 The Novikov–Adian Theorem

Here we present a "road map" for Olshanskii's proof from [258] of a version of Novikov–Adian's theorem for sufficiently large odd exponents [3, 251].

Note that this is only a road map: we wanted to describe main ideas, methods and "points of interest" of the proof. For the actual proof the reader is referred to [258, 260].

Theorem 5.2.4 (Novikov–Adian, [3, 251], Olshanskii [258, 260]). *For every* $m \geq 2$, *the* m-*generated free Burnside group* $B_{m,n}$, *i.e., the free group in the variety of groups* $\text{var}\{x^n = 1\}$, *is infinite for every sufficiently large odd* n *(say,* $n > 10^{10}$*).*[5]

In preparation of this section, we essentially used the text sent to us by Victor Guba [131] , and further explanations by Alexander Olshanskii.

5.2.3.1 The Basic Rough Idea

Fix a big enough odd number n. It suffices to show that $B(2, n)$ is infinite because $B(2, n)$ is a homomorphic image of $B(m, n)$ for every $m \geq 2$.

Order the cyclically reduced words in the free group $F_2 = \text{gp}\langle a, b\rangle$ in an almost arbitrary way: $u_1 < u_2 < \ldots$ The only requirement is: if u is shorter than v, then $u < v$. Consider the following sequence of groups G_i with group presentations

$$\mathcal{PB}_i = \text{gp}\langle a, b \mid C_1^n = 1, C_2^n = 1, \ldots, C_i^n = 1\rangle$$

where G_0 is the free group F_2, and C_i is the smallest word which has infinite order in G_{i-1}, for every $i \geq 1$. Note that then C_i is not a proper power in G_{i-1} of a smaller word because of our choice of the order $<$ (prove that!). The main fact to prove about the group G_i is this

Theorem 5.2.5. *The group* G_i *is infinite for every* $i \geq 0$. *In fact all Prouhet cube-free words* p_m *(see Section 2.1) in the alphabet* $\{a, b\}$ *are pairwise different in* G_i.

It is proved in [258] that the group $B_{2,n}$ is given by the group presentation

$$\mathcal{PB} = \text{gp}\langle a, b \mid C_i^n = 1, i \geq 1\rangle.$$

i.e. it is an *inductive limit* of groups G_i (this is not completely obvious, it is only obvious that the inductive limit is periodic). Suppose that $B_{2,n}$ is finite. Then it is given by a finite presentation $\text{gp}\langle a, b \mid R\rangle$ (the multiplication table, for example). Since R follows from $\{C_i^n = 1 \mid i \geq 1\}$, it must follow from a

---

[5] Although the formulations in [3, 251] and [258, 260] are similar, the book [3] has estimate $n \geq 665$ instead of $n > 10^{10}$. This is a very important distinction. Even from our road map, it will be clear that proving the theorem for much smaller n would require significant additional effort. Currently there are no methods of lowering the estimate to "below 50", for example.

finite number of these relations $\{C_1^n = 1, \ldots, C_m^n = 1\}$. But this means G_m is a homomorphic image of $B_{2,n}$ which is impossible since G_m is infinite. Thus Theorem 5.2.4 follows from Theorem 5.2.5.

It is in fact proved in [258] that the groups G_i are infinite and hyperbolic (similar facts are also basically proved in [251], see also Section 5.2.3.15 below), moreover its presentation satisfies a natural generalization of the small cancelation condition $C'(\lambda)$ where contiguity subdiagrams are used to define "pieces of relations" (as in Section 5.1.4.6). To show that all Prouhet words p_m are different in G_i, suppose, by contradiction, that $p_k = p_m$ in G_i, $k < m$. Since p_k is a prefix of p_m, $w = p_k^{-1} p_m$ is a subword of p_m, hence cube-free. Since $w = 1$ in G_i, there exists a van Kampen diagram Δ over the presentation of G_i with boundary label w. We need to prove an analog of Theorem 5.1.17: if Δ is minimal in some sense, then one of the cells shares large enough part of its boundary with the boundary of Δ. Indeed, a big enough subword of C_i^n contains cubes, so w cannot be cube-free.

The index i of G_i and C_i^n is called the *rank*. Thus C_i is called a *period of rank i*. Accordingly we call a cell in a diagram over $\mathcal{PB}$ a cell of *rank i* if it corresponds to the relation C_i^n, by definition 0-cells have rank 0. The *rank of a diagram* is the maximal rank of its cells.

We also define the *type* of a van Kampen (or annular) diagram Δ over $\mathcal{PB}$ as the sequence of ranks of its cells arranged in the non-increasing order $(s_1, s_2, \ldots)$. We order types lexicographically.

5.2.3.2 j-Pairs

We say that two cells π, π' of rank j form a *j-pair* if there is a path p without self-intersections (i.e., *simple*) connecting these cells whose label is equal to 1 in G_{j-1}, and the boundary labels of π, π' starting at p_- and p_+ respectively are mutually inverse. In that case we can cut off the two cells together with the path p, and insert a diagram of rank at most $j - 1$ in the resulting hole (indeed, the boundary label of the hole is equal to 1 in G_{j-1}). The new diagram will have the same boundary label but smaller type, so we shall always assume that our diagrams over $\mathcal{PB}$ do not contain j-pairs of cells, see Figure 5.18. Note that the path p can be empty as on Figure 5.7.

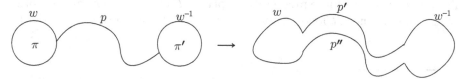

Figure 5.18 Removing a j-pair

We call a van Kampen or annular diagram of rank i *reduced* if it does not contain j-pairs for any $j \leq i$.

5.2.3.3 Parameters

Now let us turn to the proof in [258]. It consists of several lemmas proved by a simultaneous induction on the type of a diagram over $\mathcal{PB}$. This means that proving every lemma, we can assume that all other lemmas are already proved for diagrams of smaller type. One can view all these lemmas as one large lemma or as a "crown of lemmas" (which is similar to a "crown of sonnets" [199]). Our goal is to explain the semantic meaning of these lemmas and give some ideas of their proofs.

In the proof, we often use the phrase that some quantity a (length of a path, or weight of a cell, etc.) is "much smaller" than the other quantity b. This usually means that $a \leq \nu b$ for some very small parameter ν. There are several concrete parameters in [258] (and even better chosen parameters in [260]). They are denoted by small Greek letters $\alpha, \beta, \eta, \theta, \dots c$. All inequalities involving these parameters become obvious if we take into account that the ratio between a "very small number" and an "even smaller number" can be assumed arbitrary large but fixed in advance, hence do not depend on the rank. This leads to a very large value of n, which is the reciprocal of the smallest parameter, but we do not care here about how large n is as long as it is finite.

5.2.3.4 Contiguity Subdiagrams. The Definition and the Spirit of Zimin Words

As in Section 5.1.4.6, contiguity subdiagrams will connect a cell π with another cell π' or a boundary arc p in a van Kampen or annular diagram. Let us define only contiguity subdiagrams connecting a cell π_1 with another cell π_2 in a van Kampen diagram. Other definitions are completely similar.

Every contiguity subdiagram Ψ will have boundary subdivided into four arcs

$$p_1 q_1 p_2^{-1} q_2^{-1}$$

where q_1 is an arc from $\partial(\pi_1)$, q_2 is an arc in $\partial(\pi_2)$ (hence the need to consider diagrams with boundary subdivided into at most four arcs!). The arcs p_1, p_2 are called the *sides of a contiguity subdiagram*. The quotient $\frac{|q_1|}{|\partial(\pi_1)|}$ is called the *degree of contiguity of π_1 to π_2 via Ψ*. The role of this quantity is similar to the λ in $C'(\lambda)$, i.e., labels of p_1 and p_2 play the role of "pieces" of relations. See Figure 5.19.

The contiguity subdiagrams of a diagram Δ are defined by induction. The 0-contiguity subdiagrams are defined using 0-cells as in Section 5.1.4.6.

Suppose that j-contiguity subdiagrams for every $j < i$ are defined already. Consider a cell π of rank k and two contiguity subdiagrams: a j_1-contiguity subdiagram Γ_1 connecting π with some cell Π_1 of rank $> k$ and a j_2-contiguity subdiagram Γ_2 connecting π with a cell Π_2 of rank $> k$. Assume that $j_1, j_2 < k$, $j_1, j_2 < i$, and Γ_1, Γ_2 do not have common cells. Assume also that the degree of contiguity of π via Γ_s is at least β, $s = 1, 2$ where β is one of the small but not extremely small parameters fixed in advance, that is the degrees of contiguity of π to Π_1 and Π_2 are large enough. Then the smallest subdiagram of Δ containing π, Γ_1, Γ_2 is called a k-bond between Π_1 and Π_2 with the principal cell π, see Figure 5.20.

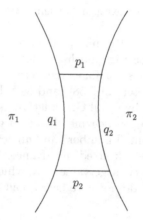

Figure 5.19 A rough sketch of a contiguity subdiagram

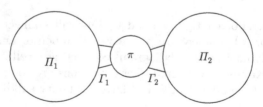

Figure 5.20 A bond

If there are two bonds, a j-bond $\mathcal{B}_1$ and a k-bond $\mathcal{B}_2$ without common cells, connecting a cell Π_1 with a cell Π_2, then the subdiagram bounded by $\mathcal{B}_1, \mathcal{B}_2, \partial(\Pi_1), \partial(\Pi_2)$, containing the bonds and not containing Π_1, Π_2 is called a $\max(j, k)$-contiguity subdiagram connecting Π_1 and Π_2. Its sides are sides of the bonds, its contiguity arcs are the obvious subarcs of $\partial(\Pi_1)$, $\partial(\Pi_2)$, see Figure 5.21.

The next exercise provides a little more detailed structure of a contiguity subdiagram. It can be easily proved by induction. Let us define a *band of bonds* connecting a cell π and a cell Π as a sequence of bonds with principal cells $\pi_1, \pi_2, \ldots, \pi_s$ of ranks $r_1, \ldots, r_s$ and 0-contiguity subdiagrams $\Gamma_0, \Gamma_1, \ldots, \Gamma_s$, Γ_0 connects π_1 with π, for $i = 1, \ldots, s-1$, Γ_i connects π_i with π_{i+1}, Γ_s connects π_s with Π, and the ranks $r_1, r_2, \ldots$ of these cells $\pi_1, \ldots, \pi_s$ form a sequence satisfying the following properties (provided it is not empty which can happen if a bond defining the contiguity subdiagram is a 0-bond):

(Z)   The sequence $r_1, r_2, \ldots$, if not empty, contains exactly one maximal number r; exactly one maximal r' among the numbers to the left of r,[6] exactly one maximal r'' among the numbers to the right of r; exactly one maximal number to the left of r', exactly one maximal number between r' and r, etc.

Exercise 5.2.6. Every contiguity subdiagram has the form depicted on Figure 5.22 where $\mathcal{U}$ and $\mathcal{V}$ are bands of bonds.

Note that the indices of letters in a Zimin word obviously satisfy properties (Z). So Zimin words appear, in spirit, in the definition of a contiguity subdiagram.

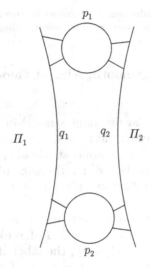

Figure 5.21 A contiguity subdiagram determined by two bonds

---

[6] The set of numbers to the right (left) of r may be empty, still among the numbers in that set there is a maximal one!

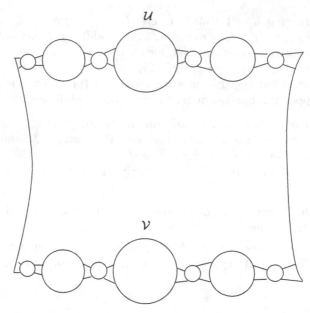

Figure 5.22 A contiguity subdiagram determined by two bands of bonds (the picture is not up to scale: in reality, the sides are much shorter than the contiguity arcs)

5.2.3.5 Boundary Arcs: Smooth, Almost Geodesic, Compatible with a Cell

The van Kampen and annular diagrams studied in [258] usually come with boundary subdivided into several (at most 4) arcs. For instance, the boundary of a contiguity subdiagram Ψ is naturally subdivided into 4 arcs: the two sides and the two contiguity arcs. Hence, for the sake of induction, if we study contiguity subdiagrams in a diagram Δ, we need to consider a subdivision of the boundary of Δ as well.

Note that the labels of the contiguity arcs q_1, q_2 are periodic words with periods C_i and C_k (where C_i is the period of rank i, C_k is the period of rank k, so that the label of $\partial(\pi_1)$ is C_i^n, the label of $\partial(\pi_2)$ is C_k^n). Thus we can assign ranks i, k to the paths q_1 and q_2. So we can assume that some boundary arcs of a diagram have ranks and periodic labels. Note that since the whole diagram from where the contiguity subdiagram Ψ was taken, does not have j-pairs for any j, we can assume that there is no cell in Ψ which is *compatible* with any boundary arc of rank j where compatibility is defined naturally: a cell π is compatible with a boundary arc q if when we attach a cell π of rank j to the diagram Ψ along the boundary arc q, we get a j-pair of cells π' and π, see Figure 5.23.

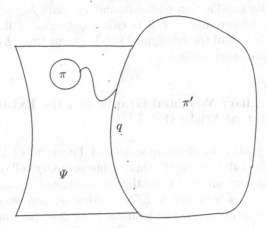

Figure 5.23 A cell π j-compatible with a boundary arc q of $\partial(\Psi)$

Thus a boundary arc q of rank j of a diagram Δ is called *smooth* if the label of q is a periodic word with period C which is not conjugate in G_j to a power of a shorter word, and the diagram does not contain a cell that is compatible with q.

A boundary arc of rank j is called *almost geodesic*, if it is not homotopic in the diagram Δ to a "substantially shorter" path. Here "substantially shorter" means "whose length is at most the length of the boundary arc times a fixed parameter < 1 that is close to 1 but not too close".

5.2.3.6 A Good System of Contiguity Subdiagrams

Now consider all possible collections of pairwise disjoint contiguity subdiagrams between different cells in Δ and between cells and boundary arcs. By definition every non-zero edge belongs to a 0-bond, hence a 0-contiguity subdiagram. Thus there is a collection of contiguity subdiagrams such that every non-0-edge belongs to a contiguity subdiagram of that collection and different contiguity subdiagrams do not share a cell. Let us call such collections of contiguity subdiagrams *full*.[7] Among all full collections of contiguity subdiagrams let us choose a collection with minimal possible number of contiguity subdiagrams. Let us call such collections *good*.[8] We fix one good collection $\Sigma(\Delta)$ of contiguity subdiagrams for each reduced diagram over $\mathcal{PB}$. We call a cell of Δ *special* if it is one of the non-0-cells of the two bands of bonds that define a contiguity subdiagram from $\Sigma(\Delta)$ (i.e., one of the non-0-cells

---

[7] In [258] these collections are called complete.

[8] In [258, 260] more complicated definitions of good collections of contiguity subdiagrams were used but our definition can be used instead.

from the contiguity subdiagram whose boundary share edges with the sides of a contiguity subdiagrams). A cell is called *concealed* if it is not special but is contained in one of the contiguity subdiagrams from $\Sigma(\Delta)$. All other non-0-cells of Δ are called *ordinary*.

5.2.3.7 The Auxiliary Weighted Graphs and the Existence of a Cell that Sticks Out

One of the main goals is to prove an analog of Theorem 5.1.17: in the good system of contiguity subdiagrams Σ, there is one ordinary cell that is attached to one of the boundary arcs via a contiguity subdiagram from Σ with high contiguity degree. This is proved in [258, Section 3], and we shall first talk about this section returning to the lemmas from [258, Section 2] later. We assume that every diagram we consider is *normal*. This means that the rank of a contiguity subdiagram is always less than the ranks of the cells or boundary arcs which this subdiagram connects. The fact that all reduced diagrams are in fact normal is proved later, in [258, Section 4] (but remember that we are dealing with a crown of lemmas!).

To show the existence of a cell that sticks out, we prove that the total length of contiguity arcs of contiguity subdiagrams between a cell and a boundary arc is large (these contiguity arcs will be called *external*). That would imply there exists a cell π for which the total length of the external contiguity arcs in $\partial(\pi)$ is large as well. Since the number of boundary arcs of the diagram is at most 4 by assumption, π is the cell we need.

One of the neat features of [258] is that instead of lengths of paths we consider their weights. Thus with every cell and every edge of a van Kampen diagram we associate its *weight*. The idea is that (a) the weight of a non-0-cell is the sum of weights of its boundary edges, (b) the weight of a non-concealed cell of a higher rank is bigger than the weight of a cell of a smaller rank, and (c) the weight of an edge on the boundary of a cell of a higher rank is smaller than the weight of an edge on the boundary of a cell of a smaller rank. The easiest way to achieve that is to assign to every non-concealed non-0-cell π the weight $|\partial(\pi)|^{\frac{2}{3}}$ and to every edge of the boundary $\partial(\pi)$ the weight $|\partial(\pi)|^{-\frac{1}{3}}$. We also assign weight 0 to every concealed cell and its boundary edges. There is no ambiguity in assigning weights to edges because no two non-0-cells share an edge by our assumption – they only share edges with 0-cells.

If one cell is attached to another cell or to a boundary arc via a contiguity subdiagram from Σ and the contiguity degree is $\geq \beta$, then we shall say that the cell is attached *essentially*. The minimality of Σ implies that there is no ordinary cell π which is attached essentially to two cells Π_1, Π_2. Otherwise we can view the cell and the two contiguity subdiagrams as a bond between Π_1 and Π_2 (making the ordinary cell special) thus decreasing the number of contiguity subdiagrams (recall that our good system Σ had minimal possible number of contiguity subdiagrams).

First estimate the total weight of all special cells in contiguity subdiagrams of Σ as very small. For this we use three lemmas about a contiguity
subdiagram Ψ connecting cells π and Π proved in [258, Section 2] (similar
lemmas hold for contiguity subdiagrams connecting a cell with a boundary
arc). The first lemma says that the sides of Ψ are very small compared to the
perimeters of the cells. The second lemma says that if a cell π is attached
essentially to a cell Π, then the rank of π is much smaller than the rank of
Π, and the label of the contiguity arc from $\partial(\Pi)$ contains at most one and
a little bit (say, less than $\frac{17}{15}$) of the period of that cell (recall that all cells
have boundary labels C_i^n for some periods C_i). The third lemma says that
the contiguity degree of π to Π cannot exceed a certain parameter α which
is only a little bigger than $\frac{1}{2}$.

A few words about the proofs of these lemmas. Since Ψ has smaller type
than the whole diagram Δ, we can assume that all the other lemmas from
the crown of lemmas are true for Ψ. The principal cell π' of each of the two
bonds on the sides of Ψ must have ranks less than the minimum of ranks of π
and Π by definition. It is attached to π and Π by two contiguity subdiagrams
Ψ_1, Ψ_2 of even smaller ranks. Since Ψ_1 and Ψ_2 have smaller type than Ψ, we
can assume that the sides of these contiguity subdiagrams are very small and
the contiguity arc of Ψ_i in $\partial(\pi)$ and $\partial(\Pi)$ contains at most one and a little
bit of a period of that cell. On the other hand, the contiguity arcs of these
contiguity subdiagrams that are contained in $\partial(\pi')$ contain at least $\beta|\partial(\pi')|$
edges. Since all contiguity arcs in this case are almost geodesic (it is one of
the lemmas in the crown of lemmas), and the sides of Ψ_i are small, we can
deduce that the length of the period of the cell π' is much smaller than the
lengths of periods of π and Π. Hence the perimeter of π' is much smaller
than the perimeters of π and Π. This of course implies that the part of $\partial(\pi')$
which is contained in the side of Ψ is very small. The whole side of Ψ_1 is then
the union of three very short paths, hence very small.

The second lemma is a direct consequence of the analog of Fine–Wilf
theorem proved later (see Section 5.2.3.12) which states that there are no
"rectangular" diagrams with very short sides, and smooth long sides where
one long side has more than $1 + \epsilon$ of a period, and the other long side has
more than βn periods (here ϵ is a small number, say, $\frac{1}{120}$).

The proof of the third lemma goes as follows. Suppose that the contiguity
degree of π to Π via a contiguity subdiagram Ψ is large, say, $> \frac{8}{15}$. Let u
be the label of the contiguity arc in $\partial(\pi)$. Then u is equal in rank j (where
j is smaller than the rank of Π) to the label u' of the complementary arc
$p' = \partial(\pi) \setminus p$ which has length at most $\frac{7}{15}|\partial(\pi)| \leq \frac{7}{8}|u|$. As we have seen the
contiguity arc $\partial(\Psi) \cap \partial(\Pi)$ cannot contain much more than $1 + \epsilon$ periods. Let
C be that period and the label of the contiguity arc be CC' where $|C'| < \epsilon|C|$.
Then CC' is equal to u multiplied by two short words on the left and on the
right. Since u is equal to u' in rank j, we conclude that CC' is equal to u'
multiplied by two short words in rank j which contradicts almost geodesicity
of the contiguity arc.

Now let us finish the estimate of the total sum of weights of special cells in a band of bonds $\mathcal{B}$ of a contiguity subdiagram Ψ. We know that $\mathcal{B}$ has property (Z), the Zimin type structure. The cell π_1 of maximal rank from the band of bonds has (as we have established above) weight that is much smaller than the weights of both π and Π. The two cells π_2, π_3 of the next to the maximal ranks of $\mathcal{B}$ have weights which are much smaller (same scaling constant!) than the perimeters of π_1 and π (resp. π_1 and Π). Continue in this manner we see that the weights of the non-0-cells of $\mathcal{B}$ form a geometric progression, and its sum is small compared to the weights of π and Π. Therefore the total weights of the cells of $\mathcal{B}$ is at most δ times the sum of weights of π and Π. In order to estimate the total weight of all special cells we use an auxiliary graph $\Gamma_s(\Delta)$ as in Section 5.1.4.6: put a vertex in each cell that is connected to another cell via a contiguity subdiagram from Σ, connect these cells by edges – one per each contiguity subdiagram. We assign weights to vertices and edges: the weight of a vertex is the weight of the corresponding cell, the weight of an edge is the weight of all special cells in the corresponding contiguity subdiagram.

Since the graphs have weights, we need a more "fancy" version of the old Theorem 5.1.15. That is [258, Lemma 1.4].

Lemma 5.2.7 (A weaker form of Lemma 1.4 from [258]). *Consider a planar graph Φ without parallel edges and loops where every vertex $v \in \Phi$ and every edge e are equipped with finite weights $\nu(v), \nu(e)$, so that $\nu(e) \leq a \min\{\nu(e_-), \nu(e_+)\}$ for every edge e of Φ for some number a. Then $N_2 \leq 7aN_1$ where N_1, N_2 are the sums of weights of vertices and edges of Φ respectively.*

Remark 5.2.8. In fact since the boundaries of the diagrams that we are considering are subdivided into several arcs, we also need to add to Φ several "external" vertices corresponding to the boundary arcs. The weights of these vertices are set to ∞ (and are not included into N_1). We also should have at most one edge connecting every "internal" vertex with one of the "external vertices", preserving planarity. The formulation and the proof of Lemma 5.2.7 stay almost the same.

Lemma 5.2.7 can be proved in the same way as Theorem 5.1.15.

Exercise 5.2.9. Prove Lemma 5.2.7.

Lemma 5.2.7 applied to $\Gamma_s(\Delta)$ immediately gives estimate of the total weight (denote it by A) of all special cells as very small.

Now we need to deal with ordinary cells. Every contiguity arc of an ordinary cell π is either *internal* (if the corresponding contiguity subdiagram is between π and another cell) or *external* (if the corresponding contiguity subdiagram is between π and a boundary arc). Our goal is to show that the total weight (denote it by E) of the external arcs is large, or, equivalently, that the total weight of the internal arcs (denote it by I) is small.

First we estimate the total weight of contiguity arcs of contiguity subdiagrams from Σ of contiguity degree at most β. Consider the auxiliary graph
$\Gamma_\beta(\Delta)$ where the vertices are inside cells connected by the contiguity subdiagrams from Σ and one vertex per each boundary arc, the edges are, again,
one per a contiguity arc of contiguity degree at most β. The weight of a vertex is the weight of the corresponding cell, the weight of an edge is the sum
of weights of the two contiguity arcs of the corresponding contiguity subdiagram. Then Lemma 5.2.7 gives that the sum of weights (denote it by B) of
contiguity arcs with contiguity degree at most β is very small.

It remains to estimate the total weight (denote it C) of the internal contiguity arcs with contiguity degree $> \beta$. Let Ψ be one of the corresponding
contiguity subdiagrams from Σ between a cell π and a cell Π and p, q are the
two contiguity arcs of Ψ with $p \subseteq \partial(\pi)$ and contiguity degree at least β. Then,
as we know, the label of q contains at most one plus a little bit of a period
of the cell Π, and the length of q is at most the length of p plus a little bit.
Hence the rank of Π is much bigger than the rank of π. But here we use the
nice feature of the weights: the weights of edges of a cell of higher rank are
much smaller than the weights of edges of a cell of smaller rank. Therefore
the weight of q is much smaller than the weight of p and we can simply ignore q. We also know that the contiguity degree of p is at most $\frac{1}{2}$ plus a little
bit. It is important to note that π cannot be attached to any other cell or a
boundary arc with contiguity degree $> \beta$. Indeed, that contiguity subdiagram
and Ψ together with π would then form a bond and it would contradict the
minimality of our good system of contiguity subdiagrams Σ. In particular,
the external arcs of π are very short (recall that there are at most 4 external
arcs of Δ). Therefore the weight of p is at most $\frac{1}{2}$ plus a little bit of the sum
of weights of all internal contiguity arcs from $\partial(\pi)$. Since every ordinary cell
has at most one such arc p we obtain that C is at most $\frac{1}{2}$ plus a little bit
(say, $\frac{8}{15}$) of the total weight of all internal contiguity arcs of ordinary cells.

Thus $I \leq A + B + C$ where $C \leq \frac{8}{15}I$ and A, B are very small. But this
inequality implies (move C to the left!) that I is also very small, hence E
is large, say, 1 minus a little bit (call this number θ) of the total weight of
ordinary cells in Δ. Hence, in average, external arcs of ordinary cells are large.
In particular, there exists one ordinary cell and contiguity subdiagrams from
Σ attaching π to the boundary arcs with total contiguity degree at least θ.
That is the sticking out cell we were looking for. We call this cell a θ-cell
of Δ.

5.2.3.8 Why Are All Reduced Diagrams Normal?

In [258, Section 4], it is proved that every reduced diagram Δ is normal. For
this consider any contiguity subdiagram Ψ of a cell π to a cell Π (the case
when Ψ connects π and a boundary arc is similar). As usual, we subdivide
$\partial(\Psi)$ into four boundary arcs $p_1 q_1 p_2^{-1} q_2^{-1}$ where q_1, q_2 are contiguity arcs,

$q_1 \subseteq \partial(\Pi)$, p_1, p_2 are the sides. We need to show that the rank of Ψ is small. By induction on the type of Δ, we can assume that Ψ already is normal. Hence the two bonds $\mathcal{B}_1$ and $\mathcal{B}_2$ that define the contiguity subdiagram Ψ have small rank contiguity subdiagrams. So we need to estimate the rank of the subdiagram Ψ' of Ψ between $\mathcal{B}_1$ and $\mathcal{B}_2$.

The subdiagram Ψ' between the two bonds $\mathcal{B}_1, \mathcal{B}_2$ satisfies the following properties:

- Ψ' is normal, its boundary is subdivided into four arcs: long arcs $q_1' \subseteq q$, $q_2' \subseteq q_2$, and short arcs p_1', p_2'.
- arcs q_1' and q_2' are almost geodesic.
- $|p_1| + |p_2| \le \mu(|q_1| + |q_2|)$ for some small parameter μ.

The idea is to cut Ψ' into small pieces by short bonds where the main cells (if these are not 0-bonds) are θ-cells (see Figure 5.24). In fact it is enough to find either a 0-bond between q_1' and q_2', or a θ-cell in Ψ' which is essentially attached to both q_1' and q_2', forming a bond between q_1' and q_2'. Indeed, then the bond divides Ψ' into two similar subdiagrams and we can proceed by induction on the type.

The proof in [258] of existence of the bond is by induction on the type of the diagram. We consider all subdiagrams Ψ'' of Ψ' with boundary consisting of two long arcs: $q_1'' \subseteq q_1$, $q_2'' \subseteq q_2$, and two short sides p_1'', p_2''. We can assume that the quotient of the sum P of lengths of shorter sides by the sum Q of lengths of the longer sides is the smallest possible among all these subdiagrams. In particular, p_1' and p_2' must be the shortest paths in Ψ' connecting their endpoints (i.e., these are geodesics in Ψ').

Since Ψ' is normal, it contains a θ-cell π. If π is attached via contiguity subdiagrams of degrees $\ge \beta$ to the long sides of Ψ', then we found a bond between q_1' and q_2'. If not, the cell π must be essentially attached to either p_1' or p_2' (in particular $\mu > 0$), hence $\partial(\pi)$ is very short (since p_i are geodesic).

Therefore π cannot be attached by contiguity subdiagrams to both p_1' and p_2' (the distance between p_1' and p_2' is too large). So π must be essentially attached to, say, p_1', and either only to q_1' or to both q_1' and q_2'. In both cases we cut off the subdiagram Ψ'' bounded by π, its contiguity subdiagrams and p_1. Since the contiguity degrees of π to q_1, q_2 cannot be both $\ge \beta$, and since p_1' is geodesic and q_1', q_2' are almost geodesic, we conclude that the contiguity degree of π to p_1 and to one of the paths q_1', q_2' must be almost $\frac{1}{2}$. Therefore when we cut off Ψ'', we get a new diagram where the sum P' of shorter sides can be estimated as about $P - a$ for some a, and the sum of the longer sides can be estimated as $Q - a$ for the same a. Since $(P - a) \le \mu(Q - a)$, and the new subdiagram has smaller type, we can proceed by induction on type.

Thus we can cut Ψ into small subdiagrams as in Figure 5.24.

It remains to note that diagrams of small perimeter over $\mathcal{PB}$ must have very small ranks (that is proved later, see Section 5.2.3.10 and is used here because the types of the small subdiagrams of Ψ are smaller than the type of Δ).

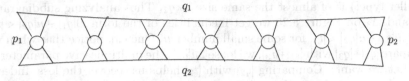

Figure 5.24 A contiguity subdiagram subdivided by bonds into small subdiagrams

5.2.3.9 Why Are Smooth Arcs Almost Geodesic?

Let us prove that smooth boundary arcs are almost geodesic [258, Lemma 4.2]. Suppose that we have a diagram Δ over $\mathcal{PB}$ with boundary subdivided into two arcs $\partial(\Delta) = qt$ where q is smooth of rank i and t is geodesic. We need to prove that q is almost geodesic, say, $.9|q| \le |t|$.

Choose a θ-cell π in Δ that is attached to t and q via contiguity subdiagrams Ψ_1, Ψ_2 with total contiguity degree at least θ. By [258, Lemma 2.3] (the third lemma from [258, Section 2] discussed in Section 5.2.3.7), the contiguity degree of π to q can only be slightly bigger than $\frac{1}{2}$. The same is true for the contiguity degree of π to t (an even easier proof since t is geodesic). Hence both contiguity degrees can be only slightly less than $\frac{1}{2}$. Then π together with Ψ_1 and Ψ_2 forms a bond $\mathcal{B}$ between q and t, the diagram Δ is divided by that bond into three smaller subdiagrams $\Delta_1, \Delta_2, \Delta_3$ as on Figure 5.25.

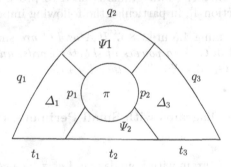

Figure 5.25 The diagram Δ is split into three subdiagrams by a bond

As before we estimate the sides p_1, p_2 of the bond $\mathcal{B}$ as very small compared to $|\partial(\pi)|$. It is important to note that the lengths of p_1 and p_2 are small also compared to $|q_2|$. This follows from the fact that almost all edges of $\partial(\pi)$ belong to the contiguity arcs of π from the bond $\mathcal{B}$, only about half of that belongs to the lower contiguity arc (a cell cannot be attached to a geodesic via a contiguity subdiagram with contiguity degree much bigger than $1/2$), so a large potion of $\partial(\pi)$ belongs to the upper contiguity arc, which by almost geodesicity (we are applying it here to a subdiagram of smaller type!) is of almost the same size as q_2. Thus analyzing subdiagrams Δ_1 and Δ_3 on Figure 5.25 we get inequalities of the form $.9|q_1| - \mu|q_2| < |t_1|$ and $.9|q_3| - \mu|q_2| < |t_3|$ for some small number μ. One can notice that when we compare $|q_1| + |q_3|$ with $|t_1| + |t_3|$, we loose $2\mu|q_2|$ (the multiplicative parameter $.9$ is what we want). Comparing $|q_2|$ with $|t_2|$ helps us recover the loss. Indeed, since the arc q is smooth, and the degree of contiguity of π to q is bigger than β, the label of q_2 contains almost one period C by Lemma [258, Lemma 2.3] (discussed in Section 5.2.3.7), more precisely, it contains at most $1 + \epsilon$ periods where ϵ is very small. Note also that C, by the definition of periods, is geodesic (without "almost"). Therefore the distance between endpoints of q_2 in Δ is not just $\geq .9|q_2|$ (as would follow from almost geodesicity) but $\geq (1 - 2\epsilon)|q_2|$. Thus, comparing q_2 with t_2 we get $(1 - 2\epsilon)|q_2| - 2\mu|q_2| < |t_2|$. Adding up the estimates for the three parts of q, we get

$$.9|q_1| + .9|q_3| + (1 - 2\epsilon)|q_2| - 4\mu|q_2| < |t|.$$

Thus

$$.9|q_1| + .9|q_2| + (1 - 2\epsilon - 4\mu)|q_2| < |t|.$$

It remains to notice that $1 - 2\epsilon - 4\mu > .9$.

Note that an almost the same proof gives that a smooth boundary component of an annular diagram cannot be much longer than the other boundary component (an "annular" version of almost geodesicity).

Similar argument of cutting a diagram by bonds whose principal cells are θ-cells into several smaller subdiagrams is used to prove also several other lemmas in [258, Section 4], in particular the following important statement:

Lemma 5.2.10 (Lemma 4.5 in [258]). *If A and C are simple words in G_i, i.e., not conjugated in G_{i-1} to powers of shorter words, and A is conjugated to a power of C in G_i, $A = X^{-1}C^l X$, then $l = \pm 1$.*

5.2.3.10 Why Do Diagrams with Small Perimeters Have Small Ranks?

Let Δ be a reduced diagram with boundary q. Let π be a cell from Δ and Δ' be the annular diagram obtained by removing π from Δ. The boundary of π is smooth because Δ is a reduced diagram. Hence by the annular version of almost geodesicity (Lemma 5.2.11 below), the length of $\partial(\pi)$ cannot be much bigger than $|q|$. Thus the rank of π cannot be large also.

5.2.3.11 Short Cuts in Annular Diagrams

The following lemma is proved in [258, Section 5].

Lemma 5.2.11. *Let Δ be a reduced annular diagram over $\mathcal{PB}$ with boundary components p, q. Then there are vertices o_1 in p and o_2 in q, and a path s connecting o_1, o_2 such that $|s|$ is much smaller than $|p| + |q|$ (say, $|s| < \frac{1}{100}(|p| + |q|)$).*

This lemma is similar to Theorem 5.1.26, and the proof is very similar too: one just needs to use a θ-cell instead of a Greendlinger cell of Δ.

5.2.3.12 A Generalization of the Fine–Wilf Theorem

Now we come to the last and the most technical part of the proof, [258, Section 6], where we consider diagrams with periodic boundary arcs (contiguity subdiagrams).

An easy consequence of the Fine–Wilf Theorem 1.2.16 is that if U and V are periodic words with periods A and C respectively, where A and C are not proper powers, $U \equiv V$, and $|U| \geq |A| + |C|$, then A and C coincide up to a cyclic shift. Olshanskii [258, Section 6] contains several generalizations of this statement for groups G_i. All these statements easily follow from Theorem 1.2.16 when $i = 0$, i.e., when we consider periodic words in the free group. In the case $i > 0$, we need to consider long and narrow diagrams with periodic long paths (e.g., contiguity subdiagrams). In Sections 5.2.3.8–5.2.3.11 we used several properties of contiguity subdiagrams. These properties follow from the next lemma.

Lemma 5.2.12. *Suppose that Δ is a reduced diagram with $\partial(\Delta) = p_1 q_1 p_2^{-1} q_2^{-1}$ where p_1 and p_2 are very short compared to q_1, q_2 (just how short p_i should be will be clear from the proof) and the labels u_1, u_2 of q_1 and q_2 are periodic words with periods A and C which are simple in G_i, $|A| \geq |C|$, and u_1 contains at least $1 + \epsilon$ periods while u_2 contains very large number of periods. Then A is a conjugate of $C^{\pm 1}$ in G_i.*

In order to prove this lemma, several similar lemmas are proved by induction on $|A| + |C|$. In each case the labels of the long paths are periodic either with different or the same periods. Thus these lemmas are of the "AC form" (when periods on the two long sides of the diagram are different) or of the "AA form" (when periods are the same). The simultaneous induction is on the sum L of the two periods (thus these lemmas form a sub-crown of lemmas). The reason we need several of these lemmas is that we should be sure that when we use an AA-lemma (resp. AC-lemma) to prove an AC-lemma (resp. AA-lemma), the number of periods on the long side of the diagram is large enough. The "large enough" in Lemma 5.2.12 is about 10^4 while in Lemmas 5.2.13, 5.2.14 below it is about 500. Here are a typical AA-lemma and an AC-lemma. We first give a sketch of proofs of these lemmas and then say (Remark 5.2.15) how the formulations should be adjusted to make the sketch work.

Lemma 5.2.13 (AA). *Suppose that Δ is a diagram of rank i with $\partial(\Delta) = p_1 q_1 p_2^{-1} q_2^{-1}$ where p_1 and p_2 are very short and the labels u_1, u_2 of q_1 and q_2 are periodic words with period A which is simple in G_i, and u_1 contains large enough number of periods. Then the boundary arcs q_1 and q_2 are compatible.*

Lemma 5.2.14 (AC). *Suppose that Δ is a diagram of rank i with $\partial(\Delta) = p_1 q_1 p_2^{-1} q_2^{-1}$ where p_1 and p_2 are very short, the label u_1 of q_1 is a periodic word with period A, the label u_2 of q_2 is a periodic word with period C, A, C are simple in G_i, and u_1, u_2 contain large enough number of periods but the number of periods in u_1 is significantly larger than the number of periods in u_2 (and so C is shorter than A). Then the boundary arcs q_1 and q_2 are compatible.*

Proof. AA → AC. Suppose the conditions of Lemma 5.2.14 hold. We can assume that the label of q_1 starts with A, the label of q_2 starts with C. Then the label of q_1 is equal to AA' and to $A'\bar{A}$ where the A' contains large enough number of periods A, $\bar{A}$ is a cyclic shift of A. Hence we can find two points o_1, o_2 on q_1, o_1 is the head of the subpath labeled by A, o_2 is the tail of the subpath labeled by $\bar{A}$. Cutting Δ into small subdiagrams as on Figure 5.24, we can find vertices o' and o'' on q_2 which are "projections" of o_1, o_2 onto q_2 i.e., the distances from o_1 to o' and from o_2 to o'' are very small. Since C is also small compared to A, we can further assume that the label of the subpath of the path q_2 between o'' and o' is a power of C (by moving if necessary o'' and o' a little bit). Let r_1 and r_2 be the shortest paths in Δ connecting o_1 with o' and o_2 with o'' respectively. These two paths divide Δ into three subdiagrams: $\Delta_1, \Delta_2, \Delta_3$ (see Figure 5.26).

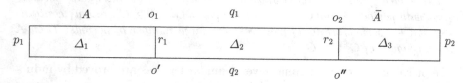

Figure 5.26 A diagram with periodic boundary arcs

Since u_2 (the label of q_2) contains very large number of periods, the labels of the bottom parts of the boundaries of Δ_1 and Δ_3 contain many periods.

Let R_1, R_2 be the labels of r_1, r_2 and P_1, P_2 be the labels of p_1, p_2.

Now let us flip the union of Δ_2 and Δ_3 and attach its top arc to the top arc of the union of Δ_1 and Δ_2 (the labels of these arcs are equal to A'). We get diagram $\bar{\Delta}$ as on Figure 5.27.

The diagram $\bar{\Delta}$ satisfies all the conditions of Lemma 5.2.13 except the sides of it, while very small, are not small enough. But by cutting off small pieces of $\bar{\Delta}$ from the left and from the right (using θ-cells again), we deduce that a very large subdiagram $\hat{\Delta}$ of $\bar{\Delta}$ satisfies all conditions of Lemma 5.2.13,

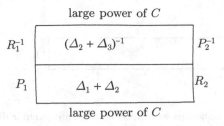

Figure 5.27 The diagram $\bar{\Delta}$ obtained by gluing the inverse of $\Delta_2 + \Delta_3$ on top of $\Delta_1 + \Delta_2$

and, moreover, $|C| + |C| < |A| + |C|$, so we can apply Lemma 5.2.13 and deduce that the top and the bottom arcs of $\bar{\Delta}$ are i-compatible, i.e., there exists a path z connecting the top and the bottom arcs whose label is 1 in G_i, and z_-, z_+ divide the arcs into subarcs containing integral number of periods, see Figure 5.28.

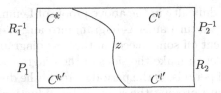

Figure 5.28 The top and the bottom arcs of $\bar{\Delta}$ are compatible

This implies that $R_1^{-1} P_1$ is equal to a power of C in G_i (on Figure 5.28 it is $C^{k-k'}$). Now consider the subdiagram Δ_1 on Figure 5.26. The top arc of that diagram is labeled by A, and the union of the other three arcs is labeled by $R_1^{-1} C^s P_2$ (for some s) which is equal to a conjugate of a power of C in G_i (why?). But this contradicts the assumption that A is not conjugate to a power of a smaller word in G_i. This completes the proof of Lemma 5.2.14.

Let us prove Lemma 5.2.13. Let Δ be a diagram satisfying the conditions of that lemma. By increasing p_1 by at most half of the length of A, and replacing A by its cyclic shift, we can assume that u_1 and u_2 start with A (note that assuming this we no longer claim that p_1 is very short, just that it does not exceed $\frac{1}{2}|A|$ plus a little bit). We will show that the label P_1 of p_1 is equal to 1 in G_i. Suppose that it is not equal to 1.

One of the words u_1 or u_2 is a prefix of the other word. Without loss of generality assume that $u_1 = u_2 v$. Note also that v is short because of almost geodesicity of periodic paths (Section 5.2.3.9). Then we can identify the two subpaths of the longer arcs of $\partial(\Delta)$ labeled by u_2 and obtain an annular diagram $\bar{\Delta}$ with inner boundary labeled by P_1 and the other boundary labeled by v times a very short word (see Figure 5.29).

By Section 5.2.3.11 there exists a very short path r connecting the boundary components of $\bar{\Delta}$. Let R be its label. There also exists a path r_1, also

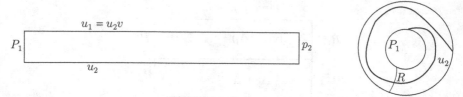

Figure 5.29 Turning diagrams with periodic arcs into annular diagrams

connecting the boundary components, whose label is u_2. By Lemma 5.1.28 then u_2 is equal in G_i to R times a power of P_1, multiplied by two short words on the left and on the right (why should we multiply by the two short words?). The word P_1 may not be simple in G_i, in which case it is conjugate (by a short word) to a power of a simple word P that is even shorter than P_1. Since $|P| + |A| < |A| + |A|$, we can then apply Lemma 5.2.14 and get a contradiction.   □

Remark 5.2.15. The difficulty in the above proofs of Lemmas 5.2.13, 5.2.14 is the following. When we turn an AA-diagram into an AC-diagram (and conversely), we need to cut off some pieces of the new diagram from the left and from the right in order to make the sides of the diagram short enough. As a result, the number of periods on the periodic arcs of the diagram can decrease and eventually (we are proving the lemmas by induction!), the numbers of periods won't be big enough any longer. The solution (used in [258, 260]) is the following. While in the AA-lemma we still require only that the periodic sides contain a big enough, say $> k$ (that number is about 500), number of periods, in the AC-lemma, we assume that the arc with period C contains, say $> 1.2k$ periods while the arc with period A has only $> .8k$ periods. Then when we convert the AA-diagram into AC-diagram and cut a number of periods A in the process, we still have at least $> .8k$ periods A on the A-side of the new diagram. On the other hand, since the length of the word C, by construction, is at most $2/3$ of the length of A, and the periodic paths are almost geodesic, the number of periods C must be $> 1.2k$, so the AC-lemma applies. Similar argument works when we convert an AC-diagram into an AA-diagram. The numbers $1.2, .8, 2/3$ are carefully chosen in such a way, that the lemmas in the crown fit together well.

5.2.3.13 The Conclusion of the Road Map

This completes our road map of the proof of the crown of lemmas from [258]. Let us deduce Theorem 5.2.4. First, why all cube-free words w are not equal to 1 in the group given by the presentation $\mathcal{PB}$? Suppose that $w = 1$ in G_i. Consider a reduced van Kampen diagram Δ with boundary label w. Then this diagram has a θ-cell π corresponding and a contiguity subdiagram Ψ connecting π with $\partial(\Delta)$ with contiguity degree at least θ. If Ψ has rank 0, then $\partial(\Delta)$ contains a subword which is equal (in the free group!) to a subword of at least θ of the length of C_j^n. Since $\theta \gg 1/n$, that subword contains C_j^6, a contradiction. So the rank of Ψ is > 1. Suppose that at least one of the two bands of bonds, say, $\mathcal{B}$, defining Ψ contains non-zero cells. Take the non-zero cell π' of $\mathcal{B}$ that is the closest to $\partial(\Delta)$. Suppose that it corresponds to a relation $C_k^n = 1$. By the definition of a bond, π' is attached to $\partial(\Delta)$ via a 0-contiguity subdiagram with contiguity degree at least β. Since $\beta \gg 1/n$, we get that w contains C_k^6 as a subword, a contradiction. Therefore both bonds defining the contiguity subdiagram are 0-bonds. Therefore the non-zero edges of $\partial(\Psi)$ belong to $\partial(\pi) \cup \partial(\Psi)$. Consider a θ-cell π_1 in Ψ. It's degree of contiguity to $\partial(\pi)$ cannot be much bigger than $1/2$. Hence it is attached to $\partial(\Delta)$ by a contiguity subdiagram Ψ_1 with contiguity degree $> \theta' > \theta - \beta > \frac{1}{2}$. Note that $\Psi_1 \subset \Psi$. Continuing in this manner, we will find a cell π_k that is attached to $\partial(\Delta)$ via a contiguity subdiagram Ψ_k which does not contain non-zero cells at all, and the contiguity degree is at least θ'. Since $\beta \gg \frac{1}{n}$, we conclude that the label of $\partial(\Delta)$ contains a big portion of the label of $\partial(\pi_k)$ and cannot be cube free.

5.2.3.14 Why Is the Group Defined by $\mathcal{PB}$ of Exponent n?

In other words, why the exponent of every element of G divides n ? It is in fact possible to prove more: every word in the generators of G is conjugate to a power of a period C_i (this gives what we need since the order of C_j divides n by definition of G). Indeed, let C be a word in the generators of G. Then $C^k = 1$ for some k, suppose that this equality is true in G_i and not true in G_{i-1}. We can assume that C is not a conjugate of a power of a shorter word D in G_i (otherwise take D instead of C).

Consider a van Kampen diagram Δ with boundary label C^k. Let i be the rank of Δ. Then $\partial(\Delta)$ cannot be smooth of rank i because otherwise it would be almost geodesic in G_i, but it is equal to 1, so has length 0 in G_i. Therefore there exists a cell π of Δ that is compatible with $\partial(\Delta)$. But that means C is equal in G_{i-1} to a cyclic shift of C_j where C_j^n is the boundary label of π. Thus C is conjugate to C_j in G.

5.2.3.15 The Groups G_i Are Hyperbolic

Indeed, consider any reduced diagram Δ over $\mathcal{PB}$ considered as a diagram with one boundary component. Proceeding as in Section 5.2.3.13, we can find a θ'-cell in Δ, i.e., a cell π and a contiguity subdiagram Ψ connecting π with $\partial(\Delta)$ with contiguity degree $> \theta' > 1/2$. Such that the rank of Ψ is smaller than the rank of π. Then the perimeter of Ψ is bounded in terms of the perimeter of π. Therefore the subdiagram $\Delta_\pi = \pi \cup \Psi$ shares more than a half of its perimeter with $\partial(\Delta)$. Now consider all (finitely many) possible diagrams of the form Δ_π for all π of rank at most i, i.e., a union of a cell π with a contiguity subdiagram (note that the contiguity subdiagram may be empty). We just proved that the set of boundary labels of these diagrams is a Dehn presentation of G_i.

5.2.3.16 Why Do We Need n to Be Odd?

We never used this assumption in our road map. In fact it is used crucially in [258, Lemma 6.3] which we skipped. It has similar formulation as Lemma 5.2.13 but with different estimates on the sizes of the boundary arcs. At some point in the proof of that lemma, we need to make sure that the period C as in Lemma 5.2.12 must be A, and cannot be A^{-1} (there are two options in Lemma 5.2.10 also), i.e., A^s cannot be conjugate to A^{-s} if $s \neq 0$. It is true if n is odd. Indeed, if $A = Z^{-1}A^{-1}Z$ in G, then $Z^{-2}AZ^2 = A$, hence $Z^{-2}A^kZ^2 = A^k$ for every k. Applying Lemma 5.2.13, we deduce that Z^2 must be equal to a power of A. Since n is odd, Z is a power of Z^2 in G (prove it!). Hence Z is a power of A, and $A = Z^{-1}A^{-1}Z$ implies $A = A^{-1}$ but that is only possible if $A = 1$ (again because n is odd).

On the other hand if $n = 2k$ is even, then $A^{2k} = 1$, so $A^k = (A^{-1})^k$. So in fact if n is even [258, Lemma 6.3] is wrong (the hardest situation is when a cell π together with a contiguity subdiagram connecting π with itself surrounds the hole in an annular diagram), and as a result the whole crown of lemmas breaks down because they are all connected to each other. It took a lot of efforts to overcome that difficulty. Finally Sergei Ivanov [156] and Igor Lysenok [213] managed to prove that the free Burnside groups of all sufficiently large exponents (odd or even) with at least 2 generators are infinite.

5.3 The Finite Basis Problem for Groups and Rings

In this section we shall present an example of R. Bryant and Yu. Kleiman of a non-finitely based variety of groups and finish the discussion of the example of Gupta and Krasilnikov of a non-finitely based variety of rings from Section 4.5.3.

5.3.1 An Example of R. Bryant and Yu. Kleiman

5.3.1.1 The Result

The existence of non-finitely based varieties of groups was shown by Olshanskii in 1969, see [256], and the very first explicit examples of such varieties were constructed by Adian [2] and Vaughan-Lee [323]. That solved one of the main problems about varieties of groups (see Hanna Neumann's book [248]). Here we present a remarkably elegant and transparent example due to Roger Bryant [56] and Yuri Kleiman [180].

Consider the variety of groups given by all identities of the form

$$(x_1^2 x_2^2 \cdots x_n^2)^4 = 1, \ n \geq 1. \tag{5.3.1}$$

Every group G satisfying these identities has a normal subgroup H of exponent 4 which is generated by all squares of elements and the factor-group G/H is of exponent 2 (prove that!). Conversely, every extension of a normal subgroup of exponent 4 by a group of exponent 2 satisfies all identities (5.3.1). That variety is denoted by $\mathcal{B}_4\mathcal{B}_2$. Note that $\mathcal{B}_4\mathcal{B}_2$ is locally finite by Corollary 3.4.2 and Theorem 5.2.3.

Theorem 5.3.1. *The variety $\mathcal{B}_4\mathcal{B}_2$ is non-finitely based.*

Proof. For every $n \geq 1$ let $u_n = (x_1^2 x_2^2 \cdots x_n^2)^4$. By definition, the infinite set of identities $u_n = 1$, $n \geq 1$, defines the variety $\mathcal{B}_4\mathcal{B}_2$. Observe that $u_n = 1$ implies $u_\ell = 1$ for all $\ell < n$ (why?). Therefore, if the variety $\mathcal{B}_4\mathcal{B}_2$ were finitely based, it would be defined by a single identity $u_n = 1$ for some n. In order to show that this is impossible, we construct a series of groups C_n, $n = 1, 2, \ldots c$, such that for each n, the group C_n satisfies $u_n = 1$ but does not satisfy $u_N = 1$ for some $N > n$.

5.3.1.2 Some Properties of Nilpotent Groups of Class 2

Recall that we denote the word $x^{-1}y^{-1}xy$ by $[x, y]$. It is convenient to denote $[[x, y], z]$ by $[x, y, z]$. We shall need some properties of nilpotent groups of class 2, that is groups G for which the factor-group $G/Z(G)$ is commutative, see Section 3.6.2 (recall that $Z(G)$ denotes the center of G, i.e., the set of elements x such that $xy = yx$ for every $y \in G$).

Exercise 5.3.2. Prove the following properties of nilpotent groups of class 2.

(1) A group is nilpotent of class 2 if and only if it satisfies the identity $[x, y, z] = 1$

(2) A group G is nilpotent of class 2 if and only if its *derived subgroup* G', i.e., the subgroup generated by all commutators $[x, y]$, $x, y \in G$, is inside $Z(G)$.

(3) If a group G is nilpotent of class 2, then for every $a \in G$, the map $x \mapsto [x,a]$ is a homomorphism from G to $Z(G)$.

(**Hint:** Use Exercise 1.4.1)

5.3.1.3 The Semigroup S and Its Semigroup Algebra $\mathbb{F}_2 S$

We start with the semigroup S with zero 0 given by the presentation $\mathrm{sg}\langle x_0, x_1, \ldots \mid x_i^2 = 0, i \geq 0 \rangle$.

Exercise 5.3.3. Prove that the presentation of S is Church–Rosser and the canonical words are words without subwords of the form x_i^2, $i \geq 0$.

We need the following

Definition 5.3.4. Let T be a semigroup with zero 0, K be a field. The *semigroup algebra* KT is the vector space over K spanned by $T \setminus \{0\}$ as a basis (so the elements of $T \setminus \{0\}$ are linearly independent over K) with multiplication naturally extending the multiplication in T:

$$\left(\sum \alpha_i g_i\right)\left(\sum \beta_j h_j\right) = \sum \alpha_i \beta_j g_i h_j.$$

Exercise 5.3.5. Prove that KT is indeed an associative algebra over K.

Let $\mathbb{F}_2$ be the field with 2 elements, i.e., the field $\{0,1\}$ of integers modulo 2. We shall need the semigroup algebra $E = \mathbb{F}_2 S$.

5.3.1.4 The Group A

For every $i \geq 0$ let a_i be the 3×3-matrix over E of the form

$$\begin{pmatrix} 1 & x_i & 0 \\ 0 & 1 & x_i \\ 0 & 0 & 1 \end{pmatrix}.$$

Then each of the matrices a_i is invertible and its square is the identity matrix (because $1 + 1 = 0$ in $\mathbb{F}_2$). Thus the semigroup $A = \langle a_i, i \geq 0 \rangle$ is a group.

Exercise 5.3.6. Show that the elements $a_i, i \geq 0$ satisfy relations

$$a_i^2 = 1, [a_i, a_j, a_m] = 1, \ i,j,m = 0,1,\ldots.$$

In fact it is possible to prove that A is isomorphic to the group given by the presentation from Exercise 5.3.6, but we shall not need it.

Let A_n be the subgroup of A generated by $a_0, \ldots, a_{2n}$.

Exercise 5.3.7. (1) Prove that in A_n, $[a_i, a_j] = (a_i a_j)^2$, in particular, every commutator of generators is a square. Deduce that every element of the derived subgroup A'_n is a product of squares.

(2) Prove that the derived subgroup A'_n is inside the center $Z(A_n)$, that is A_n is a nilpotent group of class 2.

(3) Prove that the derived subgroup A'_n satisfies the identity $x^2 = 1$

(4) Prove that the factor-group A_n/A'_n satisfies the identity $x^2 = 1$, i.e., the square of every element of A_n is in A'_n. Combining with part 1, this gives that A'_n consists of products of squares of elements of A_n. **Hint:** Use Exercise 5.3.6.

By parts (2) and (3) of Exercise 5.3.7 A'_n is a commutative group of exponent 2. If we re-denote the operation in A'_n as +, and the identity element as 0, we can view A'_n as a vector space over $\mathbb{F}_2$.

Lemma 5.3.8. *The commutators* $[a_i, a_j]$, $i < j$, *form a basis of the vector space* A'_n.

Proof. By definition, the derived subgroup of A_n viewed as a vector space over $\mathbb{F}_2$ is spanned by commutators $[a_i, a_j]$, $0 \le i < j \le 2n + 1$, each of which is equal to

$$\begin{pmatrix} 1 & 0 & x_i x_j - x_j x_i \\ 0 & 1 & 0 \\ 0 & 0 & 1 \end{pmatrix}$$

(check it! notice a nice connection between group and ring commutators). Since for any $z_1, z_2 \in R$ we have

$$\begin{pmatrix} 1 & 0 & z_1 \\ 0 & 1 & 0 \\ 0 & 0 & 1 \end{pmatrix} \begin{pmatrix} 1 & 0 & z_2 \\ 0 & 1 & 0 \\ 0 & 0 & 1 \end{pmatrix} = \begin{pmatrix} 1 & 0 & z_1 + z_2 \\ 0 & 1 & 0 \\ 0 & 0 & 1 \end{pmatrix},$$

and

$$\begin{pmatrix} 1 & 0 & z \\ 0 & 1 & 0 \\ 0 & 0 & 1 \end{pmatrix}^2 = \begin{pmatrix} 1 & 0 & 0 \\ 0 & 1 & 0 \\ 0 & 0 & 1 \end{pmatrix}.$$

These matrices are linearly independent because commutators $x_i x_j - x_j x_i$ in E are linearly independent.

Exercise 5.3.9. Prove the last statement. **Hint:** Notice that all words $x_i x_j, i \ne j$, are canonical in S. In a linear combination of commutators with non-zero coefficients look at the lexicographically maximal canonical word $x_i x_j$ and show that it appears exactly once and cannot cancel with another word in the linear combination, hence the combination is not equal to 0.

□

Exercise 5.3.10. Show that the dimension of the vector space A'_n is $\binom{2n+2}{2}$ = $\frac{(2n+2)(2n+1)}{2}$, the number of 2-element subsets of $\{a_0, \ldots, a_{2n+1}\}$.

By Exercises 5.3.10 and 1.4.16, $|A'_n| = 2^{2n^2+3n+1}$.

The key property of A_n is that while by Part (4) of Exercise 5.3.7 every element of A'_n is a product of a number of squares of elements in A_n, we have

Lemma 5.3.11. *There are elements in A'_n that are not products of n squares.*

Proof. Indeed consider any element g of A_n. It is represented by some word w in generators $a_0, \ldots, a_{2n+1}$. Then g^2 is represented by the word w^2. Note that if $w = *a_i a_j*$ where $*$ represents any words, then

$$ww = *a_i a_j * *a_i a_j* = *a_j a_i [a_i, a_j] * *a_j a_i [a_i, a_j]*$$
$$= *a_j a_i * *a_j a_i * [a_i, a_j]^2 = *a_j a_j * *a_j a_i*$$

because $[a_i, a_j]$ is in the center of A_n and its square is 1. Thus in the word ww, one can permute the letters of both copies of w as one wishes as long as this is done in a synchronous manner. Since $a_i^2 = 1$, g^2 is represented by the square of a word $a_{j_1} \ldots a_{j_s}$ where $j_1 < \ldots < j_s$. Thus the number of squares in A_n is at most the number of different subsets of the set $\{a_0, \ldots, a_{2n+1}\}$, i.e., 2^{2n+2} (Exercise 1.1.1). Hence the number of elements in A'_n that can be represented as a product of n squares does not exceed $(2^{2n+2})^n = 2^{2n^2+2n}$ (prove it!). Since $2^{2n^2+2n} < 2^{2n^2+3n+1}$, there exists an element $a \in A'_n$ which cannot be represented as a product of n squares.   □

5.3.1.5 The Group B_n

Let a be an element of A'_n that is not a product of n squares in A_n.

For $g \in A_n$, we denote by $\overline{g}$ the image of g in the (commutative) factor-group $\overline{A}_n = A_n/A'_n$.

Now for each $n = 1, 2, \ldots c$, we define another nilpotent of class 2 group B_n as the group generated by the set $\{b_g, c_k \mid g \in A_n, k \in \overline{A}_n\}$ (notice that the generators b_g are indexed by elements of A_n and the indexes of generators c_k are from $\overline{A}_n$) subject to the relations

$$(b_g)^2 = (c_k)^2 = [b_g, c_k] = 1, \quad [b_g, b_h] = \begin{cases} 1 & \text{if } gh^{-1} \neq a \\ c_{\overline{g}} & \text{if } gh^{-1} = a, \end{cases} \tag{5.3.2}$$

(Note that if $gh^{-1} = a$, then $\overline{g} = \overline{h}$.)

Exercise 5.3.12. Show that B_n is a nilpotent group of class 2 generated by $b_g, g \in A_n$, its derived subgroup $B'_n \subseteq Z(B_n)$ is a commutative group of exponent 2 (that is again can be viewed as a vector space over $\mathbb{F}_2$) and $\{c_k \mid k \in \overline{A}_n\}$ forms a basis of B'_n.

Exercise 5.3.13. Let $b = b_{g_1} \ldots b_{g_s}$, $g_i \in A_n$. Let b' be the product of the same factors b_{g_i} in a different order. Show that $b^2 = (b')^2$. **Hint:** Copy the proof of Lemma 5.3.11.

Exercise 5.3.14. Prove that for every $h, h', g \in A_n$, $b_h b_{hg}$ and $b_{h'} b_{h'g}$ commute.

5.3.1.6 The Group C_n

The group A_n acts on the set of generators $b_g, c_k, g \in A_n, k \in \overline{A}_n$ of B_n in the following natural way

$$h \cdot b_g = b_{gh}, h \cdot c_k = c_{k\overline{h}}.$$

Exercise 5.3.15. Show that for every $h \in A_n$, the map $x \mapsto h \cdot x$ preserves the defining relations of B_n, and hence extends to an automorphism of B_n.

Consider the *semidirect product* C_n of A_n and B_n, that is the set of pairs $(x, y), x \in B_n, y \in A_n$ with operation $(x, y)(x', y') = (x(y \cdot x'), yy')$.

Exercise 5.3.16. Show that C_n is a group that is an extension of a normal subgroup $\hat{B}_n = \{(x, 1) \mid x \in B_n\}$ which is isomorphic to B_n by the group $C_n/\hat{B}_n$ which is isomorphic to A_n.

Each element of C_n can be written as $(b_{g_1} b_{g_2} \cdots b_{g_s}, g)$ where $g, g_1, g_2, \ldots, g_s \in A_n$.

Our goal is to show that C_n satisfies the first n identities of the form

$$(x_1^2 \ldots x_t^2)^4 = 1 \tag{5.3.3}$$

but does not satisfy some identities of this form (that would complete the proof of the theorem).

The following lemma describes elements g from C_n with $g^4 = 1$.

Lemma 5.3.17. *Suppose that*

$$d = (b_{g_1} b_{g_2} \cdots b_{g_s}, g) \in C_n \tag{5.3.4}$$

is such that $g \in A'_n$. Then $d^4 \neq 1$ if and only if $g = a$ and for some $k \in \overline{A}_n$, the number of g_i in (5.3.4) such that $\overline{g}_i = k$ is odd.

Proof. First consider the case when d is such that all elements g_i in (5.3.4) have the same image $k = \overline{g}_i$ in $\overline{A}_n$. We have

$$d^2 = (b_{g_1 g} b_{g_2 g} \cdots b_{g_s g} b_{g_1} b_{g_2} \cdots b_{g_s}, 1) \in \hat{B}_n$$

since $g^2 = 1$. By Exercise 5.3.13, we can write

$$d^4 = ((b_{g_1} b_{g_1 g} b_{g_2} b_{g_2 g} \cdots b_{g_s} b_{g_s g})(b_{g_1} b_{g_1 g} b_{g_2} b_{g_2 g} \cdots b_{g_s} b_{g_s g}), 1). \tag{5.3.5}$$

The products $b_{g_i} b_{g_i g}$ and $b_{g_j} b_{g_j g}$ commute by Exercise 5.3.14 for every i, j. Therefore we can regroup the right-hand side of (5.3.5) as follows:

$$d^4 = ((b_{g_1} b_{g_1 g})^2 (b_{g_2} b_{g_2 g})^2 \cdots (b_{g_s} b_{g_s g})^2, 1).$$

Note that

$$(b_{g_i} b_{g_i g})^2 = b_{g_i} b_{g_i g} b_{g_i} b_{g_i g} = [b_{g_i g}, b_{g_i}] = (c_k)^\epsilon,$$

where $\epsilon = 1$ if $g = a$ and $\epsilon = 0$ otherwise. Therefore $d^4 = (c_k)^{s\epsilon}$ hence $d^4 \neq 1$ if and only if $g = a$ and s is odd.

In the general case, since $[b_g, b_h] = 1$ whenever $\overline{g} \neq \overline{h}$ (see (5.3.2)), we can collect b_{g_i}'s into blocks with the same $\overline{g}_i$. The action of the element g preserves these blocks since g belongs to A'_n. When we calculate the 4-th power of each block, we get $(c_k)^{s_k \epsilon}$ where $\overline{g}_i = k$ for each g_i in the block and s_k is the size of the group. Since the elements c_k are linearly independent (Exercise 5.3.12), $d^4 \neq 1$ if and only if $g = a$ and some s_k is odd.   □

Now let us show that the identity $(x_1^2 \ldots x_t^2)^4 = 1$ holds in C_n for all $t \leq n$. If we calculate the product $x_1^2 x_2^2 \cdots x_n^2$ with $x_1, x_2, \ldots, x_n$ being arbitrary elements of C_n, the second element of the pair has the form $g_1^2 g_2^2 \cdots g_n^2$ for some $g_1, g_2, \ldots, g_n \in A_n$ and cannot be equal to a since a is not a product of n squares. Lemma 5.3.17 then implies that $(x_1^2 x^2 \cdots x_n^2)^4 = 1$.

On the other hand, the element a is a product of some $m > n$ squares in A_n, so let $g_1, g_2, \ldots, g_m \in A_n$ be such that $a = g_1^2 g_2^2 \cdots g_m^2$. At least one of these elements does not belong to A'_n since $a \neq 1$. Since squares commute in A_n (Exercise 5.3.7, Parts (4) and (1)), we may assume that $g_m \notin A'_n$. Now let $x_i = (1, g_i)$ if $i < m$ and $x_m = (b_a, g_m)$. Then $(x_1^2 \ldots x_m^2)^4$ is not equal to 1 in C_n by Lemma 5.3.17 (check it!). Thus one of the identities of the form (5.3.3) fails in C_n.   □

5.3.2 Construction of the Algebra R Used in Section 4.5.3

Recall that the proof in Section 4.5.3 relied on the existence of an associative algebra R with identity element over the 2-element field $\mathbb{F}_2$ satisfying the following two properties:

(R$_1$) R satisfies the identities $(x, y^2) = ((x, y), z) = 0$.

(R$_2$) for every $n > 1$, there exist $s_1, \ldots, s_{n+1} \in R$ such that the product $s_1^2 \cdots s_{n+1}^2$ does not lie in the linear subspace M spanned by the set of all products of at most n squares in R;

Here we construct such an algebra R.

5.3.2.1 The Subgroup G of the Group A

We start with a subgroup G of the group A from Section 5.3.1.4 generated by all elements of the form $g_i = a_0 a_i, i > 0$, i.e., matrices

$$\begin{pmatrix} 1 & x_0 + x_i & x_0 x_i \\ 0 & 1 & x_0 + x_i \\ 0 & 0 & 1 \end{pmatrix}.$$

over the semigroup algebra $E = \mathbb{F}_2 S$.

Exercise 5.3.18. Show that G satisfies the following relations:

$$[g_i, g_j, g_k] = 1, \quad [g_i, g_j]^2 = 1, \quad g_i^4 = 1, \quad i, j, k \in \mathbb{N}.$$

We denote by G^2 the subgroup of G generated by all squares, that is, elements of the form g^2 ($g \in G$).

Observe that for all $i, j \in \mathbb{N}$

$$[g_i, g_j] = g_i^{-1} g_j^{-1} g_i g_j = g_i^3 g_j^3 g_i g_j = g_i^2 \cdot (g_i g_j^3)^2 \cdot g_j^2 \in G^2$$

(since $g_i^4 = 1$, we have $g_i^{-1} = g_i^3$).

This and the fact that the commutators $[g_i, g_j]$ commute with all elements in G (because $[g_i, g_j, g_k] = 1$ in G for all $i, j, k \in \mathbb{N}$) allow us to rewrite each element $y \in G$ as

$$y = g_{i_1} \cdots g_{i_q} h, \qquad (5.3.6)$$

where $i_1 < \ldots < i_q$ and $h \in G^2$.

Exercise 5.3.19. (1) Using the definition of G as a group of matrices, show that every $g \in G$ has a unique representation (5.3.6).

(2) Show that the elements g_i^2 and $[g_i, g_j]$, $i, j \in \mathbb{N}, i > j$ form a basis of the vector space G^2. **Hint:** Use the proof of Lemma 5.3.8 and a hint for Exercise 5.3.9.

The product $g_{i_1} \cdots g_{i_q}$ in the representation (5.3.6) is called the *trunk* of y.[9]

5.3.2.2 The Algebra R

Next we take the group algebra of the group G.

Definition 5.3.20. If H is a group, then the *group algebra* of H over a field F is the vector space FG over F spanned by G as a basis (so the elements of

---

[9] Indeed, $g_{i_1} \cdots g_{i_q}$ is reminiscent of the trunk of a car which is normally in the back. On the other hand, since G^2 is in the center of G, we can rewrite (5.3.6) in the form $h g_{i_1} \cdots g_{i_q}$, in which case $g_{i_1} \cdots g_{i_q}$ is reminiscent of the trunk of an elephant which is normally in front.

G are linearly independent over F, elements of FG are linear combinations of elements of G) with multiplication naturally extending the multiplication in G:

$$\left(\sum \alpha_i g_i\right)\left(\sum \beta_j h_j\right) = \sum \alpha_i \beta_j g_i h_j$$

(compare with Definition 5.3.4 of semigroup algebras).

The associative algebra R we are after will be a factor-algebra of $\mathbb{F}_2 G$ over some ideal V. Note that if V is such that $\mathbb{F}_2 G/V$ satisfies identities (R$_1$), then V must contain the verbal ideal W corresponding to these identities. On the other hand if (R$_2$) holds in $\mathbb{F}_2 G/V$, it also holds in $\mathbb{F}_2 G/W$ (prove it!). Thus it would be natural to define R as $\mathbb{F}_2 G/W$. In fact this R does satisfy (R$_1$) and (R$_2$). But it turns out that it is computationally easier to consider the following bigger ideal V.

Consider the map $f : G \times G \to \mathbb{F}_2 G$: $f(y,z) = [y,z] + y^2 z^2 + y^2 + z^2$. Let V be the ideal of $\mathbb{F}_2 G$ generated by all elements $f(y,z)$, $y, z \in G$. The factor-ring $\mathbb{F}_2 G/V$ is the desired algebra R.

5.3.2.3 Property (R$_1$)

First let us show that indeed $V \ge W$, that is R satisfies the identities $((x,y),z) = 0 = (x,y^2)$. The next exercise shows that the second identity is redundant.

Exercise 5.3.21. Show that in every associative algebra over a field of characteristic 2 we have

$$((x,y),y) = (xy+yx)y + y(xy+yx) = xy^2 + y^2 x = (x,y^2),$$

Thus it suffices to prove only the Lie nilpotency of class 2.

Proposition 5.3.22. *The algebra* $R = \mathbb{F}_2 G/V$ *satisfies* $((x,y),z) = 0$.

Proof. We start with the following exercise.

Exercise 5.3.23. For all $y, z \in G$,

$$(yz)^2 + 1 \equiv (y^2 + 1) + (z^2 + 1) \pmod{V}.$$

Hint:

$$
\begin{aligned}
(yz)^2 + 1 &= y^2 z^2 [y,z] + 1 \\
&= y^2 z^2 \left([y,z] + y^2 z^2 + y^2 + z^2\right) + y^2 z^2 \left(y^2 z^2 + y^2 + z^2\right) + 1 \\
&\equiv y^2 z^2 \left(y^2 z^2 + y^2 + z^2\right) + 1 \pmod{V} \\
&\equiv (y^2 + 1) + (z^2 + 1) \pmod{V}
\end{aligned}
$$

Now let $r_k = \sum_{j_k} \alpha_{kj_k} h_{kj_k}$ for $k = 1, 2, 3$ be three arbitrary elements of $\mathbb{F}_2 G$ (here $\alpha_{kj_k} \in \mathbb{F}_2$, $h_{kj_k} \in G$). We need to show that $((r_1, r_2), r_3) \in V$.

Since

$$((r_1, r_2), r_3) = \sum_{j_1, j_2, j_3} \alpha_{1j_1} \alpha_{2j_2} \alpha_{3j_3} ((h_{1j_1}, h_{2j_2}), h_{3j_3}),$$

we only need to check that $((h_1, h_2), h_3) \in V$ for all $h_1, h_2, h_3 \in G$. We expand $((h_1, h_2), h_3)$ as follows:

$$
\begin{aligned}
((h_1, h_2), h_3) &= h_1 h_2 h_3 + h_2 h_1 h_3 + h_3 h_1 h_2 + h_3 h_2 h_1 \\
&= h_1 h_2 h_3 \big(1 + [h_2, h_1] + [h_3, h_1 h_2] + [h_3, h_1 h_2][h_2, h_1]\big) \quad (5.3.7) \\
&= h_1 h_2 h_3 \big(1 + [h_2, h_1]\big)\big(1 + [h_3, h_1 h_2]\big).
\end{aligned}
$$

By the definition of V, we have $1 + [h_2, h_1] \equiv (1 + h_2^2)(1 + h_1^2)$ and $1 + [h_3, h_1 h_2] \equiv (1 + h_3^2)\,(1 + (h_1 h_2)^2) \pmod V$. By Exercise 5.3.23, $1 + (h_1 h_2)^2 \equiv (1 + h_1^2) + (1 + h_2^2) \pmod V$. Substituting these in (5.3.7), we obtain

$$((h_1, h_2), h_3) \equiv h_1 h_2 h_3 (1 + h_2^2)(1 + h_1^2)(1 + h_3^2)\big((1 + h_1^2) + (1 + h_2^2)\big) \pmod V. \tag{5.3.8}$$

Since expressions of the form $1 + h_k^2$ commute with each other and $(1 + h_k^2)^2 = 1 + h_k^4 = 0$, the right-hand side in (5.3.8) is equal to 0. Thus, $((h_1, h_2), h_3) \in V$. $\qquad \square$

5.3.2.4 Property (R$_2$)

In order to prove (R$_2$) we shall first prove three statements about G and $\mathbb{F}_2 G$.

Lemma 5.3.24. *For all $h_1, \ldots, h_{2s} \in G$,*

$$(h_1, h_2) \cdots (h_{2s-1}, h_{2s}) \equiv h_1 \cdots h_{2s} (1 + h_1^2) \cdots (1 + h_{2s}^2) \pmod V$$

Proof. Note that $f(y, z)$ can be rewritten as $(1 + [y, z]) + (1 + y^2)(1 + z^2)$ because $1 + 1 = 0$ in $\mathbb{F}_2$ and squares commute in G. Therefore by the definition of V, $1 + [h_i, h_j] \equiv (1 + h_i^2)(1 + h_j^2) \pmod V$ hence $(h_i, h_j) = h_i h_j (1 + [h_i, h_j]) \equiv h_i h_j (1 + h_i^2)(1 + h_j^2)$. It remains to use the fact that $1 + h_i^2$ commutes with all elements of $\mathbb{F}_2 G$. $\qquad \square$

Lemma 5.3.25. *For all $h_1, \ldots, h_{2s} \in G$, if the product $c = h_1 \cdots h_{2s}$ belongs to G^2, then the product $u = (h_1, h_2) \cdots (h_{2s-1}, h_{2s})$ belongs to V.*

Proof. By Lemma 5.3.24 $u \equiv h_1 \cdots h_{2s} (1 + h_1^2) \cdots (1 + h_{2s}^2) \pmod V$. We write h_{2s} as $h_{2s} = h_{2s-1}^{-1} \cdots h_1^{-1} c$. Then, using Exercise 5.3.23 and taking into account the fact that $c^2 = 1$ (since $c \in G^2$), we get

$$1 + h_{2s}^2 = 1 + (h_{2s-1}^{-1} \cdots h_1^{-1} c)^2 \equiv (1 + h_1^2) + \cdots + (1 + h_{2s-1}^2) \pmod V.$$

Since $(1 + h_i^2)^2 = 1 + h_i^4 = 0$ for all i, we have

$$u \equiv h_1 \cdots h_{2s}(1 + h_1^2) \cdots (1 + h_{2s-1}^2)\big((1 + h_1^2) + \cdots + (1 + h_{2s-1}^2)\big) \equiv 0 \pmod{V}$$

as required.   □

Let r be an element of $\mathbb{F}_2 G$, i.e., a linear combination of elements of G. We can combine elements with the same trunk together, and get a (unique) *trunk decomposition* of $r = \sum r_t$ where each *trunk component* $r_t = t h_i$ where t is the common trunk of the summands of r_t, $h_i \in \mathbb{F}_2 G^2$, and different r_t have different trunks.

Lemma 5.3.26. *If $r \in V$, then all trunk components of r are in V.*

Proof. Let V' be the ideal of the commutative algebra $\mathbb{F}_2 G^2$ spanned by all $f(y, z)$, $y, z \in G$. Then $V = \mathbb{F}_2 G \cdot V'$. Since $\mathbb{F}_2 G = \sum t \mathbb{F}_2 G^2$ where t runs over all possible trunks, we get

$$V = \sum (t \mathbb{F}_2 G^2) \cdot V' = \sum t V' = \sum (V \cap t \mathbb{F}_2 G^2). \tag{5.3.9}$$

Now suppose that $r = \sum r_t \in V$ where $r_t \in t \mathbb{F}_2 G^2$ is a trunk component of r corresponding to the trunk t. Then (5.3.9) implies that $r = \sum r_t'$ where $r_t' \in V \cap t \mathbb{F}_2 G^2$. Since a decomposition into a sum of trunk components is unique, we deduce that $r_t = r_t'$ for every t, so every trunk component of r is in V.   □

Observe that for each subspace $t \mathbb{F}_2 G^2$ (where t is a trunk), we have

$$V \cap t \mathbb{F}_2 G^2 = \Big(\sum_{t'} t' V'\Big) \cap t \mathbb{F}_2 G^2 = t V'.$$

The first equality here is true because of (5.3.9), and the second equality is a consequence of Exercise 1.4.14.

Thus we deduce

Lemma 5.3.27. $V' = V \cap \mathbb{F}_2 G^2$.

Lemma 5.3.28. *The set of products $\mathcal{A} = \{(g_{j_1}^2 + 1) \cdots (g_{j_m}^2 + 1) \mid m \geq 0,\ j_1 < \ldots < j_m\}$ from $\mathbb{F}_2 G$ is linearly independent modulo V.*

Proof. As observed before the formulation of the lemma, $\mathbb{F}_2 G^2 \cap V = V'$. Since $\mathcal{A} \subset \mathbb{F}_2 G^2$, it suffices to prove that $\mathcal{A}$ is linearly independent modulo V'.

Recall that G^2 is a group of exponent 2 with a basis consisting of all the elements g_i^2 ($i \in \mathbb{N}$) and $[g_i, g_j]$ ($i, j \in \mathbb{N}$, $i > j$). Suppose we have a homomorphism ϕ from $\mathbb{F}_2 G^2$ to some associative algebra Y over $\mathbb{F}_2$ such that $\phi(V') = \{0\}$ and $\phi(B)$ is linearly independent in Y for some subset $B \subset \mathbb{F}_2 G^2$. Then B would be linearly independent modulo V' and hence modulo V by Lemma 5.3.27.

Thus we need to construct such an algebra Y and a homomorphism ϕ for the set $\mathcal{A}$. For this we need the following property of group algebras (a similar property for semigroup algebras also holds).

Exercise 5.3.29. Suppose that K is a field, P is a group, and Q is an associative algebra over K. Let ϕ be a homomorphism from P to the multiplicative semigroup of Q. Prove that ϕ extends to a homomorphism $\bar{\phi}\colon KP \to Q$ which takes every linear combination $\sum \alpha_i p_i$ to $\sum \alpha_i \phi(p_i)$.

Recall that G^2 is a commutative group of exponent 2, and, viewed as a vector space over $\mathbb{F}_2$, G^2 has a basis consisting of all elements $[g_i, g_j]$, $i < j$ and g_i^2 (Exercise 5.3.19). Let $Q = \mathbb{F}_2[t_i]$ ($i \in \mathbb{N}$) be the algebra of (commutative) polynomials over $\mathbb{F}_2$, Z be the subset of Q consisting of all polynomials with constant term 1, and J the ideal of Q generated by the polynomials t_i^2, $i \in \mathbb{N}$. Then the set $(Z + J)/J$ is a commutative subgroup of exponent 2 in the multiplicative semigroup of the algebra Q/J (prove it!). It also can be viewed as a vector space over $\mathbb{F}_2$. By Part (3) of Exercise 1.4.13, any map ϕ from the basis of G^2 to $(Z + J)/J$ extends to a homomorphism from G^2 to $(Z + J)/J$. By Exercise 5.3.29, it will then extend to a homomorphism $\bar{\phi}$ from the group algebra $\mathbb{F}_2 G^2$ to Q/J. It is not that difficult to choose ϕ such that $\bar{\phi}$ "kills" V'. For example, we can take $\phi([g_i, g_j]) = t_i t_j + 1 + J$, and $\phi(g_i^2) = t_i + 1 + J$.

Exercise 5.3.30. Show that $\bar{\phi}(V') = \{0\}$. **Hint:** It is a straightforward computation that $\bar{\phi}$ "kills" elements $f(g_i, g_j)$ for every i, j. To prove that it "kills" $f(x, y)$ for every x, y (and hence the whole V') use induction on the sum of lengths of the trunks of x, y.

The image of the set $\mathcal{A}$ under this homomorphism is equal to the set

$$\{t_{j_1} \cdots t_{j_m} + J \mid m \geq 0 \mid j_1 < \ldots < j_m\}$$

which is linearly independent in Q/J (prove that!). Thus, $\mathcal{A}$ is linearly independent modulo V'.   □

We are ready to prove that the algebra $R = \mathbb{F}_2 G/V$ satisfies property (R_2). In fact, we will prove the following more precise statement.

Proposition 5.3.31. *Let n be a fixed positive integer and M be the subspace of $R = \mathbb{F}_2 G/V$ spanned by all products of at most n squares in R. Set $s_i = g_i + V \in R$ for $i \in \mathbb{N}$. Then $s_1^2 \cdots s_{n+1}^2 \notin M$.*

Proof. Let L be the subspace of $\mathbb{F}_2 G$ spanned by all products of at most n squares in $\mathbb{F}_2 G$. It suffices to prove that $g_1^2 \cdots g_{n+1}^2 \notin L + V$.

First we describe the squares in $\mathbb{F}_2 G$. For any $r = \sum_i \alpha_i h_i \in \mathbb{F}_2 G$ ($\alpha_i \in \mathbb{F}_2$, $h_i \in G$), we have

$$r^2 = \left(\sum_i \alpha_i h_i\right)^2 = \left(\alpha + \sum_i \alpha_i (h_i + 1)\right)^2 \qquad \text{where } \alpha = \sum_i \alpha_i$$

$$= \alpha^2 + \sum_i \alpha_i^2 (h_i + 1)^2 + \sum_{i<j} \alpha_i \alpha_j (h_i, h_j).$$

Therefore, for every $r_1, \ldots, r_\ell \in \mathbb{F}_2$, the product $r_1^2 \cdots r_\ell^2$ is a linear combination (with coefficients from $\mathbb{F}_2$) of 1 and elements of the forms $(h_1^2 + 1) \cdots (h_k^2 + 1)$ $(k \le \ell)$ and $(h_1^2 + 1) \cdots (h_p^2 + 1)(u_1, u_2) \cdots (u_{2q-1}, u_{2q})$ $(q > 0, p + q \le \ell)$ where $h_i, u_j \in G$ for all i, j. We conclude that L consists of linear combinations of 1 and elements of the forms

$$(h_1^2 + 1) \cdots (h_k^2 + 1) \quad (0 < k \le n) \tag{5.3.10}$$

and

$$(h_1^2 + 1) \cdots (h_p^2 + 1)(u_1, u_2) \cdots (u_{2q-1}, u_{2q}) \quad (q > 0, p + q \le n). \tag{5.3.11}$$

By contradiction, suppose that $g_1^2 \cdots g_{n+1}^2 \in L + V$. Then $(g_1^2 + 1) \cdots (g_{n+1}^2 + 1) \in L + V$. Indeed, this element is a sum of $g_1^2 \cdots g_{n+1}^2$ and several products of fewer squares of g_i. Hence we have

$$(g_1^2 + 1) \cdots (g_{n+1}^2 + 1) \equiv \beta + \sum_i \gamma_i c_i + \sum_j \delta_j d_j \pmod{V}$$

for some $\beta, \gamma_i, \delta_j \in \mathbb{F}_2$, some elements c_i of the form (5.3.10) and some elements d_j of the form (5.3.11). Notice that $(h_1^2 + 1) \cdots (h_{n+1}^2 + 1)$ and $\beta + \sum_i \gamma_i c_i$ lie in $\mathbb{F}_2 G^2$. Lemma 5.3.24 implies that every element

$$d_j = (h_{j1}^2 + 1) \cdots (h_{jp_j}^2 + 1)(u_1^{(j)}, u_2^{(j)}) \cdots (u_{2q_j - 1}^{(j)}, u_{2q_j}^{(j)})$$

of the form (5.3.11) belongs to

$$u_1^{(j)} \cdots u_{2q_j}^{(j)} \mathbb{F}_2 G^2.$$

By Lemma 5.3.26, we may assume that $u_1^{(j)} \cdots u_{2q_j}^{(j)} \in G^2$ for all j. Then, by Lemma 5.3.25 $(u_1^{(j)}, u_2^{(j)}) \cdots (u_{2q_j - 1}^{(j)}, u_{2q_j}^{(j)}) \in V$ hence $d_j \in V$ for all j. Therefore,

$$(g_1^2 + 1) \cdots (g_{n+1}^2 + 1) \equiv \beta + \sum_i \gamma_i c_i \pmod{V},$$

where $\beta, \gamma_i \in \mathbb{F}_2$ and each c_i is of the form (5.3.10). By Exercise 5.3.23 each c_i of the form (5.3.10) is (modulo V) a linear combination of products $(g_{i_1}^2 + 1) \cdots (g_{i_k}^2 + 1)$ with the same number k of factors as in (5.3.10).

Thus, the element $(g_1^2 + 1) \cdots (g_{n+1}^2 + 1)$ is, modulo V, a linear combination of 1 and elements $(g_{i_1}^2 + 1) \cdots (g_{i_k}^2 + 1)$ with $k \le n$. This contradicts Lemma 5.3.28. $\qquad \square$

5.4 Groups and Identities, Abért's Criterium

In this section we will give a nice and very general criterion due to Miklos Abért [1] for a group to not satisfy any group identity.

Let G be a group acting on an infinite set X on the right. We say that G *separates* X if for any finite subset $Y \subseteq X$, the *pointwise stabilizer* $G_Y = \{g \in G \mid y \cdot g = y$ for all $y \in Y\}$ does not fix any point outside Y (i.e., for every $y \notin Y$ there exists $g \in G_Y$ such that $y \cdot g \neq y$).

Theorem 5.4.1 (Abért [1]). *If G separates X, then G does not satisfy any nontrivial group identity.*

Proof. Let F_k be the free group with k free generators $x_1, \ldots, x_k$. Let $w \in F_k$ be a reduced word of length $n > 0$; that is, $w \equiv v_1 v_2 \cdots v_n$ where $v_i \in \{x_1, \ldots, x_k\}^{\pm 1}$ with $v_i \neq v_{i+1}^{-1}$ for $1 \leq i \leq n-1$. Let $w(j)$ denote the j-th prefix of w:

$$w(j) \equiv v_1 v_2 \ldots v_j, \quad 0 \leq j \leq n.$$

Given a k-tuple $\mathfrak{g} = (g_1, \ldots, g_k) \in G^k$, we can define the action $\cdot_{\mathfrak{g}}$ of F_k on X by $x \cdot_{\mathfrak{g}} w = x \cdot w(g_1, \ldots, g_k)$ for every word w in F_k. We say that a tuple $\mathfrak{g}$ is *distinctive* for a word $w \in F_k$ of length n and a point $f_0 \in X$, if the points

$$f_j = f_0 \cdot_{\mathfrak{g}} w(j), \quad 0 \leq j \leq n$$

are all distinct. If there is a point f_0 such that $(g_1, g_2, \ldots, g_k)$ is distinctive for w and f_0, then, in particular,

$$f_0 \cdot_{\mathfrak{g}} w = f_n \neq f_0,$$

and so $w \neq 1$ in G.

We claim that for all $f_0 \in X$ and $1 \neq w \in F_k$, there exists $\mathfrak{h} \in G^k$ that is distinctive for w and f_0. This implies the statement of the theorem. We prove our claim by induction on n.

For $n = 1$, the word w consists of one letter, and the claim is trivial.

Using the induction assumption, we can find $\mathfrak{g} = (g_1, \ldots, g_k) \in G^k$ such that $f_0 \cdot_{\mathfrak{g}} w(i)$, $0 \leq i < n$, are all distinct. If $f_n = f_0 \cdot_{\mathfrak{g}} w \notin \{f_0, f_1, \ldots, f_{n-1}\}$, then we have found the right $\mathfrak{h} = \mathfrak{g}$. So we assume that $f_n = f_j$ for some $j < n$. Let

$$I = \{i < n \mid v_i \equiv v_n \text{ or } v_{i+1} \equiv v_n^{-1}\}.$$

Then $j \notin I$. Indeed, if $v_j \equiv v_n$, then $f_j = f_{j-1} \cdot_{\mathfrak{g}} v_j = f_{j-1} \cdot v_n$ and $f_j = f_n = f_{n-1} \cdot_{\mathfrak{g}} v_n$. Hence $f_{j-1} = f_{n-1}$, a contradiction. If $v_{j+1} \equiv v_n^{-1}$, then $f_j = f_{j+1} \cdot_{\mathfrak{g}} v_{j+1}^{-1} = f_{j+1} \cdot_{\mathfrak{g}} v_n$, $f_j = f_n = f_{n-1} \cdot_{\mathfrak{g}} v_n$ and $f_{j+1} = f_{n-1}$, a contradiction (because $j + 1 \neq n - 1$ in this case since w is reduced).

We have $v_n \equiv x_\ell^{\pm 1}$ for some $\ell \leq k$. Put $Y = \{f_i \mid i \in I\}$ and define $\mathfrak{h} = (h_1, h_2, \ldots, h_k) \in G^k$ by $h_i = g_i$ if $i \neq \ell$, $h_\ell = g_\ell c$ if $v_n = x_\ell$ and $h_\ell = c g_\ell$ if $v_n = x_\ell^{-1}$, where $c \in G_Y$ is to be chosen later.

Then

$$f_i = f_{i-1} \cdot_{\mathfrak{g}} v_i = f_{i-1} \cdot_{\mathfrak{h}} v_i, \quad i = 1, \dots, n-1$$

and $f_{n-1} \cdot_{\mathfrak{h}} v_n = f_n \cdot c$ since $c \in G_Y$ (prove that!).

Since $f_j \notin Y$, the set $Y' = \{f_1, \dots, f_n\} \smallsetminus \{f_j\}$ contains Y. Therefore we can choose an element $c \in G_{Y'}$ which fixes Y' pointwise and does not fix f_j hence $f_j \cdot c \notin \{f_1, \dots, f_n\}$. Then $(h_1, \dots, h_k)$ is distinctive for w and f_0. □

5.5 Subgroups of Free Groups

5.5.1 The Definition of a 2-Complex and Its Fundamental Group

5.5.1.1 2-Complexes

Recall that a 2-*complex* is a graph in the sense of Serre with a set of distinguished cycles (also called *closed paths*) called *cells*. A 2-complex is called *connected* if its underlying graph is connected. For example, any van Kampen diagram can be viewed as a connected 2-complex drawn on the plane.

The *fundamental group* of a connected 2-complex X, denoted by $\pi_1(X)$, can be (syntactically) defined as follows. Take a maximal (called *spanning*) subtree T of X. Let E be the set of positive edges of X not in T. Label every edge in T by 1. Then every cell of X (i.e., one of the distinguished closed paths in X) has a label which is a word over E. Let R be the set of all these labels. The fundamental group $\pi_1(X)$ has the presentation

$$\mathrm{gp}\langle E \mid R \rangle.$$

Exercise 5.5.1. Prove that every spanning tree in X contains all vertices of X, and the number of positive edges in a spanning tree is the number of vertices minus 1. Prove that a different choice of a spanning tree T gives an isomorphic group. **Hint:** To prove the last statement, use the next Exercise 5.5.2.

An alternative definition of the fundamental group is the following. Let u be a vertex in X, $L(X, u)$ be the set of all cycles starting and ending at u. The cycles can be multiplied in the natural way (concatenation), and the product is associative, so $L(X, u)$ is a semigroup. Now we define an equivalence relation $\sim$ on $L(X, u)$ generated by the following two kinds of moves:

- If a cycle p contains two consecutive mutually inverse edges ee^{-1}, we can eliminate these two edges, producing a new cycle.
- If a cycle p contains a closed subpath $e_1 e_2 \dots e_n$ which is a cyclic shift of one of the distinguished closed paths of X, we can eliminate that subpath.

Exercise 5.5.2. Show that $\sim$ is a congruence relation on the semigroup $L(X,u)$ and $L(X,u)/\sim$ is a group. Show that $\pi_1(X)$ is isomorphic to $L(X,u)/\sim$. **Hint:** The isomorphism takes an edge from E to the closed path $p_{e_-}ep_{e_+}^{-1}$ where for every vertex x in X, p_x is the unique simple path on the spanning tree T connecting u and x.

Thus one can view elements of $\pi_1(X)$ as cycles in X starting and ending at u.

Exercise 5.5.3. Let $G = \mathrm{gp}\langle X \mid R \rangle$. Consider the following *presentation 2-complex* $C(\mathcal{P})$: it has one vertex, one edge for each letter from X labeled by the letter from X, and one cell for each $r \in R$: the closed path that reads the word r. Show that $\pi_1(C(G))$ is isomorphic to G.

We see that 2-complexes and presentations are basically the same things. But presentations are syntactic objects while 2-complexes can be viewed as topological spaces if we glue in discs whose boundaries are the distinguished cycles. The resulting topological space is called a 2-dimensional *CW-complex*.

Exercise 5.5.4. Find a connected 2-complex with ten vertices whose fundamental group is isomorphic to the surface group $\pi_1(S_g)$, $g \geq 1$.

Exercise 5.5.5. Let C_1 and C_2 be two disjoint connected 2-complexes. Let a_1, a_2 be vertices in C_1, C_2. Consider the complex $C_1 \cup C_2$, add a vertex a connected by edges to a_1 and a_2. Show that the resulting 2-complex is connected, and its fundamental group is the free product $\pi_1(C_1) * \pi_1(C_2)$.

5.5.1.2 Fundamental Groups of Graphs

If a 2-complex does not contain cells, i.e., is simply a graph (in the sense of Serre), then the fundamental group does not have relations, hence it is free. The next theorem now follows from the second part of Exercise 1.3.11.

Theorem 5.5.6. *(1) The fundamental group of every connected finite graph Γ is free.*

(2) The rank of the free group is the number of positive edges of Γ minus the number of vertices plus 1.

Exercise 5.5.7. Let Γ be a connected finite graph in the sense of Serre with the set V of vertices. For every $x \in V$ let d_x be the out-degree of the vertex x. Prove that the rank of the fundamental group of Γ is equal to $\frac{1}{2}\sum(d_x - 2) + 1$. **Hint:** Use induction on the number of edges. First assume that the graph is a tree, reduce the problem to a smaller tree obtained by removing a leaf and its edges. Next suppose that the graph is not a tree, and so there exists an edge e such that removing edges $e^{\pm 1}$ (but leaving vertices e_- and e_+) we obtains a connected graph again. Then use the induction assumption for this smaller graph.

Remark 5.5.8. An analog of Theorem 5.5.6 is true for countable graphs as well: the fundamental group of a graph is always free. The rank of the free group is finite if and only if the ranks of fundamental groups of all finite subgraphs of the graph are bounded from above. For example, the rank of the fundamental group of the graph on Figure 5.30 is equal to 1 (for every base vertex u) while the rank of the fundamental group of the graph on Figure 5.31 is infinite.

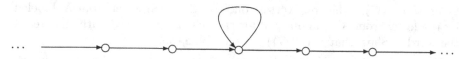

Figure 5.30 The rank of the fundamental group of this graph is 1

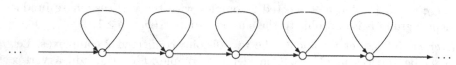

Figure 5.31 The rank of the fundamental group of this graph is ∞

5.5.2 Subgroups of Free Groups and Inverse Automata

5.5.2.1 Constructing an Inverse Automaton from a Subgroup of a Free Group

Let H be a subgroup of the free group $F = \mathrm{gp}\langle A \rangle$ generated by words $u_1, \ldots, u_n$ in $A \cup A^{-1}$ (we assume that $A \cap A^{-1} = \varnothing$ and A is finite). Consider the linear diagrams $\varepsilon(u_i), i = 1, \ldots, n$.

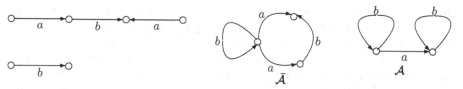

Figure 5.32 Constructing an automaton recognizing a subgroup of a free group generated by aba^{-1}, b

Identify the initial and terminal vertices of these diagrams to obtain a union of labeled circles connected at a distinguished vertex, denote it by **1**. This graph can be viewed as a graph in the sense of Serre (see Section 1.3.5) with *correct* labeling: positive edges have labels from A, negative edges have labels from A^{-1} (if e has label a, then e^{-1} has label a^{-1}). Let $\bar{A}$ be this automaton.

Let $\mathcal{A}$ be the inverse automaton obtained from $\bar{A}$ by foldings as in Section 1.3.5. Let the vertex **1** be the input and output vertex of the automaton.

Figure 5.32 shows the process of constructing $\mathcal{A}$ for the subgroup

$$H = \langle\, aba^{-1}, b\,\rangle.$$

Theorem 5.5.9. *(1) The automaton $\mathcal{A}$ recognizes those and only those reduced words that represent elements of H.*

(2) The language of all reduced words from H is rational.

(3) The group H is isomorphic to the fundamental group of the underlying graph of $\mathcal{A}$.

Proof. Let w be a reduced word from H. Then w is freely equal to a word $\bar{w}$ that is a product of words u_i and their inverses. The word $\bar{w}$ is then accepted by $\bar{A}$. The word w is obtained from $\bar{w}$ by a sequence of rewritings using rewriting rules $aa^{-1} \to 1$, $a \in A^{\pm 1}$: $\bar{w} \equiv w_1 \to w_2 \to \dots \to w_k \equiv w$. Then w_1 is accepted by an automaton obtained from $\bar{a}$ by one folding, w_2 is accepted by an automaton obtained from $\bar{A}$ by at most two foldings. By induction, we conclude that w is accepted by an automaton $\hat{A}$ obtained from $\bar{A}$ by several foldings. By Exercise 1.7.13, the rewriting system where objects are automata and rules are foldings is Church–Rosser. Therefore $\mathcal{A}$ (being a terminal object) is obtained from $\hat{A}$ by several foldings. It is easy to check that the language accepted by $\mathcal{A}$ contains the language accepted by $\hat{A}$ (check it by induction on the number of foldings needed to get $\mathcal{A}$ from $\hat{A}$). Hence w is accepted by $\mathcal{A}$.

Let

$$\bar{A} \to A_1 \to \dots \to A_m = \mathcal{A} \tag{5.5.1}$$

be a sequence of automata such that A_{i+1} is obtained from A_i by one folding of edges e_i, e_i' (with $\iota(e_i) = \iota(e_i')$ and the same label a_i). Let x_{i+1} be the head of the edge f_{i+1} in A_{i+1} that is the result of the folding of e_i, e_i'. Note that there exists a homomorphism ϕ_i from A_i to A_{i+1} which takes every edge except for e_i, e_i' and their inverses to themselves and e_i, e_i' to f_{i+1}.

Clearly every word accepted by $\bar{A}$ is a product of words u_j and their inverses, so it represents an element of H. Thus it is enough to prove (for every $i = 1, \dots, m - 1$) that if w is any word accepted by A_{i+1}, then there exists a word w' accepted by A_i and freely equal to w. The word w labels a cycle p starting and ending at the vertex **1**.

So we shall achieve our goal if we prove the following

Lemma 5.5.10. *For every path p in $\mathcal{A}_{i+1}$ starting at 1, there exists a path in $\mathcal{A}_i$ with* $\mathrm{Lab}(p') = \mathrm{Lab}(p)$ *in the free group and* $\phi_i(\tau(p')) = \tau(p)$.

Proof. Induction on the length of the path p. If p has no edges, there is nothing to prove. Suppose $p = qe$ where e is an edge. By induction, there exists a path q' in $\mathcal{A}_i$ starting at 1, and satisfying $\mathrm{Lab}(q') = \mathrm{Lab}(q), \phi_i(\tau(q')) = \tau(q)$.

Case 1. Suppose that $\tau(q) \neq \tau(f_{i+1})$, $e \neq f_{i+1}$. Then e has unique pre-image e' in $\mathcal{A}_i$, and we can set $p' = q'e'$.

Case 2. Suppose that $\tau(q) \neq \tau(f_{i+1})$, $e = f_{i+1}$. Then we can set $p' = q'e_i$.

Case 3. Suppose that $\tau(q) = \tau(f_{i+1})$. Then $\tau(q')$ must be equal to either $\tau(e_i)$ or $\tau(e_{i+1})$. Without loss of generality assume that $\tau(q') = \tau(e_i)$.

Case 3.1. Suppose that there exists an edge e' in $\mathcal{A}_i$ with $\iota(e') = \tau(q)$, $\mathrm{Lab}(e') = \mathrm{Lab}(e)$. Then we can set $p' = q'e'$.

Case 3.2. Suppose that there is no edge e' in $\mathcal{A}_i$ with $\iota(e') = \tau(q') = \tau(e_i)$, $\mathrm{Lab}(e') = \mathrm{Lab}(e)$. Then by the definition of folding there must exist an edge e' with $\iota(e') = \tau(e'_i)$, $\mathrm{Lab}(e') = \mathrm{Lab}(e)$. Then we can set $p = q'e_i^{-1}e'_ie'$.

□

This completes the proof of Part (1) of Theorem 5.5.9.

To prove Part (2) note that the language of reduced words from H is the intersection of the language of words recognized by the automaton $\mathcal{A}$ and the language of reduced words which is rational by Exercise 1.3.5. Since the intersection of two rational languages is rational by Lemma 1.3.4, the second part of Theorem 5.5.9 follows.

To prove Part (3) notice that if two cycles in the underlying graph Γ of $\mathcal{A}$ are equivalent (the equivalence is defined in Section 5.5.1.1), then their labels are equal in the free group. Therefore there exists a homomorphism Lab from the fundamental group of the underlying graph of $\mathcal{A}$ to H. That homomorphism takes every circle starting and ending at 1 to its label.

Clearly the homomorphism is surjective. It remains to show that if $\mathrm{Lab}(p)$ is equal to 1 in H, then the cycle p is equivalent to 1 (the path with no edges). Let p be a cycle with $\mathrm{Lab}(p) = 1$ in the free group and a shortest cycle in its equivalence class. By contradiction, suppose that p is not empty. Since $\mathrm{Lab}(p) = 1$, the word $\mathrm{Lab}(p)$ must contain a subword aa^{-1} for some $a \in A \cup A^{-1}$. Since $\mathcal{A}$ is an inverse automaton, that means p must contain a subpath consisting of two consecutive mutually inverse edges ee^{-1}. Removing that subpath, we get a path p' whose label is equal to $\mathrm{Lab}(p)$ in the free group. But p' is shorter than p, a contradiction. □

Remark 5.5.11. If a subgroup H was not finitely generated, then we still could construct an (infinite) automaton $\mathcal{A}$ that recognizes words from H by the same procedure as above.

Exercise 5.5.12. Show that the automaton $\mathcal{A}$ depends only on the subgroup H and does not depend on the generating set $u_1, \ldots, u_n$. **Hint:** Consider all words u in H and the corresponding linear diagrams $\varepsilon(u)$. Construct the (infinite) automaton $\hat{\mathcal{A}}$ identifying $\iota(\varepsilon(u))$ and $\tau(\varepsilon(u))$ for all u. Show that the automaton obtained by all foldings from $\hat{\mathcal{A}}$ is equal to the automaton $\mathcal{A}$ described above.

Remark 5.5.13. Recall that inverse bi-rooted automata where input and output vertices coincide are idempotents in E-unitary inverse semigroups (see Section 3.8.3). Thus Theorem 5.5.9 gives a connection between idempotents of inverse semigroups and finitely generated subgroups of free groups. For more on this see Margolis–Meakin–Sapir [221].

5.5.2.2 Some Properties of Subgroups of Free Groups

Theorem 5.5.9 gives many important results about subgroups of free groups almost "for free". Here we present only three of these results.

Theorem 5.5.14. *Every free group of at most countable rank embeds into the free group of rank 2.*

Proof. Consider the following subgroup of the free group $F_2 = \mathrm{gp}\langle a, b \rangle$ of rank 2: H is generated by words $a^{-i} b a^i$, $i \geq 0$. Then the automaton $\mathcal{A}$ looks as on Figure 5.33 (check that!). Hence by Theorem 5.5.9 and Remark 5.5.8 the rank of the free group H is infinite.

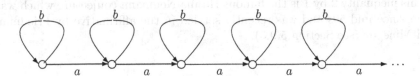

Figure 5.33 An infinite inverse automaton recognizing a free subgroup of F_2 of infinite rank

□

Theorem 5.5.15 (Schreier property). *Every subgroup of the free group is free.*

Proof. Indeed, it follows from Theorems 5.5.9, 5.5.6 and Remark 5.5.8.   □

Theorem 5.5.16 (P. Hall's LERF property). *For every finitely generated subgroup H of a free group F and every element $w \notin H$, there exists a homomorphism from F onto a finite group such that $\phi(w) \notin \phi(H)$.*

Proof. Let $\mathcal{A} = (Q, A)$ be the finite inverse automaton recognizing H with input=output vertex 1. Let $\varepsilon(w)$ be the linear diagram for w. Let us identify $\iota(\varepsilon(w))$ with vertex 1, and do all foldings in the resulting automaton to produce an inverse automaton $\mathcal{A}' = (Q', A)$. Note that $\mathcal{A}'$ recognizes all reduced words from H but does not recognize w. Note that every $a \in A$ induces a partial injective map $\cdot a: Q' \to Q'$ (the map is injective because $\mathcal{A}'$ is an inverse automaton). Extend each map $\cdot a, a \in A$ to a permutation $\alpha_a, a \in A$ of the set Q. The substitution $a \mapsto \alpha_a$ extends to a homomorphism ϕ from the free group $gp\langle A \rangle$ onto the finite group S generated by all the permutations $\alpha_a, a \in A$. Note that for every $g \in H$, we have $\phi(g): 1 \mapsto 1$ while $\phi(w): 1 \mapsto 1 \cdot w \neq 1$. Hence $\phi(w) \notin H$.

Theorem 5.5.17 (Howson property). *The intersection of any two finitely generated subgroups of a free group is finitely generated.*

Proof. Let $\mathcal{A}_i = (Q_i, A)$ be an inverse automaton recognizing the subgroup $H_i, i = 1, 2$. Then by Lemma 1.3.4 the intersection $H_1 \cap H_2$ is recognized by the equalizer automaton $\mathcal{A}$. It is easy to see that $\mathcal{A}$ is also an inverse automaton. Hence by Part (3) of Theorem 5.5.9 and Theorem 5.5.6 the subgroup $H_1 \cap H_2$ is finitely generated. $\square$

Remark 5.5.18. A more careful analysis of the equalizer automaton $\mathcal{A}$ gives that the rank r of $H_1 \cap H_2$ and ranks r_i of subgroups H_i satisfy the following inequality

$$r - 1 \le 2(r_1 - 1)(r_2 - 1)$$

(see Hanna Neumann [246, 247], and proofs involving inverse automata by Wilfried Imrich [155], Walter Neumann [249]). The claim that one can replace in this inequality 2 by 1 is the famous Hanna Neumann conjecture which was open since mid 50s and was recently settled in the affirmative by Friedman and Mineyev (see Section 5.9.5).

5.6 Diagram Groups

5.6.1 The Squier Complex of a String Rewriting System

With every string rewriting system $\mathcal{P} = \mathrm{sr}\langle X \mid \mathcal{R} \rangle$ we can associate a 2-complex $S(\mathcal{P})$, called the *Squier complex* [307]. Its vertices are all words from X^+, positive edges are the elementary diagrams $(p, u \to v, q)$ with tail puq and head pvq, and cells correspond to *independent rewritings*: if $u_1 \to v_1$ and $u_2 \to v_2$ are from R and a vertex of $S(\mathcal{P})$ (i.e., a word in X) has the form pu_1qu_2r where p, q, r are words, then the path with edges

$$(p, u_1 \to v_1, qu_2r)(pv_1q, u_2 \to v_2, r)(p, u_1 \to v_1, qv_2r)^{-1}(pu_1q, u_2 \to v_2, r)^{-1}$$

is a cell of $S(\mathcal{P})$. Note that this path is the trivial way to complete the diamond with two edges $(p, u_1 \to v_1, qu_2r)$ and $(pu_1q, u_2 \to v_2, r)$.

5.6.2 Diagrams as 2-Dimensional Words, Diagram Groups

5.6.2.1 Diagram Monoids and Groups. The Definition

First let us give a formal definition of a diagram introduced in Section 1.7.4. In order to distinguish these diagrams from van Kampen diagrams, let us called these *semigroup diagrams*.

An informal definition of semigroup diagrams has been essentially given in Section 1.7.4: a semigroup *diagram* is a planar directed labeled graph tesselated by cells. Each diagram Δ has the *top path* $\mathbf{top}(\Delta)$, the *bottom path*, $\mathbf{bot}(\Delta)$ the initial and terminal vertices $\iota(\Delta)$ and $\tau(\Delta)$. These are common vertices of $\mathbf{top}(\Delta)$ and $\mathbf{bot}(\Delta)$. The whole diagram is situated between the top and the bottom paths, and every edge of Δ belongs to a (directed) path in Δ between $\iota(\Delta)$ and $\tau(\Delta)$.

More formally, let X be an alphabet. For every $x \in X$ we define the *trivial diagram* $\varepsilon(x)$ which is just an edge labeled by x. The top and bottom paths of $\varepsilon(x)$ are equal to $\varepsilon(x)$, $\iota(\varepsilon(x))$ and $\tau(\varepsilon(x))$ are the tail and the head vertices of the edge. If u and v are words in X, a *cell* $(u \to v)$ is a planar graph consisting of two directed labeled paths, the top path labeled by u and the bottom path labeled by v, connecting the same points $\iota(u \to v)$ and $\tau(u \to v)$. There are three operations that can be applied to diagrams in order to obtain new diagrams (see Figure 5.34).

(1) **Addition.** Given two diagrams Δ_1 and Δ_2, one can identify $\tau(\Delta_1)$ with $\iota(\Delta_2)$. The resulting planar graph is again a diagram denoted by $\Delta_1 + \Delta_2$, whose top (bottom) path is the concatenation of the top (bottom) paths of Δ_1 and Δ_2. If $u = x_1 x_2 \ldots x_n$ is a word in X, then we denote $\varepsilon(x_1) + \varepsilon(x_2) + \cdots + \varepsilon(x_n)$ (i.e., a simple path labeled by u) by $\varepsilon(u)$ and call this diagram also *trivial*.

(2) **Multiplication.** If the label of the bottom path of Δ_1 coincides with the label of the top path of Δ_2, then we can *multiply* Δ_1 and Δ_2, identifying $\mathbf{bot}(\Delta_1)$ with $\mathbf{top}(\Delta_2)$. The new diagram is denoted by $\Delta_1 \circ \Delta_2$. The vertices $\iota(\Delta_1 \circ \Delta_2)$ and $\tau(\Delta_1 \circ \Delta_2)$ coincide with the corresponding vertices of Δ_1, Δ_2, $\mathbf{top}(\Delta_1 \circ \Delta_2) = \mathbf{top}(\Delta_1), \mathbf{bot}(\Delta_1 \circ \Delta_2) = \mathbf{bot}(\Delta_2)$.

(3) **Inversion.** Given a semigroup diagram Δ, we can flip it about a horizontal line obtaining a new semigroup diagram Δ^{-1} whose top (bottom) path coincides with the bottom (top) path of Δ.

Definition 5.6.1. A *semigroup diagram* over a string rewriting system $\mathcal{P} = \mathrm{sr}\langle X \mid u_i \to v_i, i \in I \rangle$ is any labeled graph obtained from the trivial diagrams and cells $u_i \to v_i$, $i \in I$, by the operations of addition, multiplication and inversion. If the top path of a diagram Δ is labeled by a word u and the bottom path is labeled by a word v, then we call Δ a (u, v)-*diagram* over $\mathcal{P}$.

One can also give a slightly different definition of a semigroup diagram using the elementary diagrams from Section 1.7.4.

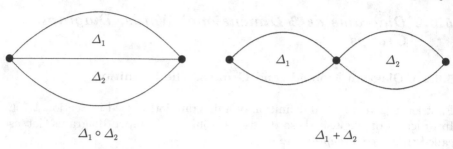

Figure 5.34 Operations on diagrams

Definition 5.6.2. A *diagram* over a string rewriting system $\mathcal{P}$ is any labeled graph obtained as a product of elementary diagrams corresponding to the rules of $\mathcal{P}$ and their inverses.

Exercise 5.6.3. Prove that the Definitions 5.6.1 and 5.6.2 are equivalent.

Exercise 5.6.4. Show that the diagram x_0 on Figure 5.38 is equal to

$$(x \to x^2) \circ (\varepsilon(x) + (x \to x^2)) \circ ((x \to x^2)^{-1} + \varepsilon(x)) \circ ((x \to x^2)^{-1}).$$

Represent x_1 from the same Figure as a product of sums of cells and trivial diagrams.

As for planar trees, two semigroup diagrams are called *equal* if there exists a continuous deformation of the plane that takes one of them to another. We do not distinguish equal diagrams (as we do not distinguish identically equal words or equal trees). In order to avoid confusion of different kinds of equality of diagrams, words, etc., equal diagrams will be also called *isotopic* and we will use the symbol $\equiv$ to denote isotopy of diagrams.

Recall that an (u, u)-diagram is called spherical. For every word u, the set of all (u, u)-diagrams with operation $\circ$ is a monoid whose identity element is the linear diagram $\varepsilon(u)$ corresponding to the word u. That monoid is denoted by $\mathrm{DM}(\mathcal{P}, u)$ and is called the *diagram monoid* corresponding to the presentation $\mathcal{P}$ and the word u.

Two cells in a semigroup diagram form a *dipole* if the bottom path of the first cell coincides with the top path of the second cell, and the cells are inverses of each other. In this case, we can obtain a new diagram removing the two cells and replacing them by the top path of the first cell. This operation is called *elimination of dipoles*. The new diagram is called *equivalent* to the initial one. A semigroup diagram is called *reduced* if it does not contain dipoles.

Exercise 5.6.5. For every string rewriting system Γ consider the following rewriting system. The objects are all semigroup diagrams over Γ. Positive moves are removing dipoles. Prove that the rewriting system is Church–Rosser, hence there exists exactly one reduced semigroup diagram in every class of equivalent semigroup diagrams over Γ.

Now let $\mathcal{P}$ be a semigroup presentation and $P = \{c_1, c_2, \ldots\}$ be the corresponding collection of cells. The diagram group $\mathrm{DG}(\mathcal{P}, u)$ corresponding to $\mathcal{P}$ and a word u consists of all reduced spherical (u, u)-diagrams obtained from the cells of P and linear diagrams by using the three operations mentioned above. The product $\Delta_1 \Delta_2$ of two diagrams Δ_1 and Δ_2 is the reduced diagram obtained by removing all dipoles from $\Delta_1 \circ \Delta_2$. By Exercise 5.6.5 the multiplication is well defined.

Exercise 5.6.6. Prove that $\mathrm{DG}(P, u)$ is a group which is a quotient of $\mathrm{DM}(P, u)$ by the equivalence relation define above. That equivalence relation is in fact a congruence.

5.6.3 Diagram Groups and Squier Complexes

There is a close connection between diagram groups of semigroup presentations and the Squier complexes of the corresponding rewriting systems.

Exercise 5.6.7 (Guba, Sapir [133]). Show that the diagram group $\mathrm{DG}(P, u)$ is the fundamental group of the connected component of the Squier complex $S(P)$ containing u. **Hint:** Consider semigroup diagrams as paths in the Squier complex, spherical diagrams being cycles. Show that equivalent diagrams correspond to cycles that are equal in the fundamental group (using the solution of Exercise 5.6.5.)

5.6.4 Diagram Groups. Examples

The following five exercises can be solved by finding fundamental groups of the corresponding Squier complexes (Exercise 5.6.7).

Exercise 5.6.8. Let $\mathcal{P} = \mathrm{sr}\langle a \mid a \to a \rangle$. Show that

(1) The diagram group $\mathrm{DG}(\mathcal{P}, a)$ is isomorphic to $\mathbb{Z}$.
(2) The diagram group $\mathrm{DG}(\mathcal{P}, a^n)$ is isomorphic to the free commutative group $\mathbb{Z}^n$.

Exercise 5.6.9. Let $\mathcal{P} = \mathrm{sr}\langle a \mid a \to a, a \to a \rangle$ (yes, two same rules). Show that $\mathrm{DG}(\mathcal{P}, a)$ is the free group with two generators.

Exercise 5.6.10. Let $\mathcal{P} = \mathrm{sr}\langle ab \to a, bc \to c, b \to b \rangle$. Show that $\mathrm{DG}(\mathcal{P}, ac)$ is isomorphic to $\mathbb{Z} \wr \mathbb{Z}$ (see Exercise 1.8.26).

Exercise 5.6.11. Let $\mathcal{P}_i$, $i = 1, 2$, be a presentation $\mathrm{sg}\langle X_i \mid \mathcal{R}_i \rangle$, where X_1, X_2 are disjoint. Let $\mathcal{P} = \mathcal{P}_1 \cup \mathcal{P}_2$. Show that for every two words $u_1 \in X_1^+, u_2 \in X_2^+$, $\mathrm{DG}(\mathcal{P}, u_1 u_2)$ is isomorphic to the direct product $\mathrm{DG}(\mathcal{P}_1, u_1) \times \mathrm{DG}(\mathcal{P}_2, u_2)$.

Exercise 5.6.12. Let $\mathcal{P}_i$, $i = 1,2$, be a string rewriting system sr$\langle X_i \mid \mathcal{R}_i \rangle$, where X_1, X_2 are disjoint. Let $u_i, i = 1,2$ be a word in X_i, a a letter not in $X_1 \cup X_2$. Consider a rewriting system $\mathcal{P} = $ sr$\langle X_1 \cup X_2 \cup \{a\} \mid \mathcal{R}_1 \cup \mathcal{R}_2 \cup \{a \to u_1, a \to u_2\}\rangle$. Show that DG$(\mathcal{P}, a)$ is isomorphic to the free product DG$(\mathcal{P}_1, u_1) * $ DG$(\mathcal{P}_2, u_2)$. **Hint:** Use Exercise 5.5.5.

5.6.5 Combinatorics on Diagrams

Semigroup diagrams can be considered as 2-dimensional words. Indeed, as we saw in Section 1.2.2, words are linear (1-dimensional) diagrams. Many facts about combinatorics on words have 2-dimensional analogs (see [133]). Here we present a description of commuting diagrams which is similar, in a sense, to the description of commuting words (Theorem 1.2.9). Note, though, that the 2-dimensionality makes a difference. For example, if a spherical diagram Δ is a sum of two spherical diagrams Δ_1, Δ_2, then every diagram $\Delta_1^k + \Delta_2^m$, $m, n \geq 0$ commutes with Δ. Hence it is not true that every two commuting diagrams have common powers (as in Theorem 1.2.9). Our goal is to show that in fact this is basically the only difference between 2-dimensional and 1-dimensional diagrams (so the situation here is somewhat similar to the situation in free associative algebras, see Section 4.3). Moreover, the proof we are giving (following [133]) is very similar to Guba's proof of Theorem 1.2.9.

In what follows, we fix a semigroup presentation $\mathcal{P}$, and consider the diagram monoid DM$(\mathcal{P})$ (which is a 2-dimensional analog of a free monoid rather than the free group).

5.6.5.1 Some Basic Properties of Semigroup Diagrams

The following lemmas contain the main properties of semigroup diagrams.

Lemma 5.6.13. *Every positive path in a diagram is a simple arc (we consider diagrams as graphs in the sense of Serre as in Section 1.3.4).*

Lemma 5.6.14. *(a) For every vertex v of a diagram Δ there exists a positive path from $\iota(\Delta)$ to $\tau(\Delta)$ passing through v.*

(b) Every diagram Δ has exactly one source-vertex $\iota(\Delta)$ (which has no incoming edges) and exactly one sink-vertex $\tau(\Delta)$ (which has no outgoing edges).

(c) For any vertex v of Δ that does not coincide with $\iota(\Delta)$ and $\tau(\Delta)$ there exists an enumeration of positive edges incident to v in the clockwise order: $e_1, \ldots, e_n$ such that for some k, $1 \leq k < n$, edges $e_1, \ldots, e_k$ are outgoing and edges $e_{k+1}, \ldots, e_n$ are incoming, see Figure 5.35.

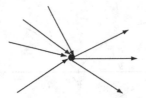

Figure 5.35 Outgoing and incoming edges incident to a vertex in a semigroup diagram do not mix

Lemma 5.6.15. *Every positive path p in a diagram Δ from $\iota(\Delta)$ to $\tau(\Delta)$ divides Δ into two diagrams Δ_1 and Δ_2 such that $\mathbf{bot}(\Delta_1) = \mathbf{top}(\Delta_2) = p$ and $\Delta = \Delta_1 \circ \Delta_2$.*

All three of these lemmas can be easily proved by induction on the number of cells.

Exercise 5.6.16. Prove Lemmas 5.6.13–5.6.15.

5.6.5.2 Nice and Simple Diagrams

A diagram is called *simple* if it is not a linear diagram and not a sum of nontrivial diagrams, that is its top and bottom paths have exactly two vertices in common. A not necessarily spherical diagram Δ is called *nice* if for every decomposition $\Delta \equiv \Delta_1 + \Delta_2$ at least one of the diagrams Δ_1 and Δ_2 is spherical. In other words a diagram is called nice if every common vertex of the paths $\mathbf{top}(\Delta)$ and $\mathbf{bot}(\Delta)$ determines decompositions $\mathbf{top}(\Delta) = p'p''$, $\mathbf{bot}(\Delta) = q'q''$ such that $\mathrm{Lab}(p') \equiv \mathrm{Lab}(q')$ or $\mathrm{Lab}(p'') \equiv \mathrm{Lab}(q'')$.

Lemma 5.6.17. *For every spherical diagram Δ there exists $n > 0$ such that Δ^m is nice for every $m \geq n$.*

Proof. We shall use an induction on the number of possible decompositions $\Delta \equiv \Delta_1 + \Delta_2$ where Δ_i are not spherical. This number is equal to the number of common vertices v of $\mathbf{top}(\Delta)$ and $\mathbf{bot}(\Delta)$ which correspond to decompositions $\mathbf{top}(\Delta) = p'p''$, $\mathbf{bot}(\Delta) = q'q''$ such that $\mathrm{Lab}(p') \not\equiv \mathrm{Lab}(q')$, $\mathrm{Lab}(p'') \not\equiv \mathrm{Lab}(q'')$ (why?).

Let us choose such a decomposition with the shortest possible path $\mathbf{top}(\Delta_1)$. In this case the path $\mathbf{bot}(\Delta_1)$ will also be the shortest possible. Then Δ_1 is nice. Let v be the common vertex of $\mathbf{top}(\Delta)$ and $\mathbf{bot}(\Delta)$ which determines this decomposition. Consider the diagram $\Delta^2 \equiv \Delta' \circ \Delta''$ where Δ' and Δ'' are isotopic copies of Δ. Let v' and v'' be the representatives of v in Δ' and Δ'' respectively. As above we can assume that the bottom path of Δ_1 is longer than the top path of this diagram, so Δ^2 has the form on Figure 5.36.

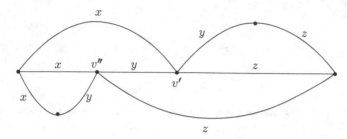

Figure 5.36 The square of a diagram which is a sum of non-spherical subdiagrams

Let w be an arbitrary common vertex of $\mathbf{top}(\Delta^2)$ and $\mathbf{bot}(\Delta^2)$ distinct from the initial and the terminal vertices of Δ^2. Every such vertex belongs also to $\mathbf{bot}(\Delta') = \mathbf{top}(\Delta'')$. Therefore w divides both Δ' and Δ'' into a sum of two diagrams. Thus the representative of w in Δ cuts Δ into a sum of two subdiagrams. If both of these subdiagrams are spherical, then w divides Δ^2 into a sum of spherical diagrams.

Consider all vertices on $\mathbf{bot}(\Delta)$ whose representatives in $\mathbf{bot}(\Delta'')$ determine decompositions of Δ^2 into sums of two non-spherical diagrams. As we have established in the previous paragraph, these vertices belong to the set of vertices of $\mathbf{bot}(\Delta)$ which determine decompositions of Δ with the similar property. But the vertex v which belongs to the second set does not belong to the first set. Indeed, its representative v'' belongs to $\mathbf{bot}(\Delta_1)$ and Δ_1 is a nice diagram. Therefore the number of decompositions of Δ^2 into a sum of non-spherical diagrams is smaller than the number of similar decompositions of Δ. Applying the induction hypothesis we deduce that Δ^n is nice for some n. Finally, if Δ^n is nice, then Δ^m is nice as well for every $m > n$ (why?). $\square$

Every spherical diagram Δ can be uniquely decomposed into a sum of spherical diagrams $\Delta_1 + \Delta_2 + \ldots + \Delta_m$ where each summand is either a linear diagram or indecomposable into a sum of spherical diagrams, and there are no consecutive linear summands. We call these summands *components* of the diagram Δ.

Exercise 5.6.18. Prove that a diagram Δ is nice if and only if each component of Δ is either simple or linear.

5.6.5.3 Automorphisms of Spherical Graphs

Let Δ be a simple spherical diagram. Identifying $\mathbf{top}(\Delta)$ and $\mathbf{bot}(\Delta)$ we obtain a graph Γ on a sphere S^2. We denote the images of the vertices $\iota(\Delta)$ and $\tau(\Delta)$ by s_1 and s_2 respectively. Consider the group of homeomorphisms of S^2 (i.e., bijective continuous maps from S^2 onto itself) which preserve the orientation of the sphere and induces automorphisms of the graph Γ (this means that the homeomorphisms must take vertices of Γ to vertices, edges to edges and preserve the incidence relation). Notice that such homeomorphisms must stabilize s_1 and s_2 since s_1 is the only "source" in Γ (there are no positive edges coming into s_1) and s_2 is the only "sink" in Γ (there are no positive edges coming out of s_2).

Consider all automorphisms of Γ induced by these homeomorphisms of S^2. We call these automorphisms *spherical*. Clearly these automorphisms form a group which we shall denote by G. Since Γ is a finite graph the group G is finite. Our goal is to describe this group.

Lemma 5.6.19. *Let $\psi : \Gamma \to \Gamma$ be a spherical automorphism. If ψ stabilizes a vertex v of Γ which is different from s_1 and s_2 or ψ stabilizes an edge, then ψ is the identity.*

Proof. Let $\psi : \Gamma \to \Gamma$ be a spherical automorphism induced by a homeomorphism $\phi : S^2 \to S^2$. Let $e_1, \ldots, e_n$ be the positive edges of Γ coming out of s_1 and numbered in, say, the clockwise order. Since ϕ preserves the orientation there must be a number k such that ϕ takes e_i to e_{i+k} (here $1 \le i \le n$ and the addition is considered modulo n). Therefore it is clear that if $\psi(e_i) = e_i$ for some i then ψ stabilizes all edges $e_1, \ldots, e_n$.

Suppose that $\psi(v) = v$ for some vertex v different from s_1 and s_2. By Lemma 5.6.14c there exists a clockwise enumeration of the positive edges incident to v in the clockwise order: $e_1, \ldots, e_n$ such that for some k, $1 \le k < n$, edges $e_1, \ldots, e_k$ are outgoing and edges $e_{k+1}, \ldots, e_n$ are incoming (we use the fact that $v \neq s_1$ and $v \neq s_2$). Since $\phi(v) = v$, and ϕ is a homeomorphism, ψ cyclically shifts the sequence of edges $e_1, \ldots, e_n$. This cyclic shift must be trivial since ψ cannot map edges coming out of v to edges coming into v or vice versa.

Applying the same argument to the initial vertices of the edges incident to v, to the initial vertices of the edges incident to those vertices, etc., we shall finally find an edge e_i incident to s_1 which is stabilized by ψ, because v is connected with s_1. Then as we mentioned before ψ stabilizes all edges e_i. Since every vertex of Γ is connected with the head vertex of one of these edges by a positive path, ψ stabilizes all the vertices of Γ. Hence ψ is an identity automorphism.

If $\psi(e) = e$ for some edge e then ψ stabilizes the tail and the head of e. Therefore it is enough to consider the case $\imath(e) = s_1$, $\tau(e) = s_2$. In this case, again, ψ must stabilize all edges coming out of s_1, which implies that ψ is the identity automorphism.   □

Lemma 5.6.20. *The group G of spherical automorphisms of the spherical graph Γ is cyclic. If m is the order of G then there exists a generator ζ of G and a positive path p on Γ from s_1 to s_2 such that the following properties hold. The paths $p, \zeta(p), \ldots, \zeta^{m-1}(p)$ subdivide S^2 into m simple diagrams Δ_i, $(1 \le i \le m)$, where $\mathbf{top}(\Delta_i) = \zeta^{i-1}(p)$ and $\mathbf{bot}(\Delta_i) = \zeta^i(p)$. All diagrams Δ_i, $1 \le i \le m$, are isotopic to the same diagram Δ.*

Proof. Denote by $e_1, e_2, \ldots, e_n$ all edges of Γ with the tail vertex s_1. Let the cyclic order on them correspond to, say, the clockwise direction. With each spherical automorphism $\psi : \Gamma \to \Gamma$ we can associate an integer $k = k(\psi)$ such that $\psi(e_i) = e_{i+k}$ for all i (the addition here is modulo n). This defines a homomorphism $\psi \mapsto k(\psi)$ from G to Z_n. Lemma 5.6.19 implies that this homomorphism is injective (prove it!). Thus G embeds in Z_n, so it is cyclic by Exercise 1.8.11. Denote by m the order of G.

Now let p be a shortest positive path on S from s_1 to s_2. It is obvious that p does not intersect itself. Let us check that if $\psi : \Gamma \to \Gamma$ is a spherical nontrivial automorphism, then p and $\psi(p)$ does not have common vertices except s_1 and s_2. We say that a path q in Γ is *geodesic* if any subpath in q cannot be replaced by a shortest path with the same initial and terminal vertices. It is clear that p and $\psi(p)$ are geodesic paths. If they have a common vertex v which is not s_1 or s_2 then the subpaths in p and $\psi(p)$ from s_1 to v have the same length since both paths are geodesic. This implies that $\psi(v) = v$ which is impossible since ψ is nontrivial.

Let f be an edge with the initial vertex s_1. Its G-orbit has exactly m elements since all edges have trivial stabilizers. Denote by $f_1, \ldots, f_m$ all edges of this orbit cyclically ordered with respect to a given direction of going around s_1. Denote by ζ the element of G which takes f_1 to f_2. It is clear that ζ takes f_i to f_{i+1} for all i (the addition is modulo m). Then the order of ζ equals m, so ζ generates G. Without loss of generality it can be assumed that f_1 is the first edge of p. Then the m paths $p, \zeta(p), \ldots, \zeta^{m-1}(p)$ have first edges $f_1, \ldots, f_m$, respectively. We know that any two of our m paths have no common points except s_1, s_2. So they subdivide S into m parts such that i-th part $(1 \le i \le m)$ is homeomorphic to a simple diagram Δ_i with the top and the bottom paths $\zeta^{i-1}(p)$ and $\zeta(p)$ respectively. Since ζ takes Δ_i to Δ_{i+1} for all i, we have that all Δ_i are isotopic to the same diagram.   □

Let $\Delta^n \equiv \Delta_1 \circ \ldots \circ \Delta_n$ be the concatenation of n isotopic copies of a diagram Δ. Consider the mapping from $\Delta_1 \circ \ldots \circ \Delta_{n-1}$ to $\Delta_2 \circ \ldots \circ \Delta_n$ which takes each vertex (edge) from Δ_i $(i = 1, \ldots, n-1)$ to the corresponding vertex (edge) of Δ_{i+1}. This partial map will be called the Δ-*shift*.

5.6.5.4 Commuting Diagrams

Lemma 5.6.21. *Let Δ_1 and Δ_2 be spherical diagrams. If $\Delta \equiv \Delta_1 \circ \Delta_2 \equiv \Delta_2 \circ \Delta_1$ is a simple diagram, then $\Delta_1^k \equiv \Delta_2^\ell$ for some $k, \ell > 0$.*

Proof. Since Δ is simple, we can identify the top path of Δ with its bottom path and obtain a graph Γ on the sphere S^2 (as we have done several times before). Consider the following spherical automorphism α of the graph Γ. Since $\Delta \equiv \Delta_1 \circ \Delta_2 \equiv \Delta_2 \circ \Delta_1$ we have two copies of Δ_1, $\Delta_{1,1}$ and $\Delta_{1,2}$, and two copies of Δ_2, $\Delta_{2,1}$ and $\Delta_{2,2}$, inside Δ. Let α take the first copy of Δ_i $(i = 1, 2)$ identically to the second copy of this diagram, that is α takes $\Delta_{i,1}$ to $\Delta_{i,2}$. It is clear that α is a spherical automorphism.

As we know (Lemma 5.6.20) the group G of spherical automorphisms of Γ is cyclic. Let $m = |G|$. Let a generator ζ and a path p connecting the initial vertex s_1 with the terminal vertex s_2 satisfy the conditions of Lemma 5.6.20. Then there exists a unique number s such that $\zeta^s = \alpha^{-1}$, $0 < s < m$. Then $\zeta^{m-s} = \alpha$.

Let q be the path on Γ which is obtained by identifying the top and the bottom paths of Δ. By the definition of α, ζ^s takes q to $\mathbf{bot}(\Delta_{1,1}) = \mathbf{top}(\Delta_{2,1})$ and ζ^{m-s} takes q to $\mathbf{bot}(\Delta_{2,2}) = \mathbf{top}(\Delta_{1,2})$.

By Lemma 5.6.20 the images of the path p under automorphisms from G divide the spherical graph Γ into several copies of some diagram Θ. For every natural number t there exists a natural map (homomorphism) from the graph Θ^t to Γ taking the top path of Θ^t to p. In fact we can take the union Θ^∞ of all Θ^t, $t \geq 0$, and obtain a natural ("covering") map δ from Θ^∞ to Γ which wraps around Γ infinitely many times. For every finite path p' in Γ, one can find a finite path p'' in Θ^∞ that maps onto p'. That path is contained in Θ^t for some t, and will be called a *lift* of p' in Θ^t. Also the Θ-shift of Θ^∞ composed with δ gives ζ (why?). Let us denote the Θ-shift on Θ^t by $\bar{\zeta}$ on Θ^t (recall that it maps the $\circ$-product of the first $t-1$ copies of Θ to the $\circ$-product of the last $t-1$ copies of Θ).

Let us take a sufficiently large power Θ^t of Θ such that the path q can be lifted to a path q' on Θ^t. Consider the corresponding partial map $\bar{\zeta}$ (which is the Θ-shift on Θ^t). Consider the path $\bar{\zeta}^{s(m-s)}(q')$. This path has only two vertices in common with the path q' (the tail and the head), so these two paths bound a subdiagram. One can check that this subdiagram is the $(m-s)$-th power of the subdiagram contained between q' and $\bar{\zeta}^s(q')$ which is isotopic to Δ_1 (prove it!). At the same time this diagram is the s-th power of the diagram contained between q' and $\bar{\zeta}^{m-s}(q')$ which is isotopic to Δ_2 (prove it!). Hence $\Delta_1^{m-s} \equiv \Delta_2^s$ as required.   □

Theorem 5.6.22. *Let Δ_1 and Δ_2 be spherical diagrams. If $\Delta \equiv \Delta_1 \circ \Delta_2 \equiv \Delta_2 \circ \Delta_1$, then $\Delta_1 = \Psi_1 + \ldots + \Psi_k$, $\Delta_2 = \Pi_1 + \ldots + \Pi_k$ where Ψ_i, Π_i are spherical diagrams such that for every i some powers of Ψ_i and Π_i are isotopic.*

Proof. Let $\Delta = \Delta_1 \circ \Delta_2 = \Delta_2 \circ \Delta_1$. By Lemma 5.6.17, there exists a natural number n such that Δ^n, Δ_1^n, Δ_2^n are nice. Let $\Delta^n = \Psi_1 + \ldots + \Psi_s$ be the sum of simple and linear components. Since $\Delta^n = \Delta_1^n \circ \Delta_2^n$, Δ_1^n decomposes as the sum of spherical subdiagrams $\Lambda_{1,1} + \ldots + \Lambda_{1,s}$ and Δ_2^n decomposes into the sum of spherical subdiagrams $\Lambda_{2,1} + \ldots + \Lambda_{2,s}$ (the subdiagrams are spherical because Δ_i^n is nice, $i = 1, 2$). Therefore $\mathbf{top}(\Delta_1^n)$ has $s + 1$ vertices in common with $\mathbf{bot}(\Delta_1^n)$ subdividing the diagram into the sum of subdiagrams $\Lambda_{1,j}$. But then $\mathbf{top}(\Delta_1)$ and $\mathbf{bot}(\Delta_1)$ also must have $s + 1$ vertices in common subdividing Δ_1 into a sum of s spherical subdiagarms $\Omega_{1,1} + \ldots + \Omega_{1,s}$. Moreover, then $\Lambda_{1,j} = \Omega_{1,j}^n$ for $j = 1, \ldots, s$. Similarly Δ_2 is a sum of spherical subdiagrams $\Omega_{2,1} + \ldots + \Omega_{2,s}$, and $\Omega_{2,j}^n = \Lambda_{2,j}$, $j = 1, \ldots, s$. Furthermore, for every j from 1 to s, Ψ_j is a simple diagrams that is equal to $\Omega_{1,j}^n \Omega_{2,j}^n = \Omega_{2,j}^n \Omega_{1,j}^n$. Applying Lemma 5.6.21, we obtain that some powers of $\Omega_{1,j}$ and $\Omega_{2,j}$ are isotopic for every j, as required.   □

5.6.6 The R. Thompson Group F

5.6.6.1 The Definitions of F

R. Thompson's group F is remarkable because it does not contain free non-cyclic subgroups (Theorem 5.6.39) and does not satisfy any nontrivial identity (Theorem 5.6.37). More importantly, it is a "universal" counterexample to many group theoretic conjectures, and stubbornly resists many attempts to answer some basic questions about it (one of these questions is Problem 5.6.47). The group F has several equivalent definitions. First of all it is the group of all piecewise linear increasing continuous maps from the unit interval $[0,1]$ onto itself with finite number of linear pieces each having slope of the form 2^k, $k \in \mathbb{Z}$ and the breakpoints of the derivative occurring at points of the form $\frac{a}{2^b}$, $a, b \in \mathbb{N}$.

The second, syntactic, definition is the following surprisingly simple presentation of F:

$$\mathrm{gp}\langle x_0, x_1 \mid x_1^{x_0^2} = x_1^{x_0 x_1}, x_1^{x_0^3} = x_1^{x_0^2 x_1} \rangle \tag{5.6.1}$$

where, as usual, a^b denotes $b^{-1}ab$. We shall prove that below.

Note that even the fact that F is finitely generated can be considered very surprising.

The third definition of F is in terms of diagram groups.

Theorem 5.6.23. *If X consists of one letter x and P consists of one rule $x \to x^2$, then the group $G = \mathrm{DG}(P, x)$ is isomorphic to the R. Thompson group F.*

Proof. Indeed, the Squier complex for the rewriting system P is easy to draw (see Figure 5.37).

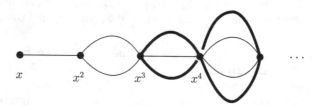

Figure 5.37 The Squier complex for $\langle x \mid x \to x^2 \rangle$. The thick line bounds a 2-cell in the complex

Exercise 5.6.24 (Guba, Sapir [133]). Use the definition of the fundamental group of a 2-complex (Section 5.5.1) and find the following infinite presentation of the fundamental group of the Squier complex on Figure 5.37:

$$\mathrm{gp}\langle x_0, x_1, x_2, \dots \mid x_i^{x_j} = x_{i+1}, i > j \in \mathbb{N} \rangle. \tag{5.6.2}$$

Hint: Choose the spanning tree in the Squier complex consisting of edges $(1, x \to x^2, x^n)$ (the tree is just the top infinite path in Figure 5.37). The relations corresponding to the cells of the Squier complex then allow one to express all edges in terms of edges of the form $(x^n, x \to x^2, 1)$. Denote this edge by x_n and prove that all relations of the fundamental group follow from the relations (5.6.2).

Exercise 5.6.25 (Guba, Sapir [133]). Show that the fundamental group of the Squier complex of P is generated by the closed paths represented by the two diagrams on Figure 5.38 (all edges are labeled by x and oriented from left to right, so we omit the labels and orientation of edges). Show that the diagram group G has the presentation (5.6.1). **Hint:** Note that each $x_i, i > 1$, in the presentation (5.6.2) is equal to $x_1^{x_0^{i-1}}$ in G, then show that the generators x_0, x_1 satisfy (5.6.1) and deduce all other relations from (5.6.2) from these two.

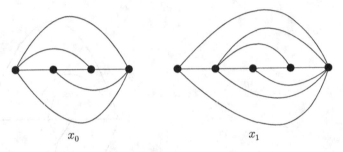

$$x_0 \qquad\qquad\qquad x_1$$

Figure 5.38 Generators of the R. Thompson group F

Now we define a homomorphism from G to F turning diagrams into functions.

5.6.6.2 From Diagrams to Functions

There are many natural homomorphisms from any diagram group to the group of continuous increasing functions from an interval onto itself (see [137]). We shall illustrate this construction using the group G (which, as we shall show, is isomorphic to the R. Thompson group F). Let f be the function $[0,1] \to [0,2]$ which takes t to $2t$ (taking any other homeomorphism $[0,1] \to [0,2]$ instead produces a different representation of G). In the cell corresponding to the rewriting rule $x \to x^2$, assume that every edge has length 1. Connect every point t on the top path to the point $f(t) = 2t$ on the bottom path by a straight interval. We obtain a *lamination* of the cell by line intervals. Now let Δ be any (x^m, x^n)-diagram over $P = \mathrm{sr}\langle x \mid x \to x^2 \rangle$. Identify each edge with a unit interval, and laminate each cell as above.

As a result, every point on the top path of the diagram Δ gets connected with unique point on the bottom path of the diagram: just follow the lamination intervals. This produces a function $f_\Delta:[0,m] \to [0,n]$ which is obviously increasing and onto. See Figure 5.39

Exercise 5.6.26. Compute the functions corresponding to the diagrams on Figure 5.38 and check that these functions belong to F (that is the functions are piecewise linear with appropriate slopes and break points).

Clearly, the function corresponding to a dipole is the identity function, hence the functions corresponding to equivalent diagrams are equal. Thus the map $\phi: \Delta \mapsto f_\Delta$ is well defined. It is also obvious that the map ϕ is a homomorphism from G into the group of all increasing functions from the unit interval onto itself. Since diagrams on Figure 5.38 generate G by Exercise 5.6.25, and the functions corresponding to these diagrams are in F by Exercise 5.6.26, ϕ is a homomorphism $G \to F$.

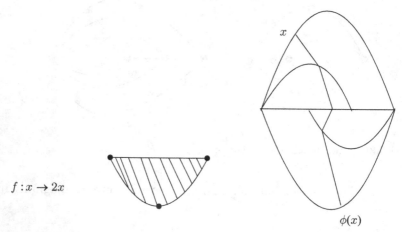

$f : x \to 2x$

Figure 5.39 Turning a diagram into a function

Let us prove that $\phi(G) = F$. Given a function h from F, let $a_0 = 0, a_1, \ldots,$ $a_k, a_{k+1} = 1$ be the increasing sequence of all the break points of the derivative of h ($k \geq 0$). Let $b_i = h(a_i)$ for all $0 \leq i \leq k + 1$. By d we denote the smallest natural number such that all the numbers $2^d a_i, 2^d b_i$ are integers ($0 \leq i \leq k+1$). Consider the function g from $[0, 2^d]$ onto $[0, 2^d]$ given by $g(t) = 2^d h(2^{-d} t)$. All breakpoints of the derivative of g occur at integer points and all slopes are the same as the corresponding slopes of h. The homeomorphism g takes $[2^d a_i, 2^d a_{i+1}]$ onto $[2^d b_i, 2^d b_{i+1}]$ for all $0 \leq i \leq k$. On each of these intervals, g has some slope of the form 2^{c_i}, where c_i is an integer.

For any $j \geq 0$, let Δ_j be an (x, x^{2^j})-diagram over P that can be defined by induction as follows. We let $\Delta_0 = \varepsilon(x)$ and $\Delta_{j+1} = \pi \circ (\Delta_j + \Delta_j)$ for all $j \geq 0$. Here π is the (x, x^2)-cell. The transition function of Δ_j is obviously

$t \mapsto 2^j t$ on $[0,1]$. Now for any number $r \geq 1$, we take the sum of r copies of the diagram Δ_j. We will denote this sum by $r \cdot \Delta_j$. Clearly $f_{r \cdot \Delta_j}$ is the linear function $t \mapsto 2^j t$ on $[0, r]$.

For any $0 \leq i \leq k$, if $c_i \geq 0$, then we let $r_i = 2^d(a_{i+1} - a_i)$. If $c_i < 0$, then we let $r_i = 2^d(b_{i+1} - b_i)$. Let us define the diagram Ψ as the sum of diagrams Ψ_i ($0 \leq i \leq k$), where $\Psi_i = r_i \cdot \Delta_{c_i}$, if $c_i \geq 0$ and $\Psi_i = (r_i \cdot \Delta_{-c_i})^{-1}$, if $c_i < 0$. The transition function of Ψ is exactly g. Now by $\Delta(h)$ we will denote the diagram $\Delta_d \circ \Psi \circ \Delta_d^{-1}$.

Exercise 5.6.27. Show that $f_{\Delta(h)} = h$.

It remains to show that ϕ is injective. For this, it is easiest to use the syntactic properties of the presentation (5.6.2) of the group G. Note that the function $\phi(x_i)$ where $x_i = x_0^{x_1^{i-1}}$, is identity on the subinterval $[0, 1 - 2^{-i}]$, $i \geq 0$, and is not identity immediately to the right of $1 - 2^{-i}$.

Exercise 5.6.28. Prove that!

Let W be a word in $x_i, i \in \mathbb{N}$, which is not equal to 1 in G. We need to show that the corresponding function in F is not the identity. Let j be the smallest index of a letter occurring in W. The relations (5.6.2) show that W is equal in G to a word $x_j^s W' x_j^{-t}$ where $s, t \geq 0$, $s + t \neq 0$, and the indices of all letters in W' are bigger than j. Indeed, the relations (5.6.2) can be rewritten as $x_j^{-1} x_i^{\pm 1} = x_{i+1}^{\pm 1} x_j^{-1}$, $j < i$, and as $x_i^{\pm 1} x_j = x_j x_{i+1}^{\pm 1}$, $j < i$, so we can move the positive powers of x_j to the left and the negative powers of x_j to the right increasing the indices of the other letters without changing the value of the word in G. If both s and t are bigger than 0, then the word $x_j^s W' x_j^{-t}$ is equal to $x_j^{s-1} W'' x_j^{-(t-1)}$ where W'' is obtained from W' by increasing all indices of letters by 1 (why?). Hence we can assume that either s or t (but not both) are equal to 0. If $s = 0$, we can replace W by W^{-1}. So we can assume that $s > 0, t = 0$. But then Exercise 5.6.28 implies that the function corresponding to the word $x^s W'$ (and hence the function corresponding to W) is not the identity function which completes the proof. □

5.6.6.3 From Diagrams to Pairs of Full Binary Trees

Let Δ be an (x^m, x^n)-diagram over $\mathcal{P} = \mathrm{sr}\langle x \mid x \to x^2 \rangle$. Then it has the longest positive path connecting $\iota(\Delta)$ and $\tau(\Delta)$ (it follows from Exercise 5.6.29 below that the longest path is unique). That path divides Δ into two subdiagrams, *positive* and *negative*, denoted by Δ^+ and Δ^-, respectively. So $\Delta = \Delta^+ \circ \Delta^-$.

Exercise 5.6.29. Prove that all cells in Δ^+ are (x, x^2)-cells, all cells in Δ^- are (x^2, x)-cells, that is Δ^+ is positive, Δ^- is negative (see Section 1.7.4). **Hint:** Use induction on the number of cells. Remove from the diagram Δ a cell that shares an edge with the top path, use the induction assumption and analyze what can happen to the longest path in the diagram.

Now let Δ be a (x,x)-diagram. As usual, consider the dual planar graph Γ of the diagram Δ: put a vertex inside every cell, connect two vertices if the cells have a common edge. If we draw the diagram in the natural way: $\mathbf{top}(\Delta)$ above $\mathbf{bot}(\Delta)$, then every vertex of Γ from Δ^+ is connected to two vertices below and to one vertex above it (unless it is the top vertex). Similarly, every vertex of Γ from Δ^- is connected to two vertices above it and to one vertex below it (unless it is the bottom vertex). Thus the longest positive path p of Δ cuts Γ into the union of two full binary trees T^+ and T^-, see Figure 5.40.

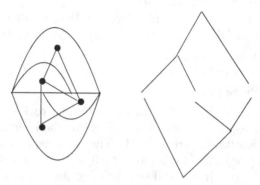

Figure 5.40 A pair of trees corresponding to the generator x_0

The number of leaves of each of the trees is the length of p. Thus T^+ and T^- have the same number of leaves. Conversely, every full binary tree corresponds to a positive derivation. Therefore every pair of full binary trees (T^+, T^-) with the same number of leaves gives rise to a diagram $\Delta = \Delta(T^+, T^-)$ whose positive part corresponds to T^+ and negative part corresponds to T^-.

Exercise 5.6.30. Show that $\Delta(T^+, T^-)$ is reduced if and only if for every $i = 1, \ldots, |p| - 1$, if leaves number i and $i + 1$ have a common parent in T^+, then leaves number i and $i + 1$ do not have a common parent in T^-.

Exercise 5.6.30 suggest defining the following *expansion* operation on a pair of trees (T^+, T^-) as above. Choose number i between 1 and the number of leaves of T^+, add two children to the leaf number i of T^+ and to the leaf number i of T^-. The result of the expansion is a new pair of trees.

Exercise 5.6.31. Consider the rewriting system where objects are pairs of full binary trees with the same number of leaves, and negative moves are the expansions. Show that this rewriting system is Church–Rosser and the terminal objects are pairs of trees corresponding to reduced diagrams.

Exercise 5.6.32. Show that for any two (finite) full binary trees T_1, T_2, there exists a tree T obtained by a series of expansions (i.e., adding two children to a leaf) from T_1 and by (another) series of expansions from T_2.

Now let (T_1^+, T_1^-) and (T_2^+, T_2^-) be two pairs of full binary trees as above. The pairs of trees corresponding to the product $\Delta(T_1^+, T_1^-)\Delta(T_2^+, T_2^-)$ can be obtained as follows. Find two series of expansions that make trees T_1^- and T_2^+ the same. Apply these series of expansions to T_1^+ and T_2^-. Obtain new pair of trees T_3^+, T_3^-. Then

$$\Delta(T_1^+, T_1^-)\Delta(T_2^+, T_2^-) = \Delta(T_3^+, T_3^-).$$

Exercise 5.6.33. Prove this formula.

Exercise 5.6.34. Prove that two diagrams $\Delta(T_1^+, T_1^-)$ and $\Delta(T_2^+, T_2^-)$ are equivalent if and only if there is a pair of trees (T_3^+, T_3^-) which can be obtained by series of expansions both from (T_1^+, T_1^-) and from (T_2^+, T_2^-).

5.6.7 Multilinear Identities of Non-associative Algebras and Elements of F

Recall that in Section 1.4.1.1 we described a correspondence between terms in the signature consisting of one binary operation and (finite) full binary trees where leaves are labeled by variables. For every full binary tree T, we can label its leaves from left to right by distinct variables $a, b, c, \ldots$. Then the tree T will correspond to a multilinear term $t(T)$ (i.e. a term where each variable occurs only once). If a pair of full binary trees has the same number of leaves, then it corresponds to a pair of multilinear terms that differ only by the arrangement of brackets. Thus there exists a correspondence between elements of F and multilinear identities $t(T_1) = t(T_2)$ where left and right sides differ only be the arrangement of brackets. Let us call such identities *bracket identities*.

Exercise 5.6.35. Suppose that a bracket identity $u = v$ is obtained from another bracket identity by a substitution. Show that the corresponding diagrams of F are equivalent.

Exercise 5.6.36. Show that the generator x_0 of F corresponds to the associativity identity $(ab)c = a(bc)$. What is the identity corresponding to x_1 ?

For more about connections between F and identities of universal algebras, and other representations of F see [80, 109].

5.6.8 F Is Lawless

Although as we saw in Section 5.6.7 the group F consists of laws, we shall prove here that F does not satisfy any group laws itself. Here we use the representation of elements of F as functions $[0, 1] \to [0, 1]$.

For $f \in F$ by supp(f) we denote the set of all $t \in I$, for which $f(t) \neq t$.

Theorem 5.6.37. *F does not satisfy any nontrivial group identity.*

Proof. Let X be the set of all diadic rational numbers, that is rational numbers of the form $\frac{a}{2^b}$, $a, b \in \mathbb{N}$, from the open interval $(0, 1)$. Clearly F acts on X: $x \cdot f = f(x)$, $f \in F, x \in X$. For every finite set $X' \subset X$ let b be any number that is not in X'. Consider any open interval $I = (b - \varepsilon, b + \varepsilon)$ containing b and disjoint from X'. There exists a function f from F that is identity outside I and does not have fixed points inside I

Exercise 5.6.38. Construct such a function. Construct a diagram of the form $(m \cdot \Delta_d) \circ \Psi \circ (n \cdot \Delta_d)^{-1}$ (see the definition before Exercise 5.6.27) corresponding to such a function.

Thus supp(f) contains b and does not intersect X'. Hence F separates X, and by Theorem 5.4.1 F does not satisfy any nontrivial identity.   □

5.6.9 F Does Not Have Non-cyclic Free Subgroups

We have seen that elements of F can be viewed as 2-dimensional words, so in this regard, F (and any other diagram group) is a 2-dimensional analog of the free group. Nevertheless here we show that F is very far from being free.

Theorem 5.6.39 (Brin, Squier [53], Guba, Sapir [135]). *F does not contain any free non-cyclic subgroups. Every subgroup of F is either commutative or contains a subgroup isomorphic to $\mathbb{Z} \wr \mathbb{Z}$ (see Exercise 1.8.26), and hence an infinitely generated free commutative group.*[10]

Proof. Let H be a non-commutative subgroup of F. Consider functions $f, g \in H$ such that $fg \neq gf$. Without loss of generality we can assume that H is generated by f, g. For every function $h \in H$ let supp(h) stand for the *support* of h that is the set $\{x \mid h(x) \neq x\}$. Then supp(h) is a union of finitely many disjoint open intervals $(a_i, b_i) \subset [0, 1]$ (prove it!). Let $J = $ supp$(f) \cup$ supp(g). Then J is a union of finitely many disjoint intervals $J_k = (a_k, b_k)$, $0 \leq k \leq m$ (why?). Moreover, for every $h \in H$, and every k, $h(a_k) = a_k, h(b_k) = b_k$ and so $h(J_k) = J_k$. Let f_1, g_1 be two non-commuting elements of H which are not identity on the least possible number of intervals J_k.

By our assumption, $[f_1, g_1] \neq 1$ in F. Then on some of intervals $J_1, \ldots, J_m$ our function $[f_1, g_1]$ is not the identity (because $[f_1, g_1]$ is not the identity function, but it must be identity on the complement $[0, 1] \setminus J$).

We claim that for any $x, y \in J_0$, where $x < y$, there exists a function $w \in H$ such that $w(x) > y$. Indeed, let z be the least upper bound for the set $X = \{h(x) \mid h \in H\}$. We need to show that $z > y$. Suppose that $z \leq y < b_0$. Note that, by definition, J is a union of supports of f and g. Hence either $f(z) \neq z$ or $g(z) \neq z$. Then one of the numbers $f^{\pm 1}(z)$ or $g^{\pm 1}(z)$ is greater

---

[10] Recall that by Corollary 1.8.30 a free group cannot contain non-cyclic commutative subgroups, so the second part of the statement of the theorem implies the first part.

than z. Therefore (by continuity of functions from H) there exists $x' \in X$ with $f^{\pm 1}(x') > z$ or $g^{\pm 1}(x') > z$ but that contradicts the definition of z as a upper bound for X since all $f^{\pm 1}(x'), g^{\pm 1}(x')$ belong to X (why?).

This implies that acting by some element of H one can make the image of every $x \in J_0$ as close to b_0 as one wishes.

Now consider the commutator $h_0 = [f_1, g_1]$. Since both f_1 and g_1 map each interval (a_i, b_i) $(1 \leq i \leq m)$ onto itself, $h_0 = [f_1, g_1]$ does the same. Let J_i be one of the intervals where h_0 acts non-identically. By renaming the intervals if necessary, we can assume that $i = 0$. Since both f_1 and g_1 fix a_0, in a little interval $[a_0, a_0 + \epsilon_1]$, $f_1(x)$ has the form $\alpha(x - a_0) + a_0$ and $g_1(x)$ has the form $\beta(x - a_0) + a_0$ (by the definition of F). Then for some $\epsilon_2 < \epsilon_1$, the function h_0 acts on the interval $[a_0, a_0 + \epsilon_2]$ as an identity (why?). Similarly for some small ϵ_3, the function $[f_1, g_1]$ is the identity on the interval $[b_0 - \epsilon_3, b_0]$. Hence the intersection of support of $[f_1, g_1]$ with the interval (a_0, b_0) is inside (c_0, d_0) where $a_0 < c_0 < d_0 < b_0$. By our claim, there exists $w \in H$ such that $d_0 < w(c_0) < b_0$. We will show that w and h_0 generate a copy of $\mathbb{Z} \wr \mathbb{Z}$. For any $n \geq 0$, let $c_n = w^n(c_0)$, $d_n = w^n(d_0)$, $h_n = h_0^{w^n}$. We have $c_0 < d_0 < c_1 < d_1 < \ldots$, and $\operatorname{supp}(h_n) \cap J_0 \subseteq [c_n, d_n]$ (since every function in F is increasing). Therefore, for any $i, j \geq 0$, the commutator $[h_i, h_j]$ is identical on J_0 (since the supports of h_i and h_j on J_0 are disjoint if $i \neq j$). By the minimality assumption in our choice of f_1, g_1, we conclude that h_i and h_j commute for every i, j.

Note that $h_n^w = h_{n+1}$ for all $n \geq 0$ by definition. Defining h_n as $h_0^{w^n}$ for negative $n \in \mathbb{Z}$ also we see that $h_n, n \in \mathbb{Z}$ and w satisfy the relations of $\mathbb{Z} \wr \mathbb{Z}$ from Exercise 1.8.26 (check it!). Thus there exists a homomorphism ϕ from $\mathbb{Z} \wr \mathbb{Z}$ onto $\operatorname{gp}\langle f, g \rangle$ taking t to w and $x_n, n \in \mathbb{Z}$ to h_n.

Exercise 5.6.40. Prove that the homomorphism ϕ is injective by showing that no nontrivial canonical word of the Church–Rosser presentation in Exercise 1.8.26 is mapped to the identity function.

□

5.6.10 Two Church–Rosser Presentations of F

The next two theorems give Church–Rosser presentations of the Thompson group F with respect to the infinite generating set (5.6.2) and another with respect to the two generators x_0, x_1 (see [134]).

Theorem 5.6.41. *The following monoid presentation with infinite set of generators $x_i, \bar{x}_i, i \in \mathbb{N}$ defines the R. Thompson group F:*

(1) $x_i \bar{x}_i = 1$ *for all* $i \geq 0$;
(2) $\bar{x}_i x_i = 1$ *for all* $i \geq 0$;
(3) $x_j x_i = x_i x_{j+1}$ *for all* $i < j$;

(4) $\bar{x}_j x_i = x_i \bar{x}_{j+1}$ for all $i < j$;

(5) $x_{j+1} \bar{x}_i = \bar{x}_i x_j$ for all $i < j$;

(6) $\bar{x}_{j+1} \bar{x}_i = \bar{x}_i \bar{x}_j$ for all $i < j$.

This rewriting system is Church–Rosser.

It is not difficult to read the canonical form of an element g (in this rewriting system) off the diagram Δ representing the element. Here is the procedure (we give it without a proof). Divide Δ into the positive and negative parts Δ^+ and Δ^- as before.

Let us number the cells of Δ^+ by numbers $1, 2, \ldots, k$ by taking every time the "rightmost" cell, that is, the cell which is to the right of any other cell attached to the bottom path of the diagram formed by the previous cells. The first cell is attached to the top path of Δ^+ (= $\mathbf{top}(\Delta)$). The ith cell in this sequence of cells corresponds to an edge of the Squier complex $\Gamma(\mathcal{P})$, which has the form $(x^{\ell_i}, x \to x^2, x^{r_i})$, where ℓ_i (r_i) is the length of the path from the initial (resp. terminal) vertex of the diagram (resp. the cell) to the initial (resp. terminal) vertex of the cell (resp. the diagram), and contained in the bottom path of the diagram formed by the first $i - 1$ cells. If $\ell_i = 0$ then we label this cell by 1. If $\ell_i \neq 0$ then we label this cell by the element x_{r_i} of F. Multiplying the labels of all cells, we get the "positive" part of the canonical form.

In order to find the "negative" part of the normal form, consider $(\Delta^-)^{-1}$, number its cells as above and label them as above. To get the canonical form of Δ, just multiply the "positive part" by the "negative part".

For example, the diagram on Figure 5.41

The "positive part" of the canonical form is equal to $x_0 x_2^2 x_4 x_5$ (cells 1 and 3 are labeled by the identity element). The negative part is $(x_1 x_3^2 x_4)^{-1}$ (cells 1, 2, 4 are labeled by 1). So the canonical form is $x_0 x_2^2 x_4 x_5 x_4^{-1} x_3^{-2} x_1^{-1}$.

Theorem 5.6.42. *The following monoid presentation with four generators* x_0, x_1, $\bar{x}_0$, $\bar{x}_1$ *defines F and is Church–Rosser.*

(1) $x_0 \bar{x}_0 = 1$,

(2) $\bar{x}_0 x_0 = 1$,

(3) $x_1 \bar{x}_1 = 1$,

(4) $\bar{x}_1 x_1 = 1$,

(5) $x_1 x_0^i x_1 = x_0^i x_1 \bar{x}_0^{i+1} x_1 x_0^{i+1}$,

(6) $\bar{x}_1 x_0^i x_1 = x_0^i x_1 \bar{x}_0^{i+1} \bar{x}_1 x_0^{i+1}$,

(7) $x_1 x_0^{i+1} \bar{x}_1 = x_0^{i+1} \bar{x}_1 \bar{x}_0^i x_1 x_0^i$,

(8) $\bar{x}_1 x_0^{i+1} \bar{x}_1 = x_0^{i+1} \bar{x}_1 \bar{x}_0^i \bar{x}_1 x_0^i$

(here i is an arbitrary positive integer).

Exercise 5.6.43. Prove Theorems 5.6.41 and 5.6.42.

Corollary 5.6.44. *Every group word in the alphabet $\{x_0, x_1, \ldots\}$ is equal in the group F to a unique word of the form*

$$x_{i_1}^{a_1} x_{i_2}^{a_2} \cdots x_{i_m}^{a_m}$$

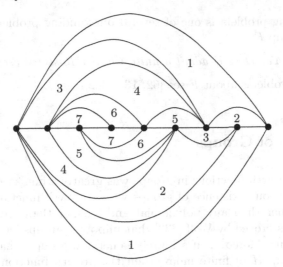

Figure 5.41 An element of F

where $m \geq 0$, $i_1, \ldots, i_m \geq 0, a_1, \ldots, a_m = \pm 1$ such that for any j, $1 \leq j < m$ one of the following three conditions hold:

- $i_j < i_{j+1}$;
- $i_j = i_{j+1}, a_j = a_{j+1}$;
- $i_j = i_{j+1} + 1, a_{j+1} = -1$.

Proof. Indeed the words of the form given in the formulation of Corollary 5.6.44 are precisely the words not containing the left parts of the rewriting rules from Theorem 5.6.41. □

Another set of canonical words in the alphabet $\{x_0, x_1, \ldots\}$ is the following (see [62]):

Exercise 5.6.45. Prove that every word in $\{x_0, x_1, \ldots\}$ is equal in F to a unique word of the form

$$x_0^{b_0} x_1^{b_1} \ldots x_n^{b_n} x_n^{-a_n} \ldots x_1^{-a_1} x_0^{-a_0}$$

where $n, a_0, \ldots, a_n, b_0, \ldots b_n$ are nonnegative integers such that (1) exactly one of a_n, b_n is not equal to 0, and (2) if $a_k > 0$ and $b_k > 0$ for some $k < n$, then $a_{k+1} > 0$ or $b_{k+1} > 0$. Find a Church–Rosser presentation of F with this set of canonical words.

Exercise 5.6.46. Check that the language of left-hand sides of the relations from Theorem 5.6.42 is rational by drawing an automaton for that language. Deduce, using Lemma 1.8.17 that the language of canonical words for this presentation is also rational.

The following problem is one of several outstanding problems about the Thompson group F.

Problem 5.6.47. *Does F admit a finite Church–Rosser presentation?*

For other problems about F see [62, 135, 282].

5.7 Growth of Groups

The study of growth functions in groups was greatly stimulated by Milnor's problem [233] about existence of groups whose growth function is intermediate, i.e., higher than any polynomial and smaller than any exponential function. It was proved by Wolf [332] that nilpotent groups have polynomial growth functions. Moreover, if a finitely generated group G has a nilpotent normal subgroup H of finite index, then the growth function of G is also polynomial. In his famous paper [126], Gromov proved that there are no other groups with growth functions bounded by polynomials. Milnor's problem was solved in the affirmative by Grigorchuk [122]. The goal of this section is to prove the theorem of Bass and Guivarc'h giving exact polynomial upper and lower bounds for the growth function of a finitely generated nilpotent group (it turned out to be the same as the lower bound in [332]), and present a short proof (due to Bartholdi) that Grigorchuk's group has intermediate growth. Proving Gromov's theorem would require much more topology than we can afford, and would take us too far into semantics. Fortunately, there are detailed proofs of his results in [91, 181, 219, 322] (the proof in [126] is not long and is not that hard to read also).

5.7.1 Similar Groups Have Similar Growth

We start with the following theorem

Theorem 5.7.1. *(1) Let G be a finitely generated group and H be a normal subgroup of G of finite index. Then H is finitely generated and the growth functions of G and H are equivalent.*

(2) Let G be a finitely generated group and H be a finite normal subgroup of G. Then the growth functions of G and G/H are equivalent.

Proof. (1) Let X be a generating set of G. For every element p of G/H add to X one $x \in G$ whose image in G/H is p. Let us denote the resulting finite generating set of G by Y. For every $a, b \in Y$, there exists an element $c(a, b) \in Y$ and an element $h(a, b) \in H$ such that $ab = c(a, b)h(a, b)$ (why?). Let Z be the union of $Y \cup H$ and the finite set of all $h(a, b)$, $a, b \in Y$. Then H is generated by Z. Indeed, let $h \in H$. Then $h = y_1 y_2 \cdots y_s, y_i \in Y$ since $h \in G$. If $s = 1$, then $h \in Z$. So assume that $s \geq 2$. Then

$$h = y_1 \cdots y_{n-2} c(y_{n-1}, y_n) h(y_{n-1}, y_n) = \ldots = h(y_1, *) h(y_2, *) \cdots h(y_{n-1}, y_n)$$
$$(5.7.1)$$

where $*$ denotes elements of Y. Thus we can represent h as a word in Z, so Z generates H and H is finitely generated.

Now we can increase Y by including Z and consider the generating set $Y \cup Z$ of G (recall that by Exercise 1.5.1 the growth function of a finitely generated group does not depend (up to the equivalence from Section 1.5) on a choice of a finite generating set). The set of elements of H represented by words of length at most n in Z is then a subset of elements of G represented by words of length at most n in $Y \cup Z$. Hence the growth function $f_H(n)$ of the group H does not exceed the growth function $f_G(n)$ of G.

To complete the proof, we need to prove the opposite inequality. Consider all elements $g_1, \ldots, g_m$ of G represented by words of length at most n in G. Then at least $\frac{m}{|G/H|}$ of them have the same image in G/H. Let it be elements $g_1, \ldots, g_s$. Then elements $1, g_1^{-1} g_2, \ldots, g_1^{-1} g_s$ belong to H (why?) and are all different. Each of these elements can be represented by a word of length at most $2n$. Therefore $f_H(2n) \geq f_G(n)/|G/H|$. Therefore the functions f_H and f_G are equivalent.

(2) Let H be a finite normal subgroup of G, X be a finite generating set of G. By Exercise 3.7.3 the growth function $f_{G/H}(n)$ does not exceed the growth function $f_G(n)$. To prove the opposite inequality, consider the set $g_1, \ldots, g_m$ of elements in G represented by words of length at most n. If the images of g_i and g_j in G/H are the same, then $g_i = g_j h$ for some $h \in H$. Therefore at least $m/|H|$ of these elements have pairwise different images in G/H (why?). Thus $f_{G/H}(n) \geq \frac{f_G(n)}{|H|}$.   □

5.7.2 Commutative Groups

In this section we shall consider the growth of finitely generated commutative groups.

5.7.2.1 Free Commutative Groups

Recall (Exercise 1.8.25) that the free commutative group A_n (i.e., the free group in the variety of commutative groups) with n free generators is isomorphic to the direct product of n copies of the infinite cyclic group $\mathbb{Z}$.

Exercise 5.7.2. Prove that the growth function $f(x)$ of the free commutative group with n free generators is equivalent to x^n. **Hint:** Compare with Example 3.7.1. Use canonical words of the Church–Rosser presentation from Exercise 1.8.25.

5.7.2.2 Arbitrary Finitely Generated Commutative Groups

The results of this subsection easily follow from the classification of finitely generated commutative groups (that every finitely generated commutative group is a direct product of finitely many cyclic groups). Nevertheless proving the classification first and then applying it here would take too long. Instead we are using some basic properties of growth functions. So those who know the classification of finitely generated commutative groups can view this subsection as an illustration of the usefulness of growth functions.

Theorem 5.7.3. *Let G be a finitely generated commutative group. Then G has a subgroup H of finite index which is a finitely generated free commutative group.*

Proof. Let us proceed by induction on the minimal number n of generators of G. If $n = 0$ there is nothing to prove. Suppose that $n > 0$. If G is free commutative, there is also nothing to prove. Otherwise suppose that $G = \mathrm{gp}\langle x_1, \dots, x_n \rangle$ and there exists a nontrivial relation $x_1^{p_1} \dots x_n^{p_n} = 1$ where some $p_i \neq 0$. Without loss of generality we can assume that $i = 1$. Then $x_1^{p_1}$ belongs to the subgroup $G_1 = \mathrm{gp}\langle x_2, \dots, x_n \rangle$. Therefore G/G_1 is a finite cyclic group (prove it!). By induction, G_1 contains a finitely generated free commutative subgroup H with G_1/H finite. Then G/H is also finite (why?).
□

Combining Theorem 5.7.3 with Exercise 5.7.2 and Theorem 5.7.1, we finally obtain.

Theorem 5.7.4. *The growth function $f(x)$ of any finitely generated commutative group G is equivalent to x^n where n is the number of free generators of a free commutative subgroup $H < G$ of finite index.*

Note that this theorem implies that all free commutative subgroups H of G of finite index have the same number of free generators. That number is called the *free rank* of the group G.

Lemma 5.7.5. *Let G be a commutative group generated by a finite set X. Then the free rank of G does not exceed $|X|$.*

Proof. Indeed, G is a homomorphic image of the free commutative group with $n = |X|$ generators. Hence by Exercises 3.7.3 and 5.7.2, the growth function of G does not exceed x^n. Hence by Theorem 5.7.4, the free rank of G does not exceed n.
□

Exercise 5.7.6. Let G be a commutative group generated by a finite number of elements of finite exponents. Then G is finite. **Hint:** If $x_1, \ldots, x_n$ are the generators, then every element is represented by a word $x_1^{m_1} \ldots x_n^{m_n}$ where each m_i is a nonnegative number not exceeding the exponent of x_i.

Lemma 5.7.7. *Let G be a finitely generated commutative group of free rank r and let T be a generating set of G. Then T contains r elements that generate a subgroup C of finite index and free rank r.*

Proof. Induction on the free rank. If G has free rank 0, there is nothing to prove. Suppose that G has free rank $r > 0$, and $x_1, \ldots, x_n$ is a generating set of G. We can assume without loss of generality that all subsets of $\{x_1, \ldots, x_n\}$ generate subgroups of free ranks $< r$. If $n = r$, there is nothing to prove. Let $n > r$. By Exercise 5.7.6, there exists a generator (without loss of generality let it be x_1) which generates an infinite cyclic subgroup $A \leq G$. The subgroup B generated by $x_2, \ldots, x_n$ has free rank $d \leq r - 1$. By the induction assumption, there are d elements generating a subgroup C of B of free rank d, and C/B is finite. Without loss of generality let these elements be $x_2, \ldots, x_{d+1}$. Then consider the subgroup C_1 generated by $x_1, x_2, \ldots, x_{d+1}$. We claim that G/C_1 is finite. Indeed, G/C_1 is generated by the images of $x_2, \ldots, x_n$, and $C_1 \geq C$ by definition. Therefore the image of each $x_i, i \geq 2$ has finite exponent in G/C_1 and by Exercise 5.7.6 G/C_1 is finite. Since the free rank of C_1 cannot be bigger than the number of generators of C_1 (by Lemma 5.7.5), and $d < r$, we conclude that $d + 1 = r$ and $x_1, x_2, \ldots, x_r$ is the required set of generators. $\quad\square$

Theorem 5.7.8. *Let G be a finitely generated commutative group of free rank r and let $T = \{t_1, \ldots, t_s\}$ be a generating set of G, so that $\{t_1, \ldots, t_r\}$ generate a subgroup C of finite index and free rank r (it exists by Lemma 5.7.7). Then T can be extended to a possibly bigger finite generating set $\{t_1, \ldots, t_s, t_{s+1}, \ldots, t_p\}$ such that for every $i, j > s$, $t_i t_j = t_k g$ where $g \in C$.*

Proof. Since G/C is finite, there exists a finite subset $t_{s+1}, \ldots, t_p \in G$ such that every element of G/C is the image of one of the t_i in G/C. It suffices to add this subset to the generating set of G. $\quad\square$

5.7.3 Nilpotent Groups

Here we prove that every finitely generated nilpotent group has polynomial growth (it was first proved by Wolf [332]), and find a precise growth function (up to the equivalence from Section 1.5) – a result obtained by Bass [25] and Guivarc'h [138, 139] (they proved that the upper bound of the growth function coincides, up to equivalence, with the lower bound found by Wolf).

5.7.3.1 Some Basic Properties of Nilpotent Groups

We have defined nilpotent groups using upper central series (see Section 3.6.2): a group G is called nilpotent of class n if G is equal to the n-th member of its upper central series. Here it is more convenient to use an alternative definition in terms of the lower central series.

Recall that the triple commutator $[x, y, z]$ is an abbreviation for $[[x, y], z]$. More generally, for each integer $n > 2$, the *iterated commutator of length* n is defined as $[x_1, \ldots, x_{n-1}, x_n] = [[x_1, \ldots, x_{n-1}], x_n]$. If G is a group, we denote by $\gamma_n(G)$ the subgroup generated by the iterated commutators $[g_1, \ldots, g_{n-1}, g_n]$ where $g_1, \ldots, g_{n-1}, g_n$ run over G. (In particular, $\gamma_2(G) = G'$, the derived subgroup of G). It is convenient to set $\gamma_1(G) = G$. The decreasing series of subgroups

$$G = \gamma_1(G) \geq \gamma_2(G) \geq \ldots \geq \gamma_n(G) \geq \ldots$$

is called *the lower central series* of G.

The following exercise can be proved by induction on the length of an iterated commutator using Exercise 1.4.1 and will be used several times later.

Exercise 5.7.9. Let $x_1, \ldots, x_n \in G$ and for some $i = 1, \ldots, n$, $x_i = y_i z_i$, where $y_i, z_i \in G$. Prove that $[x_1, \ldots, x_i, \ldots, x_n]$ is equal to

$$[x_1, \ldots, y_i, \ldots, x_n][x_1, \ldots, z_i, \ldots, x_n]$$

times a product of commutators of $x_1, \ldots, x_n, y_i, z_i$ of length $> n$, and their inverses. Using the fact that $x_i x_i^{-1} = 1$ and the iterated commutator of $x_1, \ldots, x_n$ is 1 provided one of the x_is is 1, deduce that $[x_1, \ldots, x_i, \ldots, x_n]^{-1}$ is equal to $[x_1, \ldots, x_i^{-1}, \ldots, x_n]$ times a product of commutators of bigger length.

Exercise 5.7.10. (1) Prove that every $\gamma_n(G)$ is a normal subgroup of G.

(2) Prove that $\gamma_{n-1}(G)/\gamma_n(G)$ is contained in the center of the factor-group $G/\gamma_n(G)$.

(3) Prove that if $x = [x_1, x_2, \ldots, x_n]$, then

$$x^m \equiv [x_1^m, x_2, \ldots, x_n] \equiv [x_1, x_2^m, \ldots, x_n] \equiv \ldots \equiv [x_1, x_2, \ldots, x_n^m] \pmod{\gamma_{n+1}(G)}$$

for every m, n.

(4) Prove that for every $x_1, \ldots, x_n \in G$, $a \in \gamma_2(G)$,

$$[x_1 a, x_2, \ldots, x_n] \equiv [x_1, x_2 a, \ldots, x_n] \equiv \ldots \equiv [x_1, \ldots, x_n a] \equiv [x_1, \ldots, x_n]$$

modulo $\gamma_{n+1}(G)$.

(5) Prove that $[\gamma_i(G), \gamma_j(G)] \subseteq \gamma_{i+j}(G)$ for every i, j. **Hint:** Use induction on i. Applying the group analog of the Jacobi identity and the anti-commutativity law for group commutators (Exercise 1.4.1) show that

$$[\gamma_i(G), \gamma_j(G)] = [[\gamma_{i-1}(G), \gamma_1(G)], \gamma_j(G)]$$

is inside the subgroup generated by the product of two subgroups:

$$[[\gamma_j(G), \gamma_1(G)], \gamma_{i-1}(G)]$$

and

$$[[\gamma_j(G), \gamma_{i-1}(G)], \gamma_1(G)],$$

then use the induction assumption to show that both subgroups are inside $\gamma_{i+j}(G)$.

Part (5) of Exercise 5.7.10 means that the subgroups $\gamma_i(G)$, $i = 1, 2, \ldots$, form a *filtration* of G.

Lemma 5.7.11. *Let X be a subset of G, such that the image $\bar{X}$ of X in $G/\gamma_2(G)$ generates a free commutative subgroup of finite index. Let $n \geq 2$. Then the set X_n of commutators $[x_1, x_2, \ldots, x_n]$, $x_i \in X$, belongs to $\gamma_n(G)$ (by the definition of $\gamma_n(G)$). We claim that the image $\bar{X}_n$ of X_n in $\gamma_n(G)/\gamma_{n+1}(G)$ generates a subgroup H_n of finite index.*

Proof. Let H be the subgroup of $G/\gamma_2(G)$ generated by $\bar{X}$. Let N be the size of the finite factor-group $(G/\gamma_2(G))/H$. Then by Exercise 1.8.10 for every $g \in G/\gamma_2(G)$, $g^N \in H$. Take a finite generating set $X' \supseteq X$ of G. By definition, $\gamma_n(G)$ is generated by all iterated commutators of length n of elements of X'. These commutators commute modulo $\gamma_{n+1}(G)$. Take a product p of several such commutators. Then by part 3 of Exercise 5.7.10 p^{N^n} is a product of commutators of the form $[x_1^N, \ldots, x_n^N]$ modulo $\gamma_{n+1}(G)$. Since $x_i^N \in H$ modulo $\gamma_2(G)$, each x_i^N is a product of elements of X modulo $\gamma_2(G)$. Then p^{N^n} is a product of iterated commutators of length n of elements of X. Thus every element of $\gamma_n(G)/\gamma_{n+1}(G)$ to the power N^n belongs to H_n. By Exercise 5.7.6 then H_n has finite index in $\gamma_n(G)/\gamma_{n+1}(G)$. □

The following exercise explains why nilpotency can be defined in terms of lower central series.

Exercise 5.7.12. Let G be a group and suppose a series of its normal subgroups

$$G = G_1 \geq G_2 \geq \ldots \geq G_{c+1} = \{1\}$$

has the property that G_{i-1}/G_i is contained in the center of the group G/G_i for each $i = 2, \ldots, c + 1$. Prove that then for $i = 1, \ldots, c$, the subgroup G_i is contained in the $(c - i + 1)$-th member of the upper central series of G and contains the i-th member of the lower central series of G. In particular, the s-th member of the upper central series is equal to G if and only if $s + 1$-th member of the lower central series is equal to 1. Thus G is nilpotent of class c if and only if $\gamma_{c+1}(G) = \{1\}$

5.7.3.2 Distorted Cyclic Subgroups of Nilpotent Groups

Let G be a group generated by a finite set X, and $h \in G$. Then let $|h|_X$ be the length of a shortest word in X representing h in G.

We can define the *distortion function of the element* h as

$$\delta_h(n) = |h^n|_X.$$

Exercise 5.7.13. Show that the distortion function $\delta_h(n)$, up to equivalence from Section 1.5, does not depend on the choice of the finite generating set X.

Exercise 5.7.14. Show that $\delta_g(n)$ and $\delta_{g^k}(n)$ are equivalent for every $k \neq 0$.

Exercise 5.7.15. Show that the distortion function $\delta_h(n)$ is *subadditive* that is

$$\delta_h(m+n) \leq \delta_h(m) + \delta_h(n).$$

The following exercise belongs to an elementary calculus book. Since this is not a calculus book, we provide a hint.

Exercise 5.7.16. Let $\delta: \mathbb{N} \to \mathbb{N}$ be a subadditive function and for some c, $\delta(n^c) \leq Cn$ where C is a constant. Show that there exists a constant $C' > C$ such that $\delta(m) \leq C' \sqrt[c]{m}$ for every m. **Hint:** If m is a perfect cth power, there is nothing to prove. Otherwise for some $n \approx \sqrt[c]{m}$ we have $n^c < m < (n+1)^c$. Then $m = n^c + m'$ where $m' \leq C_1 n^{c-1}$ (use the Newton binominal formula from Exercise 1.4.12). Apply the subadditivity: $\delta(m) \leq \delta(n^c) + \delta(m') \leq Cn + \delta(m')$ and use induction on m.

The following exercise is an analog of Theorem 5.7.1 (the proof is also similar).

Exercise 5.7.17. Let G be a finitely generated group and H be a normal subgroup of G of finite index. Let $g \in H$. Then the distortion function of g in H is equivalent to the distortion function of g in G.

We shall call a function $f: \mathbb{N} \to \mathbb{N}$ *linear* if it is equivalent to the identity function $n \mapsto n$.

Exercise 5.7.18. Show that in the free commutative group $\mathbb{Z}^n$ the distortion function of every non-identity element is linear.

Lemma 5.7.19. *For every element g of a finitely generated commutative group G, if g is not of finite exponent, then $\delta_g(n)$ is linear.*

Proof. By Lemma 5.7.7, G contains a finitely generated free commutative subgroup H of finite index. Let k be the size of the factor-group G/H. Then $g^k \in H$ by Exercise 1.8.10. By Exercise 5.7.18 the distortion function of g^k in H is linear. Since g is not of finite exponent $g^k \neq 1$. Hence by Exercise 5.7.17, the distortion function of g^k in G is linear. Hence by Exercise 5.7.14, the distortion function $\delta_g(n)$ in G is linear. $\square$

Exercise 5.7.20. Show that in the free group, the distortion function of every non-identity element is linear. **Hint.** Show that the distortion functions of conjugate words are equivalent. Then take a cyclically reduced word u and show that $\delta_u(n) = n|u|$.

Exercise 5.7.21. Show that the distortion function of an element in a group generating an infinite cyclic subgroup cannot be smaller than $\ln n$. **Hint:** If $\delta_u(n)$ is much smaller than $\ln n$, then the number of different powers u^k which can be represented by words of length $\leq n$ in a finite alphabet, would be super-exponential in n while the number of words of length $\leq n$ is exponential in n.

Exercise 5.7.22. Show that in the Baumslag–Solitar group $BS_{1,2} = \mathrm{gp}\langle a, b \mid b^{-1}ab = a^2 \rangle$, the distortion function of the element a is equivalent $\ln n$. **Hint:** The element a^{2^n} is represented by the word $b^{-n}ab^n$ of length $2n + 1$.

One of the key features of nilpotent groups is that the distortion functions of commutators in a nilpotent group are much smaller than linear but much bigger than logarithm.

Lemma 5.7.23. *Let G be a nilpotent group of class 2 generated by a set X, $a, b \in X$. Then $\delta_{[a,b]}(n)$ does not exceed $C\sqrt{n}$ for some constant C.[11]*

Proof. Note that by Exercise 5.7.9 and the fact that G satisfies the identity $[x, y, z] = 1$, we have $[a, b]^{m^2} = [a^m, b^m]$, and the length of the right-hand side of this equality is $4m$. Hence $\delta_{[a,b]}(m^2) \leq 4m$. Now we can apply Exercise 5.7.16.   □

The following lemma generalizes Lemma 5.7.23 to nilpotent groups of arbitrary class.

Lemma 5.7.24. *Let G be a nilpotent group of class c generated by a set X. Let $X_1 = X$, and for every $i > 1$, X_i be the union of sets $[X_j, X_{i-j}]$, $j = 1, \ldots, i - 1$ (where by definition $[A, B] = \{[a, b] \mid a \in A, b \in B\}$). Let $h \in X_c$. Then $\delta_h(n)$ does not exceed $C\sqrt[c]{n}$ for some constant C depending only on c.*

Proof. Induction on c. For $c = 2$ we can apply Lemma 5.7.23. Let $c > 2$, $h = [h_1, h_2]$ where $h_1 \in X_i$, $h_2 \in X_{c-i}$, $1 \leq i \leq c - 1$. Then by Exercise 5.7.9, $h^{m^c} = [h_1^{m^i}, h_2^{m^{c-i}}]$. By the induction hypothesis, $h_1^{m^i}$ can be represented, modulo $\gamma_{i+1}(G)$, by a word of length at most $C_1 m$, and $h_2^{m^{c-i}}$ can be represented, modulo $\gamma_{c-i+1}(G)$ by a word of length at most $C_2 m$ where C_1, C_2 are constants. Since $[\gamma_j(G), \gamma_{c+1-j}(G)] = 1$ for every $j > 0$ (by part 5 of Exercise 5.7.10), we conclude that h^{m^c} can be expressed as a word in X of length at most $2(C_1 + C_2)m$. Hence $\delta_h(m^c) \leq C_3 m$ for some constant C_3. It remains to apply Exercise 5.7.16.   □

It turns out that the distortion function $\delta_h(n)$, $h \in X_c$, is in fact equivalent to $\sqrt[c]{n}$ provided h is not of finite exponent.

---

[11] In fact one can take $C = 8$.

Theorem 5.7.25. *Let G be a finitely generated nilpotent group with filtration $G = G_1 > G_2 > \ldots > G_{c+1} = \{1\}$. Let $h \in G_c$ and h does not have a finite exponent. Then $\delta_h(n)$, up to the equivalence, exceeds $\sqrt[c]{n}$.*

Proof. We need to show that if h^n is represented by a word w of length r in the generators of G, then it is represented by a word of length at most r^c in the generators of G_c (since G_c is a finitely generated commutative group, we then can use Lemma 5.7.19).

If $c = 1$, G is commutative, so the statement follows from Lemma 5.7.19. Let $c \geq 2$. Consider the finitely generated commutative group G/G_2. Let r_1 be the free rank of this group. By Lemma 5.7.7 it contains a free commutative subgroup A_1 of free rank r_1 and finite index. Let H_1 be the preimage of A in G. Then H_1 has finite index in G (prove it!). The group H_1 has a filtration $H_1 > G_2 > \ldots > G_{c+1} = \{1\}$ and H_1/G_2 is isomorphic to A_1. By Exercise 1.8.10 if a natural number $k > 0$ is divisible by the size of G/H_1 (which depends only on G but not on g), then $g^k \in H_1$. Then by Exercise 5.7.17, the distortion functions of g^k in H_1 and in G are equivalent. So we can consider H_1 instead of G and g^k instead of g (by Exercise 5.7.14). Eventually we shall demand that k is divisible by some finitely many other numbers (all of which will not depend on g).

By Exercise 5.7.13, we can choose a finite generating set of H_1 as we want. Since all iterated commutators in H_1 of sufficiently large length are equal to 1, we can choose a finite generating set X of H_1 which is closed under taking inverses and commutators. Moreover, we can add a finite number of elements to X and assume that $X \cap G_i$ generates G_i for every i. Even moreover since commutators of any two elements of H_1 are in G_2, we can assume that the images of the first r_1 elements $x_1, \ldots, x_{r_1}$ of X in $A_1 = H_1/G_2$ form a free generating set of A_1.

Then we denote $X_i = X \cap G_i$, $i = 1, \ldots, c$, $X_i' = X \setminus X_i$. Let w be a word in X representing g^{kn} for some $n > 1$. Thus $X_1 \setminus X_2 = \{x_1^{\pm 1}, \ldots, x_{r_1}^{\pm 1}\}$. Then the *X-multilength* of w is a vector $(n_1, n_2, \ldots, n_c)$ where n_i is the number of occurrences of letters from X_i' and their inverses in w. Thus $X = X_{c+1}'$.

Now let us do some string rewriting. Pick one of the letters among $x_1, \ldots, x_{r_1}$, say, x_1. The rewriting rules are

$$x_i^\alpha x_1^\beta \to x_1^\beta x_i^\alpha x_k \qquad (5.7.2)$$

where $\alpha, \beta \in \{-1, 1\}$, x_k is a generator from X that is equal to the commutator $[x_i^\alpha, x_1^\beta]$, $j \leq r_1$. Using these rules, we can move all occurrences of letters x_1^α, $\alpha \in \{-1, 1\}$ to the left. The resulting word has the form $x_1^{q_1} w_1$ where w_1 does not have letters from $\{x_1, x_1^{-1}\}$.

Let us estimate the multilength $(n_1', n_2', \ldots, n_c')$ of w_1. It contains all the letters of w minus letters from $\{x_1, x_1^{-1}\}$ plus new letters introduced during the rewriting. Every time we apply a rule (5.7.2) with $x_i \in X_e'$, we add a new letter from X_{e+1}', keeping the letter x_i. Thus, in particular, n_1' is at most $n_1 - q$ where q is the number of occurrences of $x_1^{\pm 1}$ in w. In general we have:

Exercise 5.7.26. Prove the following "binomial inequality"

$$n_i' \le n_i + \binom{q}{1} n_{i-1} + \binom{q}{2} n_{i-2} + \ldots = \sum_{s=0}^{i} \binom{q}{s} n_{i-s}$$

Hint: Use induction on q. If $q \le 1$, the inequality was established above. If $q > 1$, then moving one occurrence of $x_1^{\pm 1}$ all the way to the left increases each n_i by at most n_{i-1}. Then use the Pascal triangle equality $\binom{q-1}{j} + \binom{q-1}{j-1} = \binom{q}{j}$ (Exercise 1.1.4).

We shall continue moving letters $x_1, \ldots, x_{r_1}$ to the left until no such letters remain in w'. Our goal is to show that the last component of the multilength of w' never exceeds some constant times r^c where r is the length of w. It is easier to show a stronger result that the i-th coordinate of the multilength never exceeds Cr^i (for some constant i). Suppose that we already have $n_i \le Cr^i$. Since $q \le r$, we have $\binom{q}{i} \le r^i$. Therefore Exercise 5.7.26 implies $n_i' \le \sum Cr^s r^{i-s} = C_1 r^i$ for some constant C_1 which does not depend on r or w. Thus by moving all letters $x_1^{\pm 1}, \ldots, x_{r_1}^{\pm 1}$ to the left, we deduce that g^{kn} is represented by a word of the form $x_1^{p_1} \ldots x_{r_1}^{p_{r_1}} w_{r_1}$ where the word w_{r_1} does not contain letters from X_1', and its multilength in H_1 does not exceed $(0, C_{r_1} r^2, \ldots, C_{r_1} r^c)$ for some constant C_{r_1}. Note that the element of G represented by w_{r_1} is in G_2. Since $g \in G_c \subseteq G_2$, the image of $x_1^{p_1} \ldots x_{r_1}^{p_{r_1}}$ is equal to 1 in A_1. Since the images of $x_1, \ldots, x_{r_1}$ in A_1 are free generators, we conclude that $p_1 = \ldots = p_{r_1} = 0$.

Now we have to consider w_{r_1} which is a word in X_2 representing g^{kn}. By the choice of X, we have that X_2 generates G_2. We will apply to G_2 the same procedure we applied to G. First find a finite index subgroup $H_2 \le G_2$ containing G_3 such that $A_2 = H_2/G_3$ is a free commutative group of free rank r_2. In order to make sure g^k belongs to H_2, we demand now that k is divisible by the product $|G/H_1||G_2/H_2|$. Then make X_2 a bit smaller so that the images of the first r_2 elements of $t_1, \ldots, t_{r_2}$ of the new generating set T of H_2 in H_2/G_3 are free generators of A_2. By both replacing G_2 with a finite index subgroup H_2 and by making X_2 smaller, we increase the multilength of the word representing g^{kn}. But the increase is only by a constant multiple (why?). Thus we can still assume that a word w in the alphabet T, represents g^{kn} and has multilength (in G_2) at most $(Cr^2, \ldots, Cr^c)$ for some constant C. As before let us move the letters from $T_1 = T \smallsetminus G_3$ to the left in w. Since $T \subset G_2$, by the definition of filtration, the "binomial inequality" now looks as follows:

$$n_i' \le n_i + \binom{q}{1} n_{i-2} + \binom{q}{2} n_{i-4} + \ldots$$

where $q < Cr^2$ is the number of occurrences of letters from T_1 in w', $(n_2, \ldots, n_c)$ is the multilength of w and $(n_2', \ldots, n_c')$ is the multilength of the word obtained by moving all occurrences of $t_1^{\pm 1}$ to the left. Now since $q < Cr^2$ we get $\binom{q}{j} \le C_1 r^{2j}$. Since $n_{i-2j} \le Cr^{i-2j}$, we get inequality $n_i' \le C_2 r^i$

for some constant C_2. As before, after moving all letters $t_1^{\pm 1}, \ldots, t_{r_2}^{\pm 1}$ to the left of w, we obtain a word of the form $t_1^{s_1} \ldots t_{r_2}^{s_{r_2}} w_{r_2}$ where w_{r_2} does not contain letters from $\{t_1^{\pm 1}, \ldots, t_{r_2}^{\pm 1}\}$. Since the images of $t_1, \ldots, t_{r_2}$ in H_2/G_3 are free generators of A_2, we conclude that $s_1 = s_2 = \ldots = s_{r_2} = 0$.

Now we can proceed with considering w_{r_2} and G_3 instead of G_2, etc.

Exercise 5.7.27. Complete the proof of Theorem 5.7.25

$\square$

5.7.3.3 The Theorem of Bass and Guivarc'h

Now we fix a finitely generated nilpotent group G with lower central series

$$G = \gamma_1(G) \ge \gamma_2(G) \ge \ldots \ge \gamma_p(G) \ge \gamma_{p+1}(G) = 1.$$

By Exercise 5.7.10, for each $h = 1, \ldots, p$, the group $\gamma_h(G)/\gamma_{h+1}(G)$ is finitely generated and commutative. Let r_h be the free rank of $\gamma_h(G)/\gamma_{h+1}(G)$ and let $d(G) = \sum_{h=1}^{p} h r_h$.

Theorem 5.7.28 (Bass [25], Guivarc'h [138, 139]). *The growth function $f(x)$ of a finitely generated nilpotent group G is equivalent to $x^{d(G)}$.*

5.7.3.4 The Lower Bound

Let $A = G/\gamma_2(G)$. Then A is a finitely generated commutative group. Hence it has a finitely generated free commutative subgroup B such that A/B is finite (by Theorem 5.7.3). Let $\bar{T}_1$ be a free generating set of B, for each element $\bar{t} \in \bar{T}_1$ pick one element $t \in G$ whose image in A is $\bar{t}$. Denote the (finite) set of all these t by T_1.

Let us define sets $T_2, T_3, \ldots$ by induction: $T_{h+1} = [T_h, T_1]$. Then $T_h \subset \gamma_h(G)$ by the definition of $\gamma_h(G)$. By Lemma 5.7.11, the image of T_h in $\gamma_h(G)/\gamma_{h+1}(G)$ generates a commutative subgroup of free rank r_h. Let $T = \cup T_h$, it is a finite subset of G. By Theorem 5.7.8, there are r_c elements $g_1, \ldots, g_{r_c}$ of T_c which generate a free commutative subgroup of free rank r_c.

Let us use induction on pairs $(c, d(G))$ ordered lexicographically. For $c = 1$ and any $d(G)$, the statement follows from Theorem 5.7.4. Let $c > 1$. If $r_c = 0$, then $\gamma_c(G)$ is finite, by Theorem 5.7.1 the growth function of G is equivalent to the growth function of $G/\gamma_c(G)$ which is nilpotent of class $c - 1$ and we can use the induction assumption.

Suppose that $r_c > 0$. Then $g = g_{c,1}$ generates an infinite cyclic subgroup H. Since H is in the center of G (why?), H is normal in G and we can consider the factor-group $G' = G/H$. The lower central series of G' is $\gamma_1(G)/H, \ldots,$ $\gamma_c(G)/H$ (prove it!). Since $\gamma_i(G')/\gamma_{i+1}(G') = (\gamma_i(G)/H)/(\gamma_{i+1}(G)/H)$ is isomorphic to $\gamma_i(G)/\gamma_{i+1}(G)$ for $i = 1, \ldots, c-1$ (why?) and $\gamma_c(G') = \gamma_c(G)/H$, we have that $d(G') = d(G) - c$. Therefore we can use the induction assumption for G' and conclude that for every n, G' contains at least $m = Cn^{d(G')} = Cn^{d(G)-c}$ elements represented by words of length at most n. Let $\bar{u}_1, \ldots, \bar{u}_m$ be these elements in $G' = G/H$ and $u_1, \ldots, u_m$ be the elements represented by the same words in G (so that $\bar{u}_i$ is the image of u_i in G/H). Note that the element g^k for k between 1 and n^c can be represented by a word of length at most $C_1 n$ for some constant C_1 by Lemma 5.7.24. Therefore the elements $u_i g^j$ where $i = 1, \ldots, m$, $j = 1, \ldots, n^c$ can be represented by words of length at most $n + C_1 n$. All these elements are different. Indeed, if $u_i g^j = u_k g^l$, then $\bar{u}_i = \bar{u}_k$ in G', hence $u_i = u_k$. Canceling u_i, we get $g^j = g^l$ which implies $j = l$ since H is infinite cyclic (i.e., isomorphic to $\mathbb{Z}$, and $\mathbb{Z}$ does not have two equal numbers). The number of these elements $u_i g^j$ is then $mn^c = n^{d(G')+c} = n^{d(G)}$, hence the growth function $f(n)$ of G satisfies $f(n + C_1 n) \geq n^{d(G)}$ which implies the required lower bound.

5.7.3.5 The Upper Bound

Consider a finite generating set T of a nilpotent group G of class c. Then find the elements $g_{i,j}$ as in Section 5.7.3.4. Again use induction on $(c, d(G))$. If $r_c = 0$, then the growth functions of G and $G/\gamma_c(G)$ are equivalent by Part (2) of Theorem 5.7.1 and we can use the induction assumption because the nilpotency class of $G/\gamma_c(G)$ is $c - 1$. So suppose that $r_c > 0$, so that the subgroup H generated by $g = g_{c,1}$ is infinite cyclic. By induction, the growth function of $G' = G/H$ does not exceed $Cn^d(G') = Cn^{d(G)-c}$. Let u be an element in G which can be represented by a word of length $\leq n$. Then the image of u in G' can be represented by a word of length at most n, and there are at most $Cn^{d(G')}$ such images. If two elements u, v which can be represented by words of length $\leq n$ have the same image in G', then $u = vg^s$ where g^s can be represented by a word of length at most $2n$ (why?). By Theorem 5.7.25 there are at most $C_1 n^c$ such numbers s for some constant C_1. Therefore the growth function $f(n)$ of G with respect to the generating set T cannot exceed $Cn^{d(G')} C_1 n^c = CC_1 n^{d(G)}$ which implies the desired upper bound.

5.7.4 Grigorchuk's Group of Intermediate Growth

We will start with a Mealy automaton $\mathcal{G}$. Its alphabet X is $\{0, 1\}$.

The automaton $\mathcal{G}$ is on Figure 5.42, it has four vertices labeled by a, b, c, d. Let us denote the transformations of the free monoid X^* corresponding to these vertices by the same letters.

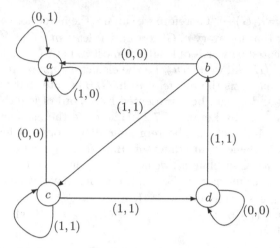

Figure 5.42 The Mealy automaton defining the Grigorchuk group

Then for every word u in the alphabet $\{0,1\}$ we have

$$
\begin{array}{ll}
a(0u) = 1u, & a(1u) = 0u, \\
b(0u) = 0a(u), & b(1u) = 1c(u), \\
c(0u) = 0a(u), & c(1u) = 1d(u), \\
d(0u) = 0u, & d(1u) = 1b(u).
\end{array}
\tag{5.7.3}
$$

Exercise 5.7.29. Prove formulas (5.7.3).

Note that G acts by automorphisms on the vertices of the infinite rooted binary tree T (see Figure 1.14), which is the Cayley graph of the free semigroup with two generators $0, 1$. Thus there exists a natural homomorphism from G into the group $\operatorname{Aut}(T)$ of all automorphisms of the tree T. That homomorphism is injective (why?), so every element of G can be viewed as an automorphism of T.

Exercise 5.7.30. Let $\mathbf{1}$ be the identity element of $\operatorname{Aut}(T)$. Prove that

(1) $a^2 = b^2 = c^2 = d^2 = \mathbf{1}$, in particular each transformation a, b, c, d is a bijection.

(2) b, c, d pairwise commute and the product of any two of these transformations is the third: $bc = d$, $cd = b$, $db = c$.

By Exercise 5.7.30, the transformations a, b, c, d are bijections, and so they generate a group which we shall denote by G. That group was first discovered by Grigorchuk [119, 122, 123] (using a different notation). The fact that it is generated by the Mealy automaton $\mathcal{G}$ was observed by Merzlyakov [230] and appeared in the third edition of [170].

It was first established by Grigorchuk [119] that G is a periodic group (later proved to have unbounded exponent) thus giving one more example of an infinite finitely generated periodic group (see Section 4.4.6).

Later, in [122, 123], Grigorchuk proved that G has intermediate growth.

Theorem 5.7.31. *The group G is infinite, periodic of unbounded exponent, and has intermediate growth.*

There are many proofs of this important theorem. Here we basically follow Bartholdi's paper [22].

Let H be the normal subgroup generated by the subgroup $V = \{1, b, c, d\}$. *Exercise* 5.7.32. Show that $H = \mathrm{gp}\langle b, c, d, b^a, c^a, d^a \rangle$ and $|G/H| = 2$. Moreover for every $u \in X^*$ and every $g \in H$, $g(u)$ and u have the same first letters. **Hint:** Show that H has index 2 in G. One coset of H is H itself. What is the representative of the other coset? The fact that in every group every subgroup of index 2 is normal is a standard abstract algebra exercise.

There is a map ψ from H into the direct square of the group $\mathrm{Aut}(T)$: $\psi(g) = (g_0, g_1)$, so that $g(0u) = 0g_0(u)$, $g(1u) = 1g_1(u)$. Note that ψ is injective (prove it!). By Exercise 5.7.32, the map ψ can be defined explicitly:

$$\psi: \begin{cases} b \mapsto (a, c), \ b^a \mapsto (c, a); \\ c \mapsto (a, d), \ c^a \mapsto (d, a); \\ d \mapsto (1, b), \ d^a \mapsto (b, 1). \end{cases} \tag{5.7.4}$$

Thus in fact $\psi(H) \subseteq G \times G$.

Note also that ψ is a homomorphism (check it!). Moreover if we compose ψ with the projection onto the first (second) coordinate, then we obtain a homomorphism $\psi_i : g \mapsto g_i$ into G ($i = 0, 1$), and its image contains all generators of G (see (5.7.4)). Thus ψ_0 is a surjective homomorphism from a proper subgroup H of G onto G. Therefore G is infinite.

Exercise 5.7.33. Show that for every $g \in H$, if $\psi(g) = (g_0, g_1)$, then $\psi(aga) = (g_1, g_0)$.

We will also consider a map from the free monoid $F = \mathrm{mn}\langle a, b, c, d \rangle$ to $F \times F$ given by the formulas (5.7.4). We shall denote that map also by ψ. Note that if a word w represents an element $g \in H$, then the components of the pair $\psi(w)$ represent the corresponding components of $\psi(g)$ (why?).

The crucial idea of Bartholdi's proof is assigning different weights to different generators of G. Let $\eta \approx .811$ be the real root of the polynomial $x^3 + x^2 + x - 2$. Let us assign the following weights ω to the generators of G:

$$\omega(a) = 1 - \eta^3, \omega(b) = \eta^3, \omega(c) = 1 - \eta^2, \omega(d) = 1 - \eta.$$

It should become clear why such (strange on the first glance) weights are assigned to the generators of G. As usual, the weight of a word is the sum of weights of its letters, and the *weight of an element* g of G, denoted $\omega(g)$, is the minimal weight of a word that represents this element (note that since every generator squared is 1, we only need to consider *positive words*), and the weighted growth function: $f_\omega(n)$ is the number of elements of G of weight at most n.

Exercise 5.7.34. Show that the weighted growth function f_ω of G is equivalent to its (usual) growth function.

Lemma 5.7.35. *Every $g \in G$ is represented by a minimal weight word of the form*

$$[*]a * a \ldots * a[*] \tag{5.7.5}$$

where $* \in \{b, c, d\}$ and the first and the last $*$'s are optional.

Proof. Indeed, if a word representing g has two consecutive letters from $\{b, c, d\}$, then their product can be replaced either by 1 (if the letters are equal) or by the third letter from $\{b, c, d\}$. The number η is chosen in such a way that for any permutation (x, y, z) of $\{b, c, d\}$ we have $\omega(x) \le \omega(y) + \omega(z)$. Thus the substitution of z for xy will not increase the weight of the word. $\square$

Note that the property of η used in the proof of Lemma 5.7.35 is not the main reason we choose this number. Many weight functions ω satisfy this property, in particular, any constant function. Thus Lemma 5.7.35 is still true if we replace "minimal weight" by "minimal length".

The next lemma shows that if $\psi(g) = (g_0, g_1)$, then the total weight of g_0 and g_1 is usually smaller than the weight of g. Thus although g_0, g_1 encode g, they weigh less. Basically this is the property that makes the growth of G intermediate. A version of it was used in most other proofs of Theorem 5.7.31 including the original proof in [123].

Lemma 5.7.36. *Let $g \in H$ with $\psi(g) = (g_0, g_1)$. Then*

$$\eta(\omega(g) + \omega(a)) \ge \omega(g_0) + \omega(g_1).$$

Proof. Let w be a minimal weight word of the form (5.7.5) representing g in G. Thus the number of $*$'s in w is at most the number of a's plus one. Let $\psi(w) = (w_0, w_1)$, so that w_0 represents g_0, w_1 represents g_1. Note that

$$\begin{aligned}
\eta(\omega(a) + \omega(b)) &= \omega(a) + \omega(c), \\
\eta(\omega(a) + \omega(c)) &= \omega(a) + \omega(d), \\
\eta(\omega(a) + \omega(d)) &= 0 + \omega(b).
\end{aligned} \tag{5.7.6}$$

(Check it! These formulas are the first significant reason for the choice of η and the weight function ω.) Since $\psi(b) = (a, c)$ and $\psi(aba) = \psi(b^a) = (c, a)$ (see Exercise 5.7.33), each b in w contributes $\omega(a) + \omega(c)$ to the total weight of w_0 and w_1; a similar argument applies to c and d. After we subdivide w into letters from $\{b, c, d\}$ and subwords from $\{aba, aca, ada\}$, we may have just one letter, a, left. Therefore $\eta\omega(g)$ is a sum of left-hand terms in (5.7.6) possibly minus $\eta\omega(a)$ while $\omega(g_0)+\omega(g_1)$ is bounded by the sum of the corresponding right-hand terms (it is not necessarily an equality because w_0, w_1 may be not minimal weight words representing g_0, g_1).   □

The idea to count the number of elements of G of weight not exceeding n is to represent elements of G by finite full binary trees introduced in Section 1.4.1.1. Recall that a *full binary tree* is a tree drawn on a plane where every vertex has out-degree either 2 or 0 (and in-degree 1 or 0).

Exercise 5.7.37. Prove that the number of full binary trees with $n+1$ leaves (i.e., the *Catalan number* C_n) does not exceed $n! \le n^n = 2^{n\log_2(n)}$. **Hint.** From every tree with n leaves one can get n trees with $n+1$ leaves: the tree number i is obtained by adding two "children" to the leaf number i.[12]

We need two more real numbers: $\alpha = \frac{\ln 2}{\ln \frac{2}{\eta}} \approx .767$ and $\zeta = \frac{\omega(a)}{\frac{2}{\eta}-1} \approx .319$.

Lemma 5.7.38. *Let $K > \zeta$. For every natural n let L_n be the maximal of two numbers: 1 and the smallest integer exceeding*

$$\left(2\frac{n-\zeta}{K-\zeta}\right)^\alpha - 1.$$

Then the weighted growth function $f_\omega(n)$ satisfies

$$f_\omega(n) \le C_{L_n-1}2^{L_n-1}f_\omega(K)^{L_n}. \tag{5.7.7}$$

Hence $f_\omega(n) \le 2^{cn^{.8}}$ for some $c > 0$.

Proof. We construct an injective map τ from G into the set of labeled binary rooted trees each of whose leaves is labeled by an element of G of weight at most K and each non-leaf vertex is labeled by a or 1. The tree $\tau(g)$ will be called the *picture* of g. It is constructed as follows. If $\omega(g) \le K$, then the picture of g is a one-vertex tree with the vertex labeled by g. Let $\omega(g) > K$. Since $|G/H| = 2$ (Exercise 5.7.32) and $a \notin H$ (why?), either $g \in H$ or $ga \in H$. Accordingly, denote by h either 1 or a. Let $\psi(g) = (g_0, g_1)$. By Lemma 5.7.36, $\omega(g_i) \le \eta\omega(g)$. So by induction on the weight, we can assume that the tree pictures T_0 and T_1 of g_0 and g_1 are already known. Connect each of these trees to a new root vertex by an edge to obtain a new binary rooted tree T with T_0 being to the left of T_1. Label the root of T by h. The resulting tree is the picture $\tau(g)$ of g.

Let us prove that τ is injective. Let T be a tree in the image of τ. If T consists of just one vertex labeled by g, then $\tau^{-1}(T) = g$. Suppose that T has more than one vertex. Let $h \in \{a, 1\}$ be the label of the root of T, and T_0, T_1 be the two disjoint subtrees obtained by removing the root and the

---

[12] Some of these trees are the same. The precise formula is $C_n = \frac{1}{n+1}\binom{2n}{n}$ which is much smaller than $n!$, see [308], but we will not use that formula.

adjacent edges. Then, by induction, we can assume that there is at most one g_i ($i = 0, 1$) such that $\tau(g_i) = T_i$. Since ψ is injective, there exists unique $g \in G$ with $gh \in H$ and $\psi(gh) = (g_0, g_1)$. Hence $\tau^{-1}(T) = g$.

We next prove by induction on n that if $\omega(g) \le n$, then its picture has at most L_n leaves. Indeed, if $\omega(g) \le K$, then the picture of g is a single vertex tree, hence has one leaf, and $L_1 = 1$. Otherwise the picture of g consists of a root connected to the pictures of g_0 and g_1. Suppose that $\omega(g_0) = l, \omega(g_1) = m$. Then by Lemma 5.7.36 we have $l + m \le \eta(n + \omega(a))$. By induction, these two pictures have at most L_l and L_m leaves. Since $\alpha < 1$, we have $L_l + L_m \le 2L_{(l+m)/2}$ for all m, l (check it!). By direct computation $L_{\frac{\eta}{2}(n+\omega(a))} = \lfloor L_n/2 \rfloor$ (prove it! – this is the place in the proof where ζ is used), so the number of leaves of the picture of g is at most

$$L_l + L_m \le 2L_{(l+m)/2} \le 2L_{\frac{\eta}{2}(n+\omega(a))} \le L_n$$

(check this!)

Thus $f_\omega(n)$ is bounded by the number of pictures with L_n leaves. There are C_{L_n-1} full binary trees with L_n leaves, 2 choices for labels of each of the (at most $L_n - 1$ by Exercise 1.3.11) non-leave vertices (the labels are a and $\mathbf{1}$) and at most $f_\omega(K)$ choices for labels of each of the leaves. This gives us (5.7.7).

It remains to note that, by Exercise 5.7.37, for some constants $c_1, c > 0$ and every $\alpha' > \alpha$, we have $C_{L_n-1} \le 2^{c_1 L_n} \ln L_n \le 2^{c_2 n^\alpha \ln n} \le 2^{cn^{\cdot 8}}$ for all n (here we used the fact that for every $\epsilon > 0$, we have $\ln n < n^\epsilon$ for all sufficiently large n).   □

Thus the growth function of G is at most (up to the equivalence from Section 1.5) $2^{n^{\cdot 8}}$. Note that by using the precise formula for Catalan numbers, we would obtain the upper estimate for the growth 2^{n^α} which is currently the best known upper estimate for the growth function of G.

Let us prove that G is periodic. Here we basically follow the book by de la Harpe [81, Chapter VIII]

Let $g \in G$. We shall prove, by induction on the length $|g|$, that $g^{2^m} = 1$ for some m. For the generators of G, the statement is obvious: all generators have exponent 2. By Lemma 5.7.35 (using length instead of weight) a minimal length word w representing g has the form $[*]a * a \ldots * a[*]$ as in that lemma. Let n be the length of that word. If w starts and ends with a or with $*$, we consider a cyclic shift of w moving the first letter to the last position. Applying one of the reductions $a^2 \to 1, bc \to d, \ldots$, we get an element of length at most $n - 1$ representing a conjugate of g. It has exponent 2^m for some m. Hence g has the same exponent. Thus we can assume that (up to a cyclic shift) $w = au_1 a \ldots u_l$, and $n = 2l$ is even, $u_i \in \{b, c, d\}$.

Suppose that l is even. Then $g \in H$. We can represent

$$\psi(w) = \psi(au_1 a)\psi(u_2)\psi(au_3 a) \ldots \psi(u_l) = (w_0, w_1)$$

where the length of each of w_0, w_1 is at most $n/2$ (check it!). By induction, the elements g_0, g_1 represented by w_0, w_1 have exponents 2^{m_0} and 2^{m_1} respectively for some m_0, m_1. Since ψ is injective on H, we have $g^{2^m} = 1$ where m is the maximum of m_0 and m_1.

Finally suppose that l is odd so that $n = 4s - 2$ for some $s \geq 2$. Then consider

$$ww = (au_1a)u_2 \ldots u_{2m-2}(au_{2m-1}a)u_1(au_2a) \ldots (au_{2m-2}a)u_{2m-1}.$$

The length of ww is $8s - 4$, so, as in the previous case $\psi(ww) = (w', w'')$ for some words w', w'' of length at most $4s - 2$ (i.e., at most the length of w). We show that the exponents of elements g', g'' are finite powers of 2.

Consider three cases.

Case 1. If one of the u_i is d then we have $\psi(au_ia) = (b, 1)$ and $\psi(u_i) = (1, b)$, thus w' and w'' are of length at most $4s - 3$, and we can apply induction on length.

Case 2. If one of u_i is c, then $\psi(au_ia) = (d, a)$, $\psi(u_i) = (a, d)$ (Exercise 5.7.33). Hence both w' and w'' contain d and the previous case applies.

Case 3. Finally if none of u_i is d or c, then $w = (ab)^l$. But ab has exponent 8 (check it!), so g has exponent dividing 8.

Since G is infinite, finitely generated and periodic, it cannot be an extension of a nilpotent group by a finite group by Corollary 3.4.2. Hence by Gromov's polynomial growth theorem [126] the growth of G is super-polynomial. In fact Gromov's polynomial growth theorem is not necessary here. Grigorchuk [123] gave a purely combinatorial proof that the growth function $f(n)$ of G is not smaller (with respect to the partial order $\prec$ from Section 3.7.1) than $2^{\sqrt{n}}$. The biggest known lower estimate is due to Bartholdi [23] and is $\approx 2^{n^{.5157}}$.

It remains to prove that G does not have bounded exponent. That easily follows from Zel'manov's solution of the restricted Burnside problem [333, 334]. Indeed, suppose that G has exponent m. For every element $g \neq 1$ in G, g must act non-identically on some vertex v_g of T. Recall that vertices of T are words in $\{0, 1\}^*$. Let n be the length of the word v_g. Consider the (finite) set D_n of all words of length n. Then G (and in fact the whole $\mathrm{Aut}(T)$) permutes D_n (why?). Thus there exists a homomorphism ψ_n from G into a finite group G_n of permutations of D_n (this homomorphism simply restricts every $g \in G$ on D_n). Note that $\psi_n(g) \neq 1$. Therefore the orders of finite groups G_n cannot be bounded independently on n (why?). But each G_n is generated by 4 elements (images of a, b, c, d) and has exponent $\leq n$. That contradicts Zel'manov's Theorem [333, 334].

Of course this is an extremely non-geodesic proof. A direct and easy, purely syntactic, proof was found by Igor Lysenok in 1996 (unpublished) and communicated to us by Laurent Bartholdi and Rostislav Grigorchuk. Denote by w_3 the word $abab$. The element of G represented by w_3 has order 8 (prove it!). Denote by σ the substitution $a \mapsto aca, b \mapsto d, d \mapsto c, c \mapsto b$. Then set $w_{n+1} = a\sigma(w_n)$. Since

$$\psi(\sigma(w_{n+1})) = (1, w_n)$$

(check it!), we have $\psi(w_{n+1}^2) = \psi(a\sigma(w_n)a)\sigma(w_n) = (w_n, w_n)$ (by Exercise 5.7.33). Thus the exponent of the element w_n in G is 2^n for every $n \geq 3$.

The proof of Theorem 5.7.31 is complete.

Remark 5.7.39. As for semigroups of intermediate growth considered in Section 3.7.2, the Grigorchuk group G has a syntactic description also. It was found by Lysenok [212]. Let $X = \{a, b, c, d\}$, and $\sigma: X \to X^+$ be the substitution $a \mapsto aca, b \mapsto d, c \mapsto b, d \mapsto c$. Then the group G has the following monoid presentation $\mathrm{mn}\langle a, b, c, d \mid a^2 = 1, b^2 = 1, c^2 = 1, d^2 = 1, bcd = 1, \sigma^n((ad)^4) = 1, \sigma^n((adacac)^4) = 1, n = 0, 1, 2, \ldots \rangle$.

5.8 Amenable Groups

Amenability and related notions are an extremely important part of group theory now, with many applications beyond algebra: in geometry, functional analysis, operator algebras, dynamical systems, probability, and so on. It all started with Hausdorff's theorem that one can cut a sphere S^2 minus a countable set of points Z into a finite number of pieces, rearrange the pieces using rotations of $\mathbb{R}^3$ and obtain two copies of $S^2 \setminus Z$. Later Banach and Tarski proved that there is no need to remove a countable set of points: one can double the sphere S^2 itself. Analysing Hausdorff's proof, von Neumann observed that the reason for Hausdorff's phenomenon is the fact that the group of rotations of $\mathbb{R}^3$, the special orthogonal group $\mathrm{SO}(3, \mathbb{R})$ (see Exercise 1.4.2), contains a free subgroup with two generators. For example, if we allow only translations $T_{\vec{p}}: \vec{v} \mapsto \vec{p} + \vec{v}$ instead of rotations, we won't be able to double the sphere (by Corollary 5.8.36 below). Thus he defined a class of groups (he called these groups *measurable*, now these groups are called *amenable*) which cannot be used to double sets on which these groups act. In this section, we give several definitions of amenable groups, examples of amenable and non-amenable groups and some properties of amenable groups. For more information see books [117, 329] or [82].

5.8.1 The Free Groups of Orthogonal Matrices and the Hausdorff–Banach–Tarski Paradox

5.8.1.1 A Free Subgroup of SO(3, ℝ)

Exercise 1.8.33 shows that the group $SL(2, \mathbb{R})$ contains an isomorphic copy of the free group F_2 with 2 generators. Now we show that $SO(3, \mathbb{R})$ also contains a copy of F_2.

Exercise 5.8.1 (Requires some knowledge of Linear Algebra). (1) Show that $SO(2, \mathbb{R})$ is commutative and so F_2 is not a subgroup of $SO(2, \mathbb{R})$.

(2) Show that every matrix $A \in SO(3, \mathbb{R})$ is similar to an (orthogonal) matrix of the form

$$\begin{pmatrix} \cos(\phi) & -\sin(\phi) & 0 \\ \sin(\phi) & \cos(\phi) & 0 \\ 0 & 0 & 1 \end{pmatrix}$$

for some ϕ. That is, A is the matrix of a rotation of $\mathbb{R}^3$ about a line containing $(0, 0, 0)$ through the angle ϕ. **Hint:** Show that A has a real eigenvalue (since the degree of the characteristic polynomial is odd). Consider an orthonormal basis $\mathcal{B}$ of $\mathbb{R}^3$ containing an eigenvector of A and the matrix B of the linear transformation $\vec{v} \mapsto A\vec{v}$ in that basis. Show that B is also an orthogonal matrix (since the transition matrix from the standard basis of $\mathbb{R}^3$ to $\mathcal{B}$ is orthogonal).

An easy proof of the fact that $SO(3, \mathbb{R})$ contains an isomorphic copy of F_2 was found by Świerczkowski [311].

Exercise 5.8.2 (see [311]). Show that the subgroup of $SO(3, \mathbb{R})$ generated by the two orthogonal matrices

$$A = \begin{pmatrix} \frac{1}{3} & -\frac{2\sqrt{2}}{3} & 0 \\ \frac{2\sqrt{2}}{3} & \frac{1}{3} & 0 \\ 0 & 0 & 1 \end{pmatrix}, B = \begin{pmatrix} 1 & 0 & 0 \\ 0 & \frac{1}{3} & -\frac{2\sqrt{2}}{3} \\ 0 & \frac{2\sqrt{2}}{3} & \frac{1}{3} \end{pmatrix}$$

is isomorphic to F_2. **Hint:** Let w be a reduced ... word in A, B. You need to show that $w(A, B)$ is not the identity matrix. Clearly if w is of length 1, the matrix $w(A, B)$ is not the identity. Let $|w| = k \geq 2$. Show that it is enough to consider the case when w starts with A or B. Consider the first column of the matrix $w(A, B)$, i.e., the vector

$$w(A, B) \begin{pmatrix} 1 \\ 0 \\ 0 \end{pmatrix}$$

and prove by induction that it has the form $(\frac{a}{3^k}, \frac{b\sqrt{2}}{3^k}, \frac{c}{3^k})$ for some integers a, b, c, moreover $b \not\equiv 0 \mod 3$ (hence the matrix $w(A, B)$ is not the identity).

To prove the second of this assertions, you need to represent the word w as uv where $u \equiv xy$ is of length 2 (here $x \in \{A, B\}, y \in \{A^{\pm 1}, B^{\pm 1}\}$), assume that the assertion is true for the word yv, and then consider all possibilities for u $(AA, AB, AB^{-1}, BA, BB, BA^{-1})$.

5.8.1.2 The Hausdorff–Banach–Tarski Paradox

Consider a sphere S^2 of radius $r > 0$ in $\mathbb{R}^3$. Since every orthogonal matrix A is the matrix of a rotation ϕ_A of $\mathbb{R}^3$ about some line l_A (see Exercise 5.8.1, Part (2)), the group $\mathrm{SO}(3, \mathbb{R})$ acts on S^2. So we will identify the group of matrices $\mathrm{SO}(3, \mathbb{R})$ with the group of rotations of S^2. Each rotation $C \in \mathrm{SO}(3, \mathbb{R})$ fixes exactly two points p_C, q_C on S^2 – these are the intersection points of the line l_C with S^2.

The Hausdorff–Banach–Tarski paradox is contained in the following

Theorem 5.8.3. *The sphere S^2 contains four disjoint subsets $U_1, \ldots, U_4$, such that by rotating these subsets we can compose two spheres of the same radius.*

To prove Theorem 5.8.3. take any free subgroup $\mathrm{gp}\langle A, B \rangle$ in $\mathrm{SO}(3, \mathbb{R})$, say, the one from Exercise 5.8.2. We shall denote this subgroup $\hat{F}_2$. Since $\hat{F}_2$ is free, its elements are both rotations of S^2 and words in A, B, A^{-1}, B^{-1}. Thus we shall refer to elements of $\hat{F}_2$ as rotations and as words.

Let us start with cutting the group $\hat{F}_2$ into disjoint pieces.

Definition 5.8.4. We say that a group G and its subset X are *equidecomposable of degree k* if X is a disjoint union of k subsets $X_1, X_2, \ldots, X_k$ and there exist k elements $g_1, \ldots, g_k \in G$ (called the *translating elements*), such that G is the union of translations $X_1 g_1, \ldots, X_k g_k$.

Remark 5.8.5. Note that since $Gg = G$ for every $g \in G$, if $\{g_1, \ldots, g_k\}$ is a set of translating elements, then $\{g_1 g, \ldots, g_k g\}$ is also a set of translating elements. Therefore we can always assume that the set of translating elements contains 1.

Notice that $\hat{F}_2$ contains 2 disjoint subsets W_a and W_b where W_a consists of reduced words that end with $A^{\pm 1}$ and W_b consists of reduced words that end with $B^{\pm 1}$. Moreover W_a is the union of disjoint subsets W_A (words that end with A) and $W_{A^{-1}}$ (words that end with A^{-1}. Similarly W_b is the union of two disjoint subsets W_B and $W_{B^{-1}}$. Note also that $\hat{F}_2$ is the disjoint union of subsets W_A and $W_{A^{-1}}A$, and also the disjoint union of W_B and $W_{B^{-1}}B$. Thus $\hat{F}_2$ is equidecomposable of degree 2 with W_a (translating elements 1 and A) and with W_b (translating elements 1 and B).

Now let us cut S^2. Let U be the set of points $p_C, q_C, C \in \hat{F}_2$.

Exercise 5.8.6. Show that $\hat{U} \cdot \hat{F}_2 = U$.

Note that U is a countable set because $\hat{F}_2$ is countable. Therefore there exists a point $z \in S^2$ which does not belong to U. Moreover S^2 is the union of

U and all orbits $z \cdot \hat{F}_2$ for $z \notin U$. Consider the set $\mathcal{Y}$ of all orbits $z \cdot \hat{F}_2$, $z \notin U$. Two distinct orbits in $\mathcal{Y}$ do not intersect (why?), hence S^2 is a disjoint union of U and orbits from $\mathcal{Y}$. Pick one point z in each orbit $Y \in \mathcal{Y}$. Let Z be the set of these points z. Note that then S^2 is a disjoint union of U and $Z \cdot F_2$.

Note that S^2 is the disjoint union of U, Z, $Z \cdot W_A$, $W_{A^{-1}} \cdot Z$, $W_B \cdot Z$ and $W_{B^{-1}} \cdot Z$. Then $S^2 \smallsetminus U$ is a disjoint union of $\bar{W}_A = Z \cdot W_A$, $\bar{W}_B = Z \cdot W_B$, $\bar{W}_{A^{-1}} = Z \cdot W_{A^{-1}}$, $\bar{W}_{B^{-1}} = Z \cdot W_{B^{-1}}$ and Z, and also a union of $Z \cdot W_A$, $Z \cdot W_{A^{-1}} A$ and of $Z \cdot W_B$, $Z \cdot W_{B^{-1}} B$ (why?). Thus the sphere S^2 without a countable set of points is equidecomposable of degree 2 with two disjoint subsets.

It remains to take care of the countable set U. We shall increase the sets $\bar{W}_A, \bar{W}_B, \bar{W}_{A^{-1}}, \bar{W}_{B^{-1}}$ by adding some points from U.

We have already cut every orbit of $\hat{F}_2$ that is not contained in U. Let us cut orbits which are contained in U. Let $x \in U$, and consider the orbit $\mathcal{O} = x \cdot \hat{F}_2$. Let w be a shortest word from $\hat{F}_2$ fixing an element in $\mathcal{O}$. We can assume that x is a point of $\mathcal{O}$ fixed by w.

The point x is on the axis ℓ of rotation w. Consider the stabilizer S_x of x in $\hat{F}_2$, i.e., the subgroup of all rotations in $\hat{F}_2$ that fix x. Let L be the big circle of the sphere S^2 that is perpendicular to ℓ. All rotations from S_x take L to L (why?) and different rotations from S_x induce different rotations of L. Therefore S_x is isomorphic to a subgroup of the group of rotations of the circle L. That group is commutative (we have used this fact in the second proof of Theorem 1.2.9). Thus S_x is a commutative subgroup of the free group $\hat{F}_2$. Let $v \in S_x$. Then $wv = vw$ in the free group $\hat{F}_2$. By Theorem 1.8.29 then w and v are powers of some other word $u \in \hat{F}_2$: $w = u^m, v = u^n$ for some m, n. Since the length of w is minimal possible, we have $n \geq m$. Therefore $n = km + r$ for some $0 \leq r < m$. Since both $x \cdot w = x, x \cdot v = x$ we have $x \cdot u^r = x$ which can happen only if $r = 0$ (again because of the minimality of w). Therefore v is a power of w. Thus S_x is the cyclic group generated by w.

Let $C \in \{A, B, A^{-1}, B^{-1}\}$ be the last letter of w. Note that w cannot start with C^{-1} because otherwise CwC^{-1} would be a shorter word fixing a point in $\mathcal{O}$, namely the point $x \cdot C^{-1}$ (why?).

For every $y \in x \cdot \hat{F}_2$ let us pick a nice word v_y such that $x \cdot v_y = y$. By "nice" we shall mean a word that does not start with C^{-1} or with w. Let v be a shortest word from $\hat{F}_2$ such that $y = x \cdot v$. Then v cannot start with w (why?). If v does not start with C^{-1}, then let $v_y \equiv v$. If v starts with C^{-1}, then let $v_y = wv$ (i.e., v is the reduced word that is freely equal to the concatenation wv). In both cases v_y starts neither with C^{-1} nor with w. Then v_y is determined uniquely by the point y. Indeed if $y = x \cdot v_1, y = x \cdot v_2$, $v_1 \neq v_2$ and both v_1 and v_2 are nice, then $x = x \cdot v_1 v_2^{-1}$. Hence $v_1 v_2^{-1} \in S_x$. Therefore $v_1 v_2^{-1} = w^n$ for some $n \neq 0$. We can assume that $n > 0$ (why?). Since $v_1 = w^n v_2$ and v_2 does not start with C^{-1}, the word $w^n v_2$ starts with w. This contradicts the assumption that v_1 is nice.

Now for every point $y = x \cdot v_y$ from $\mathcal{O}$, we add y to W_D (where $D \in \{A, B, A^{-1}, B^{-1}\}$) provided $v_y \in D$. The new sets will be denoted by $\tilde{W}_D$.

Exercise 5.8.7. Show that the sets $\tilde{W}_D$ are pairwise disjoint and $S^2 = \tilde{W}_A \cup \tilde{W}_{A^{-1}} \cdot A = \tilde{W}_B \cup \tilde{W}_{B^{-1}} \cdot B$.

□

In Section 5.8.1.2, we essentially proved the following.

Exercise 5.8.8. Suppose that G is a group which is equidecomposable with two of its disjoint subsets A, B. Suppose that G acts (on the right) on a set X so that $g \cdot x \neq x$ for every $x \in X$ (i.e., the action is *free*). Prove that then there exist disjoint subsets $X_1, \ldots, X_n, Y_1, \ldots Y_{k-n}$ of X ($n < k$) and k elements $g_1, \ldots g_n, h_1, \ldots, h_{k-n}$ of G such that $X = \bigcup X_i \cdot g_i = \bigcup Y_i \cdot h_i$. Thus we can double X by cutting it into a number of pieces and rearranging the pieces using elements of the group G. **Hint:** Replace in the proof of Theorem 5.8.3, $\hat{F}_2$ by G, $S^2 \setminus Z$ by X, and sets W_a, W_b by the two disjoint subsets of G which are equidecomposable with G.

5.8.1.3 The Full Formulation of the Banach–Tarski Paradox

Encouraged by the success of doubling a sphere, Banach and Tarski started cutting other things (see [21, 329]).

Their most impressive achievement is the following remarkable theorem

Theorem 5.8.9 (See Wagoner[329]). *Let M, M' be two bounded subsets of $\mathbb{R}^3$, each containing an open ball. Then one can cut M into a finite number of pieces, and rearrange the pieces using rotations and translations to obtain M'.*

Thus we can cut a ball from $\mathbb{R}^3$ into a finite number of pieces, rearrange the pieces and obtain two balls of the same radius. We can cut a chicken into a finite number of pieces, rearrange the pieces and obtain a dog. Or 101 dogs. Or a cow. The opportunities are endless.

Note that the main cause of all that madness is the trivial syntactic fact that two freely reduced words that end with different letters represent different elements in the free group.

Another cause, first noticed by von Neumann (see [328, 329]) is the group of transformations that we use to rearrange pieces of the partition. For example, the group of translations (consisting of maps $\phi_{\vec{a}} : \vec{v} \to \vec{v} + \vec{a}$) is commutative, hence amenable (Corollary 5.8.36 below). Therefore if we use translations instead of rotations, we would not be able to double a sphere.

And yet another cause is that we allow ourselves to consider a set Z that has exactly one element in common with each orbit $z \cdot F_2$. The ability to use this seemingly innocent operation is in fact equivalent to the Axiom of Choice (for collections of subsets of, say, $\mathbb{R}$), and is independent from the usual Zermelo–Fraenkel axioms of set theory [329]. In truth, Zorn's Lemma 1.1.7 is also equivalent to the Axiom of Choice. Thus if we declare Axiom of Choice false in order to not allow turning chicken into cows, we would also loose the

Zorn lemma, which would be a big loss for the whole mathematics (say, we have used Zorn's lemma in this book: Theorem 1.6.2, Proposition 4.4.25).

5.8.2 The First Two Definitions of Amenability

5.8.2.1 Paradoxical Decompositions and Invariant Means

Probably the easiest way to define amenable groups is by using equidecomposable sets (see Definition 5.8.4).

Definition 5.8.10. A group G is called *amenable* if G does not contain two disjoint subsets each of which is equidecomposable with G.

Definition 5.8.11. Suppose that a group G is not amenable. Then the smallest number k such that G contains two disjoint subsets X, Y such that G and X are equidecomposable of degree $n < k$, G and Y are equidecomposable of degree $k - n$ (for some n) is called the *Tarski number* of the group G.

Thus in Section 5.8.1.2, we showed that the free group with two generators is not amenable and its Tarski number is at most 4.

Exercise 5.8.12. Show that no group has Tarski number ≤ 3, so the Tarski number of the free group F_2 is equal to 4.

The following theorem gives many more examples of non-amenable groups.

Theorem 5.8.13. *Let H be a non-amenable subgroup of a group G. Then G is not amenable. Moreover, the Tarski number of G does not exceed the Tarski number of H.*

Proof. Indeed, suppose that H is equidecomposable with its disjoint subsets A, B, that is A is a disjoint union of sets $A_1, \ldots, A_n$, B is a disjoint union of sets $B_1, \ldots, B_{k-n}$ so that $H = \bigcup_{i=1}^{n} A_i g_i = \bigcup_{j=1}^{k-n} B_j h_j$. The group G is a disjoint union of left cosets $gH, g \in X$ (X is a set containing one representative of each coset of H). Then G is equidecomposable with $\bar{A} = \bigcup_{g \in X} Ag$ and with $\bar{B} = \bigcup_{g \in X} Bg$. Indeed, $\bar{A}$ is a disjoint union of subsets $\bar{A}_i = \bigcup_{g \in X} A_i g$, $i = 1, \ldots, n$, $\bar{B}$ is a disjoint union of subsets $\bar{B}_i = \bigcup_{g \in X} B_i g$, and

$$G = \bigcup_{i=1}^{n} \bar{A}_i g_i = \bigcup_{j=1}^{k-n} \bar{B}_j h_j.$$

Hence if k is the Tarski number of H, then the Tarski number of G does not exceed k.   □

Corollary 5.8.14. *Every group containing a copy of the free group F_2 is non-amenable with Tarski number 4.*

The next theorem is in some sense a converse of Theorem 5.8.13.

Theorem 5.8.15. *Suppose that a group G is non-amenable with Tarski number k. Then G contains a k-generated non-amenable subgroup.*

Proof. Let G contain a disjoint union of subsets $X_1, X_2, \ldots, X_n,\ Y_1, Y_2 \ldots Y_{k-n}$ and $g_1, g_2, \ldots, g_n,\ h_1, h_2, \ldots, h_{k-n}$ be the translating elements from G such that $G = \bigcup X_i g_i = \bigcup Y_j h_j$. Let H be the k-generated subgroup of G generated by $g_1, g_2, \ldots, g_n,\ h_1, h_2, \ldots, h_{k-n}$. Then H is equidecomposable with its two disjoint subsets $H \cap \bigcup X_i$ and $H \cap \bigcup Y_j$ (why?). $\square$

Another easy consequence of the definition of amenability in terms of decompositions is the following

Theorem 5.8.16. *If G is an amenable group, then any homomorphic image $\bar{G}$ of G is amenable. Moreover, if $\bar{G}$ is non-amenable with Tarski number k, then the Tarski number of G does not exceed k.*

Proof. Indeed, let G be an amenable group, $\phi: G \to H$ be a surjective homomorphism. Suppose that H is not amenable and A, B are disjoint subsets of H each of which is equidecomposable with H. Then the preimages $\phi^{-1}(A)$, $\phi^{-1}(B)$ are subsets of G each of which is equidecomposable with G (prove that!). $\square$

One of the main properties of non-amenable groups is that we cannot assign weights to its subsets in an invariant way so that the weight of the whole group is finite. Indeed, then the weight of each of the two disjoint subsets which are equidecomposable with the whole group should be at least as large as the weight of the whole group which is impossible if the weight satisfies some obviously natural assumptions. More precisely,

Definition 5.8.17. Let G be a group. An *invariant mean* on G is a function μ that assigns a nonnegative number to each subset of G so that

1. $\mu(G) = 1$,
2. If A and B are disjoint subsets of G, then $\mu(A \cup B) = \mu(A) + \mu(B)$.
3. For every $A \subseteq G, g \in G$, we have $\mu(Ag) = \mu(A)$.

Remark 5.8.18. It is not necessary to assume that $\mu(G) = 1$ in Definition 5.8.17. It is enough to assume that $0 < \mu(G) < \infty$. Indeed, if $\mu(G) = a > 0$, and conditions (2) and (3) are satisfied, then $\frac{1}{a}\mu$ satisfies all three conditions of Definition 5.8.17 (prove it!).

Note that an invariant mean satisfies conditions (1) and (2) of a good measure (see Definition 3.9.12) but it is additive only for finite unions of disjoint subsets.

Thus if G has an invariant mean, then G is amenable. Tarski showed that the converse is true too.

Theorem 5.8.19 (Tarski, for the proof see [329]). *A group is amenable if and only if it has an invariant mean.*

5.8.3 Følner Sets

Every finite group is amenable. Indeed, if G is finite, we can assign to each subset $X \subseteq G$ the number $\mu(X) = \frac{|X|}{|G|}$.

Exercise 5.8.20. Show that this μ is an invariant mean on G.

It is less trivial to prove the following

Theorem 5.8.21. *The cyclic group $\mathbb{Z}$ is amenable.*

Proof. Indeed, suppose that $\mathbb{Z}$ has disjoint subsets $X_1, \ldots, X_k$ and elements $g_1, \ldots, g_k$ such that $\mathbb{Z} = \bigcup_{i=1}^{n}(g_i + X_i) = \bigcup_{i=n+1}^{k}(g_i + X_i)$ for some $n < k$. Consider a very long interval of integers $F = \{1, 2, \ldots, L\}$. If L is very large, then for a very small $\epsilon > 0$ and all but at most ϵL numbers $p \in F$, we have $-g_i + p \in F$ for all i. The larger L the closer ϵ is to 0. Therefore the numbers $|(g_i + X_i) \cap F|$ and $|X_i \cap F|$ differ by at most $2\epsilon L$. Therefore the numbers $L = |F| = |\bigcup_{i=1}^{n}((g_i + X_i) \cap F)|$ and $|\bigcup_{i=1}^{n}(X_i \cap F)|$ differ by at most $2n\epsilon L$. Similarly the numbers $L = |F| = |\bigcup_{i=n+1}^{k}((g_i + X_i) \cap F)|$ and $|\bigcup_{i=n+1}^{k}(X_i \cap F)|$ differ by at most $2(k-n)\epsilon L$. Since F contains the disjoint union of $X_i \cap F$ ($i = 1, \ldots, k$), we get that $2L$ differs from L by at most $2k\epsilon L$ which is impossible for small enough ϵ. $\qquad\square$

Remark 5.8.22. To appreciate Tarski's Theorem 5.8.19, try finding an invariant mean on $\mathbb{Z}$. The reader is referred to a discussion of this on Mathoverflow [223].

One can generalize the proof of Theorem 5.8.21 to other groups.

Theorem 5.8.23 (Følner condition). *Suppose that G is a group and the following property holds.*

($\mathcal{F}$) *For every finite subset $U \subseteq G$ and every $\epsilon > 0$ there exists a finite set $F = F(\epsilon, U)$ (called a Følner set) such that*

$$|FU \smallsetminus F| < \epsilon|F|.$$

Then the group G is amenable.

Exercise 5.8.24. Prove Theorem 5.8.23. **Hint:** Follow the proof of Theorem 5.8.21.

5.8.3.1 The Marriage Lemma for Polygamists

We shall need the following classical combinatorial statement. Suppose that we have m males and a number of females. Every male likes some females and does not like others. Suppose that we want to arrange marriages so that each male marries k females which he likes and none of the females marries two different males. The next lemma by Ph. Hall gives a solution. This lemma is one of the most important statements in combinatorics and has numerous applications in many areas of mathematics (see [9, 273, 287]).

Lemma 5.8.25 (Hall's lemma). *Let $C = \{A_1, \ldots, A_m\}$ be a collection of subsets of a finite set $A = \{a_1, \ldots, a_n\}$, $k \geq 1$ be a natural number. Suppose that for every $J \subseteq \{1, \ldots, m\}$, the number of elements in $\bigcup_{i \in J} A_i$ is at least $k|J|$. Then one can choose a k-element subset B_i in each A_i such that $B_i \cap B_j = \varnothing$ if $i \neq j$.*

Remark 5.8.26. Clearly the condition of the lemma is also necessary.

Rado's proof [273]. Induction on k. Let $k = 1$. Suppose that C satisfies the conditions of the lemma. Clearly each A_i contains at least one element. If each A_i contains one elements, then $A_i \cap A_j = \varnothing$ for every $i \neq j$ and we are done. So suppose that, say, A_1 contains a_1, a_2. Let $B_1 = A_1 \setminus \{a_1\}$, $B_2 = A_1 \setminus \{a_2\}$. We show that either $\{B_1, A_2, \ldots, A_m\}$ or $\{B_2, A_2, \ldots, A_m\}$ satisfies the condition of the lemma (with $k = 1$). Indeed, otherwise there exist subsets $J, K \subseteq \{2, \ldots, m\}$ such that

$$|J| \geq |B_1 \cup A_J|,$$

$$|K| \geq |B_2 \cup A_K|.$$

where $A_J = \bigcup_{i \in J} A_i$. Let $S = B_1 \cup A_J, T = B_2 \cup A_K$. Then

$$S \cup T = A_1 \cup A_{J \cup K}.$$

By the assumptions of the lemma, we get

$$|S \cup T| \geq 1 + |J \cup K|.$$

Now consider $S \cap T$. We have

$$S \cap T \supseteq A_J \cap A_K \supseteq A_{J \cap K}.$$

Therefore (by the condition of the lemma)

$$|S \cap T| \geq |J \cap K|.$$

Combining the inequalities obtained above, yields

$$|J| + |K| \geq |S| + |T| = |S \cup T| + |S \cap T| \geq (1 + |J \cup K|) + |J \cap K| = 1 + |J| + |K|,$$

a contradiction. The proof can be finished by induction on the number $\sum_{i=1}^{m} |A_i|$.

Now suppose that $k > 1$. Since the assumption of the lemma for $k = 1$ also holds, we can choose one element b_i in each A_i, $b_i \neq b_j$ if $i \neq j$. Consider the sets $A_i' = A_i \setminus \{b_i\}$. These sets satisfy the assumptions of the lemma for $k - 1$

(why)? By induction on k we can choose a $k - 1$-element subset B_i' in each A_i' so that $B_i' \cap B_j' = \emptyset$ if $i \neq j$. It remains to take $B_i = B_i' \cup \{b_i\}$. □

Here is a useful infinite version of Hall's lemma.

Lemma 5.8.27. *Let X be a countable set, and $A_i, i \in \mathbb{N}$, be a collection of finite subsets of X and $k \in \mathbb{N}$. Suppose that for every finite subset $I \subseteq \mathbb{N}$, we have*

$$\left| \bigcup_{i \in I} S_i \right| \geq k|I|.$$

Then one can choose a k-element subset B_i in each A_i such that $B_i \cap B_j = \emptyset$ if $i \neq j$.

Proof. Notice that for every finite subset $I \subseteq \mathbb{N}$ the collection of subsets $A_i, i \in I$ of the finite set $\bigcup_{i \in I} A_i$ satisfies the conditions of Hall's Lemma 5.8.25. Notice also that if $I \subseteq J$, then every "proper" (i.e., as in Lemma 5.8.25) choice of subsets $B_i \subseteq A_i, i \in J$, gives us a "proper" choice of subsets $B_i \subseteq A_i, i \in I$. Since given a finite $I \subset \mathbb{N}$, there are only finitely many "proper" choices of subsets $B_i \subseteq A_i, i \in I$, we can use a compactness argument (Remark 3.3.10) similar to the one used in Section 3.3.1 □

Exercise 5.8.28. Complete the proof of Lemma 5.8.27.

5.8.3.2 Amenable Groups Satisfy the Følner Condition ($\mathcal{F}$)

The following theorem together with Theorem 5.8.23 shows that property ($\mathcal{F}$) is equivalent to amenability.

Theorem 5.8.29. *If a group G is amenable, then it satisfies Følner's property ($\mathcal{F}$) from Theorem 5.8.23.*

Proof. Indeed, suppose that ($\mathcal{F}$) does not hold. Then there exists a finite subset $U \subseteq G$ and a number ϵ such that for every finite subset $F \subseteq G$, $|UF \smallsetminus F| \geq \epsilon|F|$, that is

$$|F \cup FU| = |FU \smallsetminus F| + |F| \geq (1 + \epsilon)|F|. \tag{5.8.1}$$

We can always replace U by a bigger finite set (why?), so we will assume that U contains 1. Thus $F \cup FU = FU$

Using FU instead of F, we get

$$|FU^2| \geq (1 + \epsilon)|FU| \geq (1 + \epsilon)^2|F|.$$

By induction, $|FU^n| \geq (1 + \epsilon)^n|F|$. Since $\epsilon > 0$, there exists n such that $(1 + \epsilon)^n > 2$. Let V be the set U^n, that is V consists of all elements of G representable by words of length $\leq n$ in the alphabet U. Then

$$|FV| \geq 2|F| \tag{5.8.2}$$

for every finite set $F \subseteq G$.

Consider the subgroup H of G generated by V. Note that H is finitely generated, hence countable. Now for every $x \in H$ consider the finite set $S_x = xV$. Property (5.8.2) implies that the collection of sets $S_x, x \in H$, satisfies the condition of Lemma 5.8.27 for $k = 2$. Therefore we can "marry" each $x \in H$ with two elements $a_x = xg_x, b_x = xh_x$ from S_x, where $g_x, h_x \in V$ so that the sets $\{a_x, b_x\}$ and $\{a_y, b_y\}$ are disjoint provided $x \neq y$.[13] Now let $A = \{a_x, x \in H\}$, $B = \{b_x, x \in H\}$. The subsets A and B of H are disjoint (why?). Let us show that both A and B are equidecomposable with H. Indeed for every $g \in V$ let $A_g \subseteq A$ be the set of all a_x such that $ga_x = x$. Then

(1) The sets A_g are disjoint and cover the whole A.
(2) The set $\bigcup_{g \in V} A_g g$ coincides with H.

Exercise 5.8.30. Prove (1) and (2).

Thus A is equidecomposable with H. A similar argument proves that B is also equidecomposable with H. Hence H is not amenable. Therefore G is also not amenable by Theorem 5.8.13.

Using Følner sets, one can prove amenability of many groups. We shall give two such statements.

5.8.3.3 Amenability and Group Extensions

Theorem 5.8.31. *Suppose that a group G contains an amenable normal subgroup N such that G/N is amenable. Then G is amenable.*

Proof. Indeed, pick a finite subset $U = \{u_1, \ldots, u_m\} \subseteq G$ and a number $\epsilon > 0$. Since G/N is amenable, there exists a subset $F = \{Ng_1, \ldots, Ng_n\}$ such that $|FU \setminus F| < \epsilon n$. This means that "for almost every" $i \in \{1, \ldots, m\}, j \in \{1, \ldots, n\}$, $Ng_j u_i$ is in F, that is there exists $s \in \{1, \ldots, n\}$ and $n_{i,j} \in N$ such that $g_j u_i = n_{i,j} g_s$. Let $U' \subseteq N$ be the (finite) set of all $n_{i,j}$. By Theorem 5.8.23 there exists a finite subset $F' \subseteq N$ such that $|F'U' \setminus F'| < \epsilon |F'|$. Let $F'' \subseteq G$ be the set $\bigcup_{i=1}^{n} F'g_i$.

Exercise 5.8.32. Show that $|F''U \setminus F''| < \mu|F''|$ for some $\mu = \mu(\epsilon)$ such that $\lim_{\epsilon \to 0} \mu = 0$.

$\square$

Exercise 5.8.33. Suppose that a group G has an amenable subgroup H of finite index. Prove that G is amenable. **Hint:** The group G acts on the finite set X of right cosets of H (we have considered this action in Section 1.8.10): $Hx \cdot g = Hxg$. This gives a homomorphism from G to the group of permutations of X. The group G is an extension of the kernel of that homomorphism, which is amenable, by a finite group which is also amenable by Exercise 5.8.20.

---

[13] Note that in this infinite "marriage ceremony", it is not clear how to distinguish males from females. We hope, though, that the reader is progressive enough and will not be bothered by this.

Exercise 5.8.34. Let G be an amenable group, H be a subgroup of G which is isomorphic to G, and let ϕ be an isomorphism $G \to H$. Show that the HNN extension $\text{HNN}_\phi(G)$ is amenable. **Hint:** Consider the homomorphism γ from $\text{HNN}_\phi(G)$ to the cyclic group $\mathbb{Z}$ that sends G to $\{0\}$ and the free letter t to 1. Prove that the kernel N of γ is an increasing union of subgroups $t^n G t^{-n}$ which are isomorphic to G. Using Theorem 5.8.15, prove that N is amenable. Then apply Theorem 5.8.31.

5.8.3.4 Growth and Amenability

Theorem 5.8.35. *Suppose that a finitely generated group G has subexponential growth function. Then G is amenable.*

Proof. Let U be a finite subset of G. Then the subgroup H generated by U also has a subexponential growth by Exercise 1.5.1 (indeed, consider a finite generating set of G containing a generating set of H). Without loss of generality we can assume that U is symmetric, that is $U^{-1} = U$. Pick a number $\epsilon > 0$. Let B_n be the set of all elements in H represented by words of length $\leq n$ in U. Suppose that for every n we have $|B_n U \smallsetminus B_n| > \epsilon |B_n|$. Then $|B_{n+1}| = |B_n| + |U B_n \smallsetminus B_n| > (1 + \epsilon)|B_n|$ for every n. Therefore the growth function of H with respect to the generating set U is at least $(1 + \epsilon)^n$, hence exponential, a contradiction. $\square$

Theorems 5.8.31, 5.8.35 apply to many groups. For example, using Theorems 5.7.25 and 5.7.31, we deduce the following

Corollary 5.8.36. *Every nilpotent group and the Grigorchuk group from Section 5.7.4 are amenable.*

Remark 5.8.37. One can construct a finitely presented amenable group containing the Grigorchuk group G. Indeed the substitution σ from Remark 5.7.39 induces an injective homomorphism $\phi: G \to G$ [212]. The corresponding HNN extension $\text{HNN}_\phi(G)$ is amenable by Exercise 5.8.34 and has a finite presentation $\text{gp}\langle a, b, c, d, t \mid a^2 = 1, b^2 = 1, c^2 = 1, d^2 = 1, bcd = 1, (ad)^4 = 1, (adacac)^4 = 1, a^t = aca, b^t = d, c^t = b, d^t = c \rangle$ (prove that!).

Corollary 5.8.38. *Every solvable group is amenable.*

Proof. Indeed, by definition, a solvable group G has a chain of normal subgroups $G_0 = \{1\} < G_1 < \ldots < G_m = G$ such that G_{i+1}/G_i is commutative for every $i = 0, \ldots, m$. Since each G_{i+1}/G_i are amenable by Corollary 5.8.36, G is amenable by Theorem 5.8.31. $\square$

Note that the growth function of any solvable group which does not have nilpotent subgroups of finite index is exponential by a result of Wolf [332]. This is illustrated by the following exercise.

Exercise 5.8.39. Show that the groups $\mathrm{BS}_{1,2}$ from Exercise 5.7.22 and $\mathbb{Z} \wr \mathbb{Z}$ from Exercise 1.8.26 are solvable but do not have polynomial growth functions. In particular, these groups do not have nilpotent subgroups of finite index. **Hint:** In the group $\mathrm{BS}_{1,2} = \mathrm{gp}\langle\, a, b \mid b^{-1}ab = a^2 \,\rangle$ the elements a, ba generate a free subsemigroup. To prove that consider any positive word $w(x,y)$ and using the relation $ab = ba^2$ transform $w(a, ba)$ into a word of the form $b^k a^l$. Show that if $w \not\equiv w'$, then $w(a, ab)$ and $w'(a, ab)$ are different elements of $\mathrm{BS}_{1,2}$. For this use a homomorphism $\phi \colon a \mapsto \begin{pmatrix} 1 & 0 \\ 0 & 2 \end{pmatrix}, b \mapsto \begin{pmatrix} 1 & 1 \\ 0 & 1 \end{pmatrix}$ from $\mathrm{BS}_{1,2}$ into $\mathrm{GL}(2, \mathbb{R})$. For $\mathbb{Z} \wr \mathbb{Z}$, use the fact from Exercise 1.8.26 that it contains free commutative groups of arbitrary rank, and Exercise 5.7.2, or find a free subsemigroup of rank 2.

Therefore there exist amenable groups with exponential growth functions.

5.8.4 Groups with Tarski Number 4

The following Theorem was proved by Jónsson and Dekker (see [329] or [82]).

Theorem 5.8.40. *A group G has Tarski number 4 if and only if it contains a copy of the free group F_2.*

Proof. The "if" statement is Corollary 5.8.14. Suppose that the Tarski number of G is 4. Then there exist four disjoint subsets A_1, A_2, B_1, B_2 and two 2-element sets of translating elements $\{1, g\}$, $\{1, h\}$: the first set translates subsets A_1, A_2, the second set translates B_1, B_2 (recall that by Remark 5.8.5 we can always assume that each of the two sets of translating elements contains 1). We will use the ping-pong Theorem 1.8.32 to prove that the subgroup generated by g, h is isomorphic to the free group F_2. The parts of the ping-pong table are $A_1 \cup A_2$, $B_1 \cup B_2$. We have

$$G = A_1 \cup A_2 g_1.$$

Therefore

$$A_1 \supseteq G \smallsetminus A_2 g \supseteq (A_1 \cup A_2 \cup B_1 \cup B_2)g \smallsetminus A_2 g = (A_1 \cup B_1 \cup B_2)g.$$

Hence $A_1 \supset A_1 g^m \supset (B_1 \cup B_2)g^{m+1}$ for every $m \geq 0$.

Exercise 5.8.41. Show that $(B_1 \cup B_2)g^m \subseteq A_2$ for every $m < 0$, and that $(A_1 \cup A_2)h^m \subseteq B_1 \cup B_2$ for every integer $m \neq 0$. Thus, indeed, the sets $A_1 \cup A_2$ and $B_1 \cup B_2$ form a ping-pong table for g, h and by Theorem 1.8.32, the subgroup $\langle g, h \rangle$ is free.

□

5.8.5 Co-growth and the von Neumann–Day Conjecture

The goal of this subsection is to prove that existence of free non-cyclic subgroups is not equivalent to non-amenability.

5.8.5.1 Co-growth and Amenability

The following characterization of amenable groups was found by Kesten [175] in terms of *random walks*. Its purely syntactic reformulation was announced by Grigorchuk in[121]. For a full proof in English see [120], different and easier proofs can be found in [69, 312, 331].

Let N be a normal subgroup of a free group F_X, $|X| = k \geq 1$. Let L_N be the language of all (not necessarily reduced) words in $X \cup X^{-1}$ representing elements in N. Let $s(n)$ be the spherical growth function of L_N. For any word u in the alphabet $X \cup X^{-1}$ the word uu^{-1} is in L_N since N contains 1. This implies the inequality

$$s(n) \geq (2k)^{n/2}. \tag{5.8.3}$$

On the other hand, $s(n)$ does not exceed the number of all words in the alphabet $X \cup X^{-1}$ of length n. Therefore

$$s(n) \leq (2k)^n. \tag{5.8.4}$$

The *cogrowth in the sense of Kesten* $\mathcal{C}_K$ of N is the *exponential rate of growth* of the function $s(n)$. As in Section 3.7.1, it is defined as the following limit:

$$\mathcal{C}_K = \limsup_{n \to \infty} \sqrt[n]{s(n)}$$

Inequalities (5.8.3) and (5.8.4) show that

$$(2k)^{\frac{1}{2}} \leq \mathcal{C}_K \leq 2k.$$

Remark 5.8.42. It is often more convenient to consider the language L'_N of all reduced words in N. The exponential rate of growth of that language is called the *cogrowth in the sense of Grigorchuk* and will be denoted by $\mathcal{C}_G$. One can deduce from Exercise 1.8.28 that $(2k - 1)^{1/2} \leq \mathcal{C}_G \leq 2k - 1$ provided N is nontrivial.

If $\mathcal{P} = \mathrm{gp}\langle X \mid R \rangle$ is a nontrivial group presentation of a group G, X is finite, and N is the normal subgroup of the free group F_X generated (as a normal subgroup) by R, then the cogrowth of G (more precisely of the presentation $\mathcal{P}$) is, by definition, the cogrowth of N.

Now we can formulate

Theorem 5.8.43 (Kesten [175], Grigorchuk [120]). *A group $G = F_k/N$ is amenable if and only if its cogrowth in the sense of Kesten is maximal possible, i.e., $\mathcal{C}_K = 2k$, and, provided N is nontrivial, if and only if the cogrowth in the sense of Grigorchuk is maximal possible, i.e., $\mathcal{C}_G = 2k - 1$.*

We shall not prove this theorem, referring the reader to [69, 120]

Example 5.8.44. We know that the infinite cyclic group $G = \mathrm{gp}\langle a \rangle$ is amenable (Theorem 5.8.21). Let us show directly that G has the maximal possible cogrowth in the sense of Kesten for 1-generated groups, i.e., 2 (the co-growth in the sense of Grigorchuk is 0: the normal subgroup N in that case is $\{1\}$). Indeed, a word in $\{a, a^{-1}\}$ is equal to 1 in G if and only if it is of even length, say, $2n$, and contains exactly n occurrences of a and exactly n occurrences of a^{-1} (why?). Therefore the number of words of length m which are equal to 1 in G is 0 if m is odd. If $m = 2n$ is even, then the number of words of length m which are equal to 1 in G is the number of possibilities to place n occurrences of the letter a in a word of length $2n$. Thus it is equal to $\binom{2n}{n}$, the middle binomial coefficient in the expansion of $(x + y)^{2n}$. By Exercise 3.7.6

$$\binom{2n}{n} \geq \frac{4^n}{2n + 1}.$$

Therefore

$$\mathcal{C}_K = \limsup_{n \to \infty} \binom{2n}{n}^{\frac{1}{2n}} \geq \lim_{n \to \infty} \left(\frac{4^n}{2n + 1} \right)^{\frac{1}{2n}} = 2,$$

as required.

5.8.5.2 The von Neumann–Day Conjecture

Recall that von Neumann [328, 329] discovered that the cause of Hausdorff's paradox is a free non-cyclic subgroup in $SO(3, \mathbb{R})$, and introduced amenable groups. The next natural question to ask was whether every non-amenable group contains a free non-cyclic subgroup. That question was first formulated explicitly by Day in [79]. A negative answer was found by Olshanskii [257]. His example was a Tarski monster (see Section 5.9.1 below). Then Adian [4] proved that the free Burnside groups $B(m, n)$ of odd exponents ≥ 665 with at least 2 generators are non-amenable. Since every element of $B(m, n)$ has finite order, $B(m, n)$ cannot contain nontrivial free subgroups.

Below we shall explain some of the ideas of Adian's proof.

5.8.5.3 Nonamenability of Groups with Dehn Presentations

Adian's proof uses the fact that Burnside groups have nice infinite Dehn presentations (see Section 5.2.3.15). It starts with the following general result about groups admitting Dehn presentations.

Let $\mathcal{P} = \mathrm{gp}\langle X \mid R \rangle$ be a Dehn presentation of a group G, $X \cap X^{-1} = \varnothing$, $|X| = m$. We shall, as usual, assume that R is closed under cyclic shifts and taking inverses. Then by the definition of Dehn presentation every . . . reduced word w in the alphabet $X \cup X^{-1}$ that is equal to 1 in G contains more than a half of a relator $r \in R$. Thus $w \equiv p E_w q$, $r = E_w v_w \in R$ and $|E_w| > |v_w|$. We shall assume that E_w is the largest possible such subword in w. Let us denote the smallest difference $|E_w| - |v_w|$ (for all . . . w which are equal to 1 in G) by $\delta_\mathcal{P}$ and call it the *speed* of the Dehn presentation $\mathcal{P}$. The infimum of all quantities $\frac{|E_w| - |v_w|}{|r|}$ is called the *relative speed* of $\mathcal{P}$ and is denoted by $\gamma_\mathcal{P}$. We shall need also the number $\beta_\mathcal{P}$ which is the infimum of all numbers β such that the number of words of length n in R is $\le (2m - 1)^{\beta n}$ for every $n \ge 1$ (show that then $(2m - 1)^\beta$ is the exponential rate of growth of the language L by the definition from Section 5.8.5.1!).

Theorem 5.8.45 (Adian, [4]). *Suppose that a Dehn presentation $\mathcal{P} = \mathrm{gp}\langle X \mid R \rangle$ as above has speed δ, relative speed γ and the exponential rate of growth $(2m - 1)^\beta$. Let $\alpha = \frac{4}{\delta} \log_{2m-1} \left(e \left(1 + \frac{\delta}{4\gamma} \right) \right)$ where $e \approx 2.71828 \ldots$ is the base of the natural logarithm. Then the cogrowth in the sense of Grigorchuk of the group G defined by $\mathcal{P}$ does not exceed*

$$(2m - 1)^{\frac{1}{2} + \frac{\beta}{\gamma} + \alpha}.$$

Remark 5.8.46. Thus if $\frac{\beta}{\gamma}$ is small enough, and δ is big enough, then the cogrowth in the sense of Grigorchuk is smaller than $2m - 1$ and the group given by the presentation $\mathcal{P}$ is non-amenable by Theorem 5.8.43.

Proof of Theorem 5.8.45. We follow Adian's proof from [4]. Let w be any reduced word in the alphabet $X \cup X^{-1}$ that is equal to 1 in G, $|w| = x$. Then (by the definition of Dehn presentations), there exists a word $r \in R$ represented as $r \equiv E_1 v_1$ such that E_1 is the maximal subword of r which is contained in w, and $|E_1| - |v_1| \ge \delta$. Replacing E_1 by v_1^{-1} in w we get a word w_1, $|w| - |w_1| \ge \delta$. Since $w_1 = 1$ in G, we can continue the chain of words until we reach the empty word:

$$w \equiv w_0 \to w_1 \to \ldots \to w_\lambda \equiv \varnothing.$$

Then $\lambda \le \frac{x}{\delta}$. Let $E_1, v_1, \ldots, E_\lambda, v_\lambda$ be the corresponding parts of the relations from R. Note that by the maximality of E_i, if v_i is not empty, then after replacing E_i by v_i^{-1} in w_{i-1} we get a reduced word (prove it!). If v_i is empty, then after replacing E_i by 1, some cancelations may be necessary. Thus each transition from w_i to w_{i-1} can be described as follows: represent w_i as $p z_i v_i^{-1} \cdot z_i^{-1} \cdot q$ in the free group where $p z_i v_i^{-1}$ is a prefix of w_i, $z_i^{-1} q$ is a suffix of w_i, z_i is empty if v_i is not empty. Then

$$w_{i-1} \equiv p z_i E_i z_i^{-1} q.$$

We can also obtain w_{i-1} from w_i by inserting $z_i E_i v_i z_i^{-1}$ after p and then canceling v_i.

In order to minimize the number of possible substitutions, let us slightly modify our chain of words. Consider the sequence of substitutions $w'_\lambda \equiv \varnothing \to w'_{\lambda-1} \to \dots \to w'_0 \equiv w'$ where at each step $w_{i-1} \to w_i$ we insert $(z_i[E_i v_i]z_i^{-1})$ without canceling v_i. Thus we do not replace a part of a word from R by another part of that word but insert the whole word. To specify different parts of the inserted word, we use parentheses and square brackets. Thus each w'_i is equal to w_i in the free group. In particular, $w' = w$ in the free group. The word w' will be called a *companion* of w. To simplify counting, we leave the brackets in w' but we do not count the brackets when we consider the length of w'. Note that w is the unique reduced word (see Section 1.8.6) that is equal to w' in the free group, so w is uniquely determined by any of its companions.

Let us estimate from above the length of w' in terms of the length x of w. By the definition of speed and relative speed, we have that for every $i \geq 0$,

$$|w_i| - |w_{i+1}| \geq 2|z_i| + |E_i| - |v_i| \geq 2|z_i| + \delta.$$

$$|w_i| - |w_{i+1}| \geq 2|z_i| + \gamma|E_i v_i|.$$

Summing these equalities for all i gives

$$|w| \geq 2\sum_{i=1}^{\lambda} |z_i| + \lambda\delta,$$

$$|w| \geq 2\sum_{i=1}^{\lambda} |z_i| + \gamma\sum_{i=1}^{\lambda} |E_i v_i|.$$

Since by the definition of the companion $\bar{w}$ we have

$$|\bar{w}| = 2\sum_{i=1}^{\lambda} |z_i| + \sum_{i=1}^{\lambda} |E_i v_i|$$

we deduce

$$|w| \leq |\bar{w}| \leq \frac{x}{\gamma}. \tag{5.8.5}$$

Now let us estimate the number $N(x, h)$ of companion words $\bar{w}$ of length h of reduced words w of length x as above. One can obtain $\bar{w}$ as follows.

- Start with the word $123\dots h$ of length h.
- Place in that word 2λ pairs of parentheses and square brackets.
- Repeat 2λ times the following steps until all pairs of corresponding brackets are used.

  – Pick a new pair of corresponding parentheses or square brackets, and the subword V bounded by that pair.

- Remove all proper subwords from V bounded by square brackets or parentheses. The remaining word V' is a product of some letters from $\{1, 2, \ldots, h\}$.
- Replace that word by a word of the same length $|V'|$ from R (if V was bounded by square brackets) or $z_i z_i^{-1}$ (if V was bounded by parentheses).
- Insert back the removed subwords from V (there is only one natural way to do that).

The number of possible placements of 2λ pairs of square brackets and parentheses in the word $123\ldots h$ of length h can be (very roughly) estimated from above as

$$\binom{h+4\lambda}{4\lambda} \leq \frac{(h+4\lambda)^{4\lambda}}{(4\lambda)!} = \frac{(4\lambda)^{4\lambda}}{(4\lambda)!}\left(1 + \frac{h}{(4\lambda)}\right)^{4\lambda} \tag{5.8.6}$$

In order to further simplify that expression, we need the following fact from Calculus.

Lemma 5.8.47. *For every natural $n \geq 1$ we have $\frac{n^n}{n!} \leq e^n$.*

Proof. Integrating by parts the function $\ln x$ we get

$$\int_1^n \ln x = n \ln n - n + 1.$$

Since $\ln x$ is an increasing function, we can estimate the integral $\int_1^n \ln x\, dx$ from above by using the Darboux sums with step 1:

$$\int_1^n \ln x \leq \ln 2 + \ldots + \ln n = \ln n!.$$

Thus $n \ln n - n + 1 \leq \ln n!$ or

$$n \ln n - \ln n! \leq n - 1.$$

Taking exponents of both sides of this inequality yields the result. $\square$

By Lemma 5.8.47 the expression (5.8.6) does not exceed

$$e^{4\lambda}\left(1 + \frac{h}{(4\lambda)}\right)^{4\lambda}$$

which does not exceed

$$e^{4\frac{x}{\delta}}\left(1 + \frac{\frac{x}{\gamma}}{(4\frac{x}{\delta})}\right)^{4\frac{x}{\delta}} = (2m-1)^{x\alpha} \tag{5.8.7}$$

where (as in the formulation of the theorem)

$$\alpha = \frac{4}{\delta} \log_{2m-1} \left(e \left(1 + \frac{\delta}{4\gamma} \right) \right).$$

Here we used the fact that the function $\ell(p, q) = (1 + \frac{q}{p})^p$ is increasing in both p and q (check it!).

Fix the lengths $k_1, \ldots, k_\lambda$ of words from R and lengths $t_1, \ldots, t_\lambda$ of the words z_i used in the construction of $\bar{w}$. Then the number of companion words of length $h = \sum_{i=1}^{\lambda} k_i + 2 \sum_{i=1}^{\lambda} t_i$ is bounded from above by

$$\prod_{i=1}^{\lambda} (2m-1)^{\beta k_i} \prod_{i=1}^{\lambda} (2m-1)^{t_i} \leq (2m-1)^{\beta \sum_{i=1}^{\lambda} k_i + \sum_{i=1}^{\lambda} t_i} \leq (2m-1)^{\beta h + \frac{x}{2}} \leq (2m-1)^{\frac{\beta x}{\gamma} + \frac{x}{2}}.$$

$$(5.8.8)$$

Note that the last term of that sequence of inequalities does not depend on k_i or t_i or even on the way the parentheses and square brackets are placed. Note also that any placement of parentheses in the word $12 \ldots h$ determines the numbers $k_1, k_2, \ldots, t_1, t_2 \ldots$ uniquely or does not correspond to any companion word at all. Therefore multiplying the last term in (5.8.8) by $(2m-1)^{x\alpha}$ gives us an upper bound of $N(x, h)$. Multiplying that further by the number of possible values of h ($1 \leq h \leq \frac{x}{\lambda}$), we get an upper bound of all companion words of reduced words w of length x which are equal to 1 in G. That upper bound is equal to

$$\frac{x}{\gamma} (2m-1)^{x \left(\frac{1}{2} + \frac{\beta}{\gamma} + \alpha \right)}$$

hence the cogrowth in the sense of Grigorchuk does not exceed

$$(2m-1)^{\left(\frac{1}{2} + \frac{\beta}{\gamma} + \alpha \right)}$$

as required.   □

5.8.5.4 Non-cyclic Burnside Groups of Sufficiently Large Odd Exponents Are Not Amenable

Let exponent n be sufficiently large and odd. In Section 5.2.3.15 we described a Dehn presentation $\mathcal{P}_i$ of the intermediate group G_i, $i \geq 1$. By definition $\mathcal{P}_i \subseteq \mathcal{P}_{i+1}$ for every i. Therefore $\mathcal{P} = \bigcup \mathcal{P}_i$ is a Dehn presentation of the group $B(2, n)$. It is easy to see (from Section 5.2.3.15) that the speed of this presentation is at least cn for some constant c and the relative speed is close to 1. Still we cannot immediately apply Theorem 5.8.45 because the exponential rate of growth of that presentation is too large: $\mathcal{P}$ has "too many" relations. One can trim the set $\mathcal{P}_i$ from Section 5.2.3.15, for example, by considering only contiguity subdiagrams Ψ which do not contain cells that are attached to the boundary of Δ with large contiguity degree. But estimating the number of possible subdiagrams Ψ is not an easy task and would require a very detailed analysis of diagrams over $\mathcal{P}_i$. Instead we refer to Adian's paper [4] where he constructs, using the techniques from [251], a Dehn presentation of the Burnside group $B(m,n)$ for every odd $n \geq 665$ and every $m \geq 2$ with speed ≥ 228, relative speed $\frac{1}{3}$, the exponential rate of growth $\leq (2m-1)^{\frac{1}{45}}$. Plugging these numbers in the formula from Theorem 5.8.45, we conclude that the cogrowth in the sense of Grigorchuk of $B(m,n)$ does not exceed $(2m-1)^{\frac{1}{6}}$. Applying Theorem 5.8.43 we obtain

Theorem 5.8.48. *The free Burnside group $B(m,n)$ is non-amenable for every $m \geq 2$ and every odd $n \geq 665$.*

Note that $B(m,n)$ was the first counterexample to the von Neumann-Day conjecture that satisfied a nontrivial identity. As of today, the varieties of all groups of exponent n (n big enough) are still the smallest known varieties of groups containing non-amenable groups. But the free non-cyclic groups in the non-commutative variety of groups where all finite groups are commutative constructed by Olshanskii [259] and in the non-commutative variety of groups where all periodic groups commutative constructed by Ivanov and Storozhev [158] are probably non-amenable also.

5.9 Further Reading and Open Problems

5.9.1 Further Applications of Olshanskii's Method

The method outlined in Section 5.2.3, and its modifications, have been used to solve numerous other famous group theory problems. Here is a very incomplete list of these results. The first three and the fifth results can be found in [260], the fourth is in [232], and the sixth is in [265] (for more see [260] and [281]).

(1) (Tarski monster of the first kind) There exists a finitely generated group with all proper subgroups infinite cyclic.

(2) (Tarski monster of the second kind) There exists a nontrivial finitely generated group with all proper subgroups cyclic of the same prime order.

(3) (Divisible group) There exists a nontrivial finitely generated group in which every element a has a root of any degree (i.e., for every $n \geq 0$ there exists b_n such that $b_n^n = a$).

(4) (Verbally complete group) There exists a finitely generated nontrivial group G such that for every $a \in G$ and every word w which is not freely trivial, a is a value of w in G.

(5) (Group with finitely many conjugacy classes) There exists an infinite finitely generated periodic group G and an element $a \in G$ such that every element is conjugate to a power of a (in particular, G has only finitely many conjugacy classes).

(6) (Group with two conjugacy classes) There exists an infinite finitely generated group where all non-identity elements are conjugate.

Each of these results answered a long standing problem in group theory which were completely out of reach before Olshanskii's method was developed. Note that all these groups except the last one are inductive limits of hyperbolic groups. The last example is an inductive limit of the so-called *relatively hyperbolic* groups. In each case a version of small cancelation theory based on contiguity subdiagrams was developed.

5.9.2 Syntactic Properties of Hyperbolic Groups

Hyperbolicity has been studied in geometry since at least Lobachevsky, Bolyai and Poincare and, under the name of small cancelation, by combinatorial group theorists since Dehn. Gromov was first who noticed that groups that appear in hyperbolic geometry and "abstract" small cancelation groups have many common features. He formally defined and systematically studied hyperbolic groups in [127].[14] The importance of [127] was understood immediately and by now the theory of hyperbolic groups and their generalizations is one of the largest parts of the theory of infinite groups called *geometric group theory* (one of the main goals of the geometric group theory is to consider Cayley graphs of groups as geometries and classify groups up to quasi-isometry which is a much weaker equivalence than isomorphism: see the introduction to Druţu and Kapovich [91] and the references therein). One of the most accessible presentation of the foundations of the theory of hyperbolic groups can be found in [12] and [110]. Here we survey only a very few, more syntactic, results about hyperbolic groups.

5.9.2.1 Almost All Groups Are Hyperbolic

This means that if we pick a finite group presentation "at random", it defines a hyperbolic group with probability close to 1. Of course to make that

---

[14] This is a very incomplete version of the history of hyperbolic groups. For more details see [12, 110].

statement precise, we need to explain what does it mean to pick a presentation at random. There are several different ways to do that, one can read about it in my survey [281, Section 3.1.H].

5.9.2.2 Dehn Functions of Hyperbolic Groups

The *Dehn function* of a finitely presented group $G = \text{gp}\langle X \mid R \rangle$ is a function $f: \mathbb{N} \to \mathbb{N}$ such that every word $w \in (X \cup X^{-1})^*$ which is equal to 1 in G labels the boundary of a van Kampen diagram with at most $f(|w|)$ cells. The Dehn function (up to the equivalence from Section 1.5) does not depend on the presentation of a group, and is an almost as important invariant of a group as the growth function (see [281]).

Since hyperbolic groups have Dehn presentations, every hyperbolic group has linear Dehn function: in Section 5.8.5.3 we essentially proved that the Dehn function $f(n)$ is bounded by n/δ where δ is the speed of the Dehn presentation.

It turned out [127, 261] that, conversely, all groups with linear, and even subquadratic, Dehn functions are hyperbolic. Thus there is a gap between "linear" and "quadratic" in the set of Dehn functions (up to equivalence) as in the set of growth functions in semigroups, rings and groups (Section 3.10.6.1).

5.9.2.3 The Language of Geodesic Words in a Hyperbolic Group Is Rational

This is true for every finite presentation of a hyperbolic group. That fact was first proved by Cannon [61] for some groups that occur in hyperbolic geometry and then for arbitrary hyperbolic groups by Gromov [127]. This implies that the growth series for the language of geodesic words is a rational function (see Section 1.8.9). In fact the growth series of any hyperbolic group G with respect to any finite generating set X (when we count elements of G rather than geodesic words that represent them) is also a rational function (see [96], [110, Chapter 1]) because there exists an automaton $\mathcal{A} = (Q, X)$ such that for every element g of the group there exists a unique word w_g recognized by $\mathcal{A}$ which represents g in the group, the word w_g is geodesic, and the language accepted by $\mathcal{A}$ consists of the words w_g.

5.9.2.4 The Subshift of Bi-infinite Geodesics

Recall that in Section 3.3.1, with every finitely generated semigroup S we associated the subshift $D(S)$ consisting of all bi-infinite geodesics. Let $G = \text{sg}\langle A \mid \mathcal{R} \rangle$ be a hyperbolic group.

Theorem 5.9.1. *There exists a subshift of finite type* (X, T) *and a finite-to-one homomorphism* ϕ *from* (X, T) *onto* $D(G)$.

Proof. Let G be generated by a finite set A. By Section 5.9.2.3, there exists a rooted automaton $\mathcal{A} = (Q, A)$ with input and output sets of vertices Q_-, Q_+ which recognizes the language of geodesic words in G. That means (see Section 1.3.2) that the labels of paths from Q_- to Q_+ are precisely all the geodesic words in G. Let us modify $\mathcal{A}$ by removing edges and vertices that do not belong to paths in $\mathcal{A}$ from Q_- to Q_+. Clearly the new automaton $\mathcal{A}' = (Q', A)$ recognizes the same language as $\mathcal{A}$. Then the label of every path in $\mathcal{A}'$ is a geodesic in G. Indeed, every path p in $\mathcal{A}'$ can be extended to a path $p_0 p p_1$ where p_0 starts at a vertex from Q_-, p_1 ends at a vertex from Q_+. Thus the label of every path in $\mathcal{A}'$ is a subword of a geodesic word in G, hence a geodesic word itself. We conclude that the label of every bi-infinite path in $\mathcal{A}'$ belongs to the subshift $D(G)$. Let $D' \subseteq A^{\mathbb{Z}}$ be the subshift generated by all labels of bi-infinite paths in $\mathcal{A}'$. Then the sets of finite subwords of D' and $D(G)$ coincide (both are the sets of all geodesic words in G). Hence $D' = D(G)$.

Let (E, T) be the edge subshift of the underlying graph of $\mathcal{A}'$. It is a subshift of finite type (see Section 1.6.2). There exists a homomorphism from (E, T) to $D(G)$ which takes every bi-infinite path to its label. That homomorphism is finite-to-one (in fact at most k-to-one where k is the number of vertices in $\mathcal{A}'$). Indeed, every bi-infinite path in $\mathcal{A}'$ is completely determined by its label and the vertex visited at time 0 (we have used that argument in the proof of Theorem 3.9.18). □

For more on the connections between subshifts and hyperbolic groups see the book by Coornaert and Papadopoulos [73].

5.9.3 Some Semantic Properties of Small Cancelation Groups

Novikov–Adian theorem and results of Olshanskii and others (see [260] and Section 5.9.1) changed the way we think about infinite groups. A similar change is happening now with small cancelation groups (and the whole geometric group theory). One of the strongest results about small cancelation groups has been obtained recently by Ian Agol [10] (based on the work of Daniel Wise [330]). Here is one of the corollaries of the result.

Theorem 5.9.2. *Every group satisfying* $C'(\frac{1}{6})$ *embeds into the group* $\mathrm{SL}(n, \mathbb{Z})$ *of* $n \times n$*-matrices with integer entries and determinant 1 for some* n.

Not every hyperbolic group has this property.

Theorem 5.9.3 (M. Kapovich [169]). *There is a hyperbolic group that does not embed into any (multiplicative) group of matrices over a commutative ring.*

5.9.4 Open Problems About Hyperbolic Group

There are many open problems about hyperbolic (and related) groups (see Bestvina's list [47] or [282]). We mention only two of them here. Both problems were first formulated by Gromov but are still far from being solved although many attempts have been made.

Problem 5.9.4. *Suppose that a hyperbolic group G does not have a free subgroup of finite index. Is it true that G contains a subgroup isomorphic to a surface group $\pi_1(S_g)$, $g \geq 2$?*

Problem 5.9.5. *Is it true that every hyperbolic group has a homomorphism onto a finite nontrivial group?*

This is true for groups satisfying the small cancelation property $C'(1/6)$ which follows from Theorem 5.9.2.

5.9.5 The Hanna Neumann Conjecture

The Hanna Neumann conjecture formulated in Section 5.5.2.2 can be reformulated as follows. Let r be the rank of a subgroup H of a free group. Then let us denote $\bar{r} = \max(r - 1, 0)$. This number is called the *reduced rank* of H.

Conjecture 5.9.6 (Hanna Neumann [246]). *Let H_1, H_2 be subgroups of reduced ranks $\bar{r}_1, \bar{r}_2$ of a free group. Then the reduced rank $\bar{r}$ of the intersection $H_1 \cap H_2$ satisfies*

$$\bar{r} \leq \bar{r}_1 \bar{r}_2.$$

That conjecture has been strengthened by Walter Neumann in the 80s:

Conjecture 5.9.7 (Walter Neumann [249]). *Let H_1, H_2 be subgroups of reduced ranks $\bar{r}_1, \bar{r}_2$ of a free group F_k. Then the sum of reduced ranks of distinct intersections $H_1^g \cap H_2$, $g \in F_k$, does not exceed $\bar{r}_1 \bar{r}_2$.*

Note that although the number of intersections $H_1^g \cap H_2$, $g \in F_k$, is infinite it can be proved (using Theorem 5.5.9) that only finite number of these intersections are distinct.

After more than 50 years of attempts, both conjectures have been proved in 2011 almost simultaneously (papers using quite different sets of ideas appeared in the arXiv one 5 days after the other) by Joel Friedman [104] and Igor Mineyev [234]. Amazingly short modifications of both proofs were given by Warren Dicks [85, 86]. Note that both Mineyev's proof and Dicks' version of it essentially use the fact that there exists a total order $\leq$ on any free group which is compatible with the product (Theorem 1.8.37).

5.9.6 Diagram Groups

More information about diagram groups can be found in Farley [99, 100] and Guba and Sapir [136, 137]. It turned out that a diagram group can be defined as the fundamental group of the space of certain paths in a *directed 2-complex*. Farley proved, in particular, that this space of paths has nice geometric properties (its universal cover is CAT(0)). Some results proved in [136, 137] are summarized in the next theorem.

Theorem 5.9.8 (Guba, Sapir [133, 136, 137]). *(1) If a string rewriting system* $\mathcal{P} = sr\langle X \mid \mathcal{R} \rangle$ *is Church–Rosser, then presentations of the corresponding diagram groups* $\mathrm{DG}(\mathcal{P}, u)$ *can be explicitly found.*
(2) The diagram group corresponding to the rewriting system $sr\langle x \mid x^2 \to x, x \to x \rangle$ *(and many other diagram groups) is universal, i.e., contains all other countable diagram groups as subgroups. That group can be explicitly constructed as an extension of a "nice" group by the R. Thompson group* F.
(3) Every diagram group admits a total order compatible with the multiplication.

5.9.7 The R. Thompson Group F

There are several characterizations of R. Thompson group F which we have not mentioned before. One of these was first discovered by F. Galvin and R. Thompson in [106] and then used by Higman in [153]. They proved that the R. Thompson group F is the group of certain "strong order preserving" automorphisms of the finitely generated relatively free algebra in the Jónsson–Tarski variety defined in Section 1.4.4. The group V of all automorphisms of that algebra turned out to be also interesting. That was the first example of a finitely presented infinite simple group (for more on this see [54] and [62]).

Many papers are devoted to the study of the R. Thompson group F. The most difficult and unexpected result about that group was obtained by Victor Guba

Theorem 5.9.9 (Guba, [130]). *The Dehn function of the R. Thompson group* F *is quadratic.*

The growth function and growth series of F are also subject of intensive study. Although by Theorem 5.6.42 and Exercise 5.6.46, the R. Thompson group F admits a Church–Rosser presentation with a rational language of canonical words, we cannot conclude that the growth series of F (say, with respect to the generating set $\{x_0, x_1\}$) is represented by a rational function as in Section 1.8.9 because the canonical words are not geodesic. But Elder, Fusy and Rechnitzer [94] formulated the following

Conjecture 5.9.10. *The difference between the length of a "typical" canonical word from Theorem 5.6.42 and the length of the element of F represented by this word does not exceed a constant.*

The word "typical" which should be intuitively clear, can be precisely defined (see [168], for example).

Problem 5.9.11. *Is it true that the growth series of F (with respect to the generating set $\{x_0, x_1\}$ or any other finite generating set) is a rational function?*

José Burillo proved [59] that the answer is positive for the growth series of the submonoid generated by x_0, x_1. This growth series is represented by $\frac{1-z^2}{1-2z-z^2+z^3}$. The exact values of the spherical growth function $s(n)$ of F (with respect to $\{x_0, x_1\}$) are known for $n \leq 1500$ [94]. This data implies that if the growth series of F with respect to the generating set $\{x_0, x_1\}$ is represented by a rational function P/Q, then the degree of the polynomial Q must be at least 749.

The most famous open problem about R. Thompson group F is whether F is amenable. That problem was first formulated by R. Thompson in around 1973 (unpublished) and it was then independently formulated by Geoghegan[108] in 1979. Recall that F does not contain free non-cyclic subgroups and is finitely presented with two generators and two relations (see Section 5.6.6). Thus if F was non-amenable, it would give an easy finitely presented counterexample to the von Neumann–Day conjecture. At this time (the end of 2013) the only published finitely presented counterexamples have about 10^{200} relations [157, 264] (although see [204]). The best currently known result about amenability of the R. Thompson group F is the following statement showing that even if F is amenable, it is "barely so" because its Følner sets are enormously large.

Theorem 5.9.12 (J. Moore [237]). *Let U be any finite generating set of F. Then for every n the smallest finite set satisfying the Følner condition $(\mathcal{F})$ for $\epsilon = \frac{1}{2^n}$ has at least $2^{2^{2^{\cdots}}}$ elements (where the number of 2s is at least Cn for some constant $C > 0$).*

5.9.8 The Finite Basis Problem for Varieties of Groups

Many results and open problems about the finite basis problem for group varieties are surveyed by Gupta and Krasilnikov in [141]. The following is one of the main open problems in that area.

Problem 5.9.13 (Shmelkin). *Is every group of matrices over a field finitely based?*

By the *Tits alternative* [313], every finitely generated group of matrices over a field either satisfies no nontrivial identity or has a solvable subgroup of finite index (the condition "finitely generated" can be removed if the characteristic is 0). Solvable groups of matrices have been described by Lie, Kolchin and Mal'cev (see [170]). Every such group G has a finite index subgroup H which is nilpotent-by-commutative, i.e., H has a nilpotent normal subgroup N such that H/N is commutative. Thus Problem 5.9.13 reduces to groups which have finite index nilpotent-by-commutative subgroups.

The following theorem "almost" solves the problem.

Theorem 5.9.14 (Krasilnikov, [186]). *Any nilpotent-by-commutative group is finitely based.*

We also know (Theorem 1.4.33) that every finite group is finitely based, but all efforts to combine Theorems 5.9.14 and 1.4.33 failed so far.

5.9.9 Zel'manov Words and Inherently Non-finitely Based Varieties of Groups

Since every finite group has finite basis of identities by Oates and Powell (Theorem 1.4.33), there are no inherently non-finitely based finite groups. It is quite possible that there are no inherently non-finitely based locally finite group varieties at all. The main evidence for this is the following theorem of Zel'manov where Zimin words reappear again, this time "dressed up" using group commutators.

Let us define the *Zel'manov word* $\mathcal{Z}_n$ by the following rule:

$$\mathcal{Z}_1 = x_1, \ldots, \mathcal{Z}_{n+1} = [\mathcal{Z}_n, x_{n+1}, \mathcal{Z}_n]$$

where, as before, brackets denote the group commutator: $[x, y] = x^{-1}y^{-1}xy$, $[x, y, z] = [[x, y], z]$. One can easily see that the Zel'manov word is precisely the Zimin word where the multiplication is replaced by the group commutator. The following theorem is proved in [335] (it solved a long-standing problem by B.H.Neumann [248] in the case of prime exponents).

Theorem 5.9.15 (Zel'manov, [335]). *For every prime number p there exists a natural number n such that a group of exponent p is locally finite if and only if it satisfies the identity $\mathcal{Z}_n = 1$.*

As an immediate consequence of Theorem 5.9.15 and Theorem 1.4.37 we obtain the following

Corollary 5.9.16. *There are no inherently non-finitely based locally finite varieties of groups of prime exponent.*

For exponent 5, Corollary 5.9.16 was proved before by Endimioni [95].

Conjecture 5.9.17. *The statement of Corollary 5.9.16 should remain true if we replace there "prime" by "any finite".*

5.9.10 Amenable Groups

The literature concerning amenability is very large and fast growing (see, say, [82]). Here I will mention only a few recent results (see also Section 5.9.7).

5.9.10.1 More Counterexamples to the von Neumann-Day Conjecture

Currently there are several classes of counterexamples obtained by completely different methods. Here we mention two of these methods.

Golod–Shafarevich groups. In Section 5.2.1 we constructed examples of infinite periodic groups using associative algebras given by "sparse" sets of relations (from Section 4.4.6). Let us call these *Golod groups*. Similar groups can be constructed directly, by using group presentations.

Definition 5.9.18. Consider a free group F_d, $d > 1$. Let $F_d > \gamma_1(F_d) > \ldots$ be the lower central series of F_d. For every i, n let $\gamma_i^n(F_d)$ be the (normal) subgroup generated by all n-th powers of elements of $\gamma_i(F_d)$. Pick a prime number p. For every $m \geq 1$ let D_m be the subgroup of F_d generated by subgroups $\gamma_i^{p^j}(F_d)$ for $ip^j \geq m$. Then $F_d \geq D_1 \geq D_2 \geq \ldots$. It is possible to prove (using the Magnus embedding of the free group from Section 1.8.8 where K is the field of integers modulo p) that the intersection $\bigcap_{m \geq 0} D_m = \{1\}$. Then for every element $r \in F_d$ we can define its *degree* $\deg(r)$ as the maximal m such that $r \in D_m$.

For a set of elements $R \subseteq F_d$, we define the series

$$H_R(t) = \sum_{r \in R} t^{\deg(r)}.$$

Then consider the *Golod–Shafarevich function* $\mathrm{GS}_R(t) = 1 - dt + H_R(t)$. A group G is called *Golod–Shafarevich* if it has a presentation $\mathrm{gp}\langle x_1, \ldots, x_d \mid R \rangle$ such that for some $t > 0$, $\mathrm{GS}_R(t) < 0$.

The following two properties make this definition immediately useful:

- Every Golod–Shafarevich group is infinite.
- If G is a group given by a Golod–Shafarevich presentation $\mathrm{gp}\langle X \mid R \rangle$, then adding a relation $r = 1$ to R with r of high enough degree, we obtain a Golod–Shafarevich quotient of G. Indeed, if a number z is negative, $t < 1$, then we can add to z a small positive number $t^{\deg(r)}$ and keep the sum negative.

Using the second property one can construct Golod–Shafarevich groups as inductive limits of finitely presented Golod–Shafarevich groups (similar to inductive limits of hyperbolic groups used in Section 5.2.3). Golod–Shafarevich groups have been used as a source of many examples (see the excellent survey by Ershov [97]). Say, pick any prime p and let w run over all reduced group words in x, y. For each w choose i_w so that the numbers i_w grow very fast. Consider the presentations $\mathcal{P} = \mathrm{gp}\langle x, y \mid R \rangle$ where R consists of all relations $w^{p^{i_w}} = 1$. Since the degree of $w^{p^{i_w}}$ is at least i_w, the value $\mathrm{GS}_R(t)$ will be negative for, say, $t = \frac{1.5}{d}$ as in Section 4.4.6. Therefore the group given by the presentation $\mathcal{P}$ is infinite, finitely generated and periodic. The following theorem is more complicated.

Theorem 5.9.19 (See Ershov, [97]). *For every $m \geq 1$ there exists a Golod group and a Golod–Shafarevich group, each of which is $m + 1$-generated with all m-generated subgroups finite.*

For Golod groups, Theorem 5.9.19 was proved by Golod in [113], for Golod–Shafarevich groups see [97]. (Compare this theorem and Theorem 5.9.15. This shows an important difference between the bounded and unbounded versions of the Burnside problem.)

Disproving a conjecture of Lubotzky and Zel'manov, Ershov proved the following

Theorem 5.9.20 (Ershov, [97]). *Every Golod–Shafarevich group is non-amenable.*

Groups of piece-wise projective transformations of the circle. Let $\mathbb{R}^*$ be the real line with the infinity point ∞ adjoined, so that topologically $\mathbb{R}^*$ is a circle. In the hint for Exercise 1.8.33, we defined an action of $\mathrm{SL}(2, \mathbb{R})$ on the complex plane with infinity ∞ adjoined. Every invertible matrix $\begin{bmatrix} p & q \\ r & s \end{bmatrix}$ acts as the Möbius transformation

$$z \mapsto \frac{pz + q}{rz + s}.$$

Note that if z is a real number, then its image under the Möbius transformation is also real (or ∞). Thus $SL(2, \mathbb{R})$ acts on the line with infinity point ∞ adjoined, $\mathbb{R}^*$. Topologically $\mathbb{R}^*$ is a circle, so $SL(2, \mathbb{R})$ acts on a circle. Each Möbius map is a continuous bijection $\mathbb{R}^* \to \mathbb{R}^*$.

Let A be a subring of $\mathbb{R}$ containing 1. Let $G(A)$ be the group of all continuous bijections $\phi \colon \mathbb{R}^* \to \mathbb{R}^*$ which are piecewise Möbius transformations of $\mathbb{R}^*$ corresponding to matrices with coefficients from A. This means that there exists a partition of the circle $\mathbb{R}^*$ into a finite number of intervals; on each of these intervals, ϕ acts as one of the Möbius transformations corresponding to some matrix from $SL(2, A)$, and ϕ is continuous and bijective. The definition of G is similar to the definition of the R. Thompson group in terms of piecewise linear functions.

Let $H(A)$ be the subgroup of all elements ϕ of $G(A)$ that fix ∞, i.e., $\phi(\infty) = \infty$. By [62, Theorem 7.2] the R. Thompson group F is a subgroup of index 2 in $H(\mathbb{Z})$.

The following theorem of Monod is remarkable both for its strength and for the simplicity of the proof

Theorem 5.9.21 (Monod, [235]). *For every ring $A \le \mathbb{R}$ containing 1, $A \ne \mathbb{Z}$, the group $H(A)$ is non-amenable and contains no free non-cyclic subgroups.*

The fact that $H(A)$ contains no free non-cyclic subgroups is proved almost in the same way as the similar fact for F (see Theorem 5.6.39).

The group $H(A)$ is not finitely generated. Moreover it is not countable. But by Theorem 5.8.15, $H(A)$ contains a finitely generated non-amenable subgroup M which, of course, would not contain non-cyclic free subgroups. In fact such a subgroup M was found explicitly by Lodha and J. Moore [204]. Moreover they proved that the subgroup is finitely presented. It has a presentation with 3 generators and 9 relations (compare with 10^{200} relations in [264]). The group is generated by an isomorphic copy of the R. Thompson group F and one extra function. Note, though, that the group from [264] satisfies a nontrivial identity ($[x, y]^n = 1$, $n \ge 10^{70}$, odd) while the group in [204] is lawless by Theorem 5.6.37.

5.9.10.2 Tarski Numbers of Groups

The paper [82] by de la Harpe, Grigorchuk and Ceccherini-Silberstein contains one of the first detailed studies of possible Tarski numbers of non-amenable groups. Since groups containing a non-cyclic free subgroup are precisely the groups with Tarski number 4 (by Theorem 5.8.40), we need to restrict ourselves to groups without non-cyclic free subgroups. It is proved in [82] that periodic non-amenable groups cannot have Tarski number 5. Thus the Tarski number of the free Burnside group $B(m, n)$, $m \ge 2$, $n \ge 665$ odd, is at least 6. Theorem 56 in [82] relates the Tarski number and cogrowth. Using this theorem and Adian's estimate for cogrowth of $B(m, n)$ (Section 5.8.5.4), it is proved in [82] that the Tarski number of $B(m,n)$, m,n as above, is between 6 and 14. The exact value is not known. Nevertheless the following result holds (it is my answer to a question of Ozawa on Mathoverflow).

Theorem 5.9.22. *There are non-amenable periodic groups with arbitrarily large Tarski numbers.*

Proof. Indeed, by Theorem 5.9.19, for every $m \geq 1$ there exists a Golod–Shafarevich $m+1$-generated group G with all m-generated subgroups finite. By Theorem 5.9.20, G is non-amenable, and by Theorem 5.8.15 the Tarski number of G is at least $m+1$.   □

Problem 5.9.23. *Can the Tarski number of the direct product $G \times G$ be smaller than the Tarski number of G?*

Problem 5.9.24. *Is there a number ≥ 4 which is not the Tarski number of some group? Is 5 the Tarski number of a group?*

Remark 5.9.25. Gili Golan proved that if G is non-amenable, and Z is any amenable normal subgroup of G, then G and G/Z have the same Tarski numbers. This and other results about Tarski numbers of groups can be found in [98]. In particular, it is proved in [98] that 6 is the Tarski number of a group, and that for every sufficiently large n there exists a group with Tarski number between n and $2n$.

References

1. M. Abért, Group laws and free subgroups in topological groups. Bull. Lond. Math. Soc. **37**(4), 525–534 (2005) (Cited on pages 158 and 255).
2. S.I. Adian, Infinite irreducible systems of group identities. Izv. Akad. Nauk SSSR Ser. Mat. **34**, 715–734 (1970) (Cited on page 243).
3. S.I. Adian, *The Burnside Problem and Identities in Groups* (Springer, Berlin/ New York, 1979) (Cited on pages 28, 222, and 223).
4. S.I. Adian, Random walks on free periodic groups. Izv. Akad. Nauk SSSR Ser. Mat. **46**(6), 1139–1149 (1982) (Cited on pages 213, 314, 315, and 319).
5. A. Adler, S.-Y. Robert Li, Magic cubes and Prouhet sequences. Am. Math. Mon. **84**, 618–627 (1977) (Cited on page 71).
6. R.L. Adler, L.W. Goodwyn, B. Weiss, Equivalence of topological Markov shifts. Isr. J. Math **27**(1), 49–63 (1977) (Cited on pages 139, 140, 141, 148, and 150).
7. R.L. Adler, B. Marcus, Topological entropy and equivalence of dynamical systems. Mem. Am. Math. Soc. **20**(219), pp. iv+84 (1979) (Cited on page 159).
8. R.L. Adler, B. Weiss, Similarity of automorphisms of the torus. Mem. Am. Math. Soc. **98**, pp. ii+43 (1970) (Cited on page 139).
9. R. Aharoni, P. Haxell, Hall's theorem for hypergraphs. J. Graph Theory **35**(2), 83–88 (2000) (Cited on page 308).
10. I. Agol (with an appendix by Ian Agol, Daniel Groves and Jason Manning). The virtual Haken conjecture, arXiv:1204.2810 (Cited on page 322).
11. M. Aigner, G. Ziegler, *Proofs from The Book*, 4th edn. (Springer, Berlin/New York, 2010) (Cited on page 183).
12. J. Alonso, T. Brady, D. Cooper, V. Ferlini, M. Lustig, M. Mihalik, M. Shapiro, H. Short, Notes on hyperbolic groups, group theory from a geometric viewpoint, in *Proceedings of ICTP*, Trieste (World Scientific, Singapore, 1991), pp. 3–63 (Cited on page 320).
13. A.S. Amitsur, J. Levitzki, Minimal identities for algebras. Proc. Am. Math. Soc. **1**, 449–463 (1950) (Cited on page 169).
14. S.E. Arshon, A proof of the existence of n-valued infinite assymmetric sequences. Mat. Sb. **2**(44), 769–777 (1937) (Cited on page 72).
15. M. Aschbacher, The status of the classification of the finite simple groups. Not. Am. Math. Soc. **51**(7), 736–740 (2004) (Cited on page 153).
16. R. Baer, Radical ideals. Am. J. Math. **65**(4), 537–568 (1943) (Cited on page 181).
17. K.A. Baker, G.F. McNulty, W. Taylor, Growth problems for avoidable words. Theor. Comput. Sci. **69**(3), 319–345 (1989) (Cited on page 84).
18. K.A. Baker, G.F. McNulty, H. Werner, The finitely based varieties of graph algebras. Acta Sci. Math. (Szeged) **51**(1–2), 3–15 (1987) (Cited on page 32).

© Springer International Publishing Switzerland 2014
M.V. Sapir, *Combinatorial Algebra: Syntax and Semantics*,
Springer Monographs in Mathematics, DOI 10.1007/978-3-319-08031-4

19. K.A. Baker, G.F. McNulty, H. Werner, Shift-automorphism methods for inherently non-finitely based varieties of algebras. Czechoslov. Math. J. **39**(1), 53–69 (1989) (Cited on pages 32 and 36).

20. Yu.A. Bakhturin, A.Yu. Ol'shanski, Identity relations in finite Lie algebras. Mat. Sb. **96**(4), 543–559 (1975) (Cited on page 29).

21. S. Banach, A. Tarski, Sur la decomposition des ensembles de points en porties respectivement congruents. Fund. Math. **6**, 244–277 (1924) (Cited on page 304).

22. L. Bartholdi, The growth of Grigorchuk's torsion group. Int. Math. Res. Not. **20**, 1049–1054 (1998) (Cited on page 295).

23. L. Bartholdi, Lower bounds on the growth of a group acting on the binary rooted tree. Int. J. Algebra Comput. **11**(1), 73–88 (2001) (Cited on page 299).

24. L. Bartholdi, I.I. Reznykov, V.I. Sushchansky, The smallest Mealy automaton of intermediate growth. J. Algebra **295**(2), 387–414 (2006) (Cited on page 131).

25. H. Bass, The degree of polynomial growth of finitely generated nilpotent groups. Proc. Lond. Math. Soc. **25**, 603–614 (1972) (Cited on pages 285 and 292).

26. H. Bauer, *Measure and Integration Theory* (Trans. from the German by Robert B. Burckel). de Gruyter Studies in Mathematics, vol. 26 (Walter de Gruyter, Berlin, 2001) (Cited on pages 148 and 149).

27. J.A. Beachy, *Introductory Lectures on Rings and Modules*. London Mathematical Society Student Texts, vol. 47 (Cambridge University Press, Cambridge, 1999), pp. viii+238 (Cited on page 23).

28. D.R. Bean, A. Ehrenfeucht, G.F. McNulty, Avoidable patterns in strings of symbols. Pac. J. Math. **85**, 261–294 (1979) (Cited on pages 76, 77, and 84).

29. A. Belov-Kanel, Some estimations for nilpotence of nil-algebras over a field of an arbitrary characteristic and height theorem. Commun. Algebra **20**(10), 2919–2922 (1992) (Cited on page 172).

30. A. Belov-Kanel, On non-Specht varieties. Fundam. Prikl. Mat. **5**(1), 47–66 (1999) (Cited on page 191).

31. A. Belov-Kanel, Counterexamples to the Specht problem. Mat. Sb. **191**(3), 13–24 (2000); trans. in Sb. Math. **191**(3–4), 329–340 (2000) (Cited on page 191).

32. A. Belov-Kanel, The Kurosh problem, the height theorem, the nilpotency of the radical, and the algebraicity identity. Fundam. Prikl. Mat. **13**(2), 3–29 (2007); trans. in J. Math. Sci. (N. Y.) **154**(2), 125–142 (2008) (Cited on page 195).

33. A. Belov-Kanel, Local finite basis property and local representability of varieties of associative rings. Izv. Ross. Akad. Nauk Ser. Mat. **74**(1), 3–134 (2010); trans. in Izv. Math. **74**(1), 1–126 (2010) (Cited on page 195).

34. A. Belov-Kanel, V. Borisenko, V. Latyshev, Monomial algebras (Algebra, 4). J. Math. Sci. (N. Y.) **87**(3), 3463–3575 (1997) (Cited on pages 157 and 194).

35. A. Belov-Kanel, M.I. Kharitonov, Subexponential estimates in Shirshov's theorem on height. Mat. Sb. **203**(4), 81–102 (2012) (Cited on pages 174 and 176).

36. A. Belov-Kanel, L.H. Rowen, *Computational Aspects of Polynomial Identities*. Volume 9 of Research Notes in Mathematics (A K Peters, Wellesley, 2005) (Cited on pages 174, 194, and 195).

37. A. Belov-Kanel, L.H. Rowen, U. Vishne, Structure of Zariski-closed algebras. Trans. Am. Math. Soc. **362**(9), 4695–4734 (2010) (Cited on page 196).

38. A. Belov-Kanel, L.H. Rowen, U. Vishne, Full quivers of representations of algebras. Trans. Am. Math. Soc. **364**(10), 5525–5569 (2012) (Not cited).

39. A. Belov-Kanel, L.H. Rowen, U. Vishne, Specht's problem for associative affine algebras over commutative Noetherian rings. Accepted for publication in Trans. Am. Math. Soc. (Not cited).

40. A. Belov-Kanel, L.H. Rowen, U. Vishne, *PI*-varieties associated to full quivers of representations of algebras. Trans. Am. Math. Soc. (to appear) (Cited on page 196).

41. V.V. Belyaev, N.F. Sesekin, V.I. Trofimov, Growth functions of semigroups and loops. Matem. Zapiski Ural'skogo Gos. Universiteta **10**(3), 3–8 (1977) (Cited on page 131).

42. G.M. Bergman, Centralizers in free associative algebras. Trans. Am. Math. Soc. **137**, 327–344 (1969) (Cited on page 164).

43. G.M. Bergman, *A Note on Growth Functions of Algebras and Semigroups.* Mimeographed Notes (University of California, Berkeley, 1978) (Cited on page 157).

44. G.M. Bergman, The diamond lemma for ring theory. Adv. Math. **29**(2), 178–218 (1978) (Cited on page 54).

45. J. Berstel, Mots sans carré et morphismes itérés. Discret. Math. **29**, 235–244 (1979) (Cited on page 75).

46. J. Berstel, Sur les mots sans carré définis par un morphisme, in *ICALP*, Graz, ed. by A. Maurer (Springer, 1979), pp. 16–25 (Cited on pages 75 and 83).

47. Collected by M. Bestvina, Questions in geometric group theory (preprint), available at http://www.math.utah.edu/~bestvina/eprints/questions-updated.pdf (Cited on page 323).

48. O. Bogopolski, *Introduction to Group Theory* (trans., revised and expanded from the 2002 Russian original). EMS Textbooks in Mathematics (European Mathematical Society (EMS), Zürich, 2008) (Cited on page 209).

49. R.V. Book, F. Otto, *String-Rewriting Systems.* Texts and Monographs in Computer Science (Springer, New York, 1993) (Cited on page 54).

50. A. Braun, The radical in a finitely generated P. I. algebra. Bull. Am. Math. Soc. (N.S.) **7**, 385–386 (1982) (Cited on page 195).

51. M.R. Bridson, The geometry of the word problem, in *Invitations to Geometry and Topology.* Oxford Graduate Texts in Mathematics, vol. 7 (Oxford University Press, Oxford, 2002), pp. 29–91 (Cited on page 200).

52. M.R. Bridson, H. Wilton, The triviality problem for profinite completions, arXiv:1401.2273 (Cited on page 153).

53. M.G. Brin, C.C. Squier, Groups of piecewise linear homeomorphisms of the real line. Invent. Math. **79**(3), 485–498 (1985) (Cited on page 278).

54. K. Brown, Finiteness properties of groups, in *Proceedings of the Northwestern Conference on Cohomology of Groups*, Evanston, 1985. J. Pure Appl. Algebra **44**(1–3), 45–75 (1987) (Cited on page 324).

55. T.C. Brown, An interesting combinatorial method in the theory of locally finite semigroups. Pac. J. Math. **36**, 285–289 (1971) (Cited on page 106).

56. R.M. Bryant, Some infinitely based varieties of groups. Aust. J. Math. **16**(1), 29–33 (1973) (Cited on page 243).

57. R.M. Bryant, The laws of finite pointed groups. Bull. Lond. Math. Soc. **14**(1), 119–123 (1982) (Cited on page 156).

58. J. Brzozowski, Open problems about regular languages, in *Formal Language Theory: Perspectives and Open Problems* (Academic, New York, 1980), pp. 23–47 (Cited on page 154).

59. J. Burillo, Growth of positive words in Thompson's group *F*. Commun. Algebra **32**(8), 3087–3094 (2004) (Cited on page 325).

60. W. Burnside, On an unsettled question in the theory of discontinuous groups. Q. J. Pure Appl. Math. **33**, 230–238 (1902) (Cited on page 28).

61. J.W. Cannon, The combinatorial structure of cocompact discrete hyperbolic groups. Geom. Dedicata **16**(2), 123–148 (1984) (Cited on page 321).

62. J.W. Cannon, W.J. Floyd, W.R. Parry, Introductorary notes on Richard Thompson's groups. L'Enseignement Mathématique **42**(2), 215–256 (1996) (Cited on pages 281, 282, 324, and 329).

63. A. Carpi, On abelian squares and substitutions, in *WORDS*, Rouen, 1997. Theor. Comput. Sci. **218**(1), 61–81 (1999) (Cited on page 83).

64. F. Cedó, J. Okniński, Semigroups of matrices of intermediate growth. Adv. Math., **212**(2), 669–691 (2007) (Cited on page 131).

65. J. Černý, Poznámka k homogénnym eksperimentom s konečnými automatami. Matematicko-fyzikalny Časopis Slovenskej Akadémie Vied **14**(3), 208–216 (1964) (Cited on page 159).

66. A. Church, J.B. Rosser, Some properties of conversion. Trans. Am. Math. Soc. **39**(3), 472–482 (1936) (Cited on page 45).

67. R.J. Clark, The existence of a pattern which is 5-avoidable but 4-unavoidable. Int. J. Algebra Comput. **16**(2), 351–367 (2006) (Cited on page 84).

68. A.H. Clifford, G.B. Preston, *The Algebraic Theory of Semigroups. I* (American Mathematical Society, Providence, 1961) (Cited on page 23).

69. J.M. Cohen, Cogrowth and amenability of discrete groups. J. Funct. Anal. **48**(3), 301–309 (1982) (Cited on pages 313 and 314).

70. P.M. Cohn, Embedding in semigroup with one-sided division. J. Lond. Math. Soc. **31**, 169–181 (1956) (Cited on page 54).

71. P.M. Cohn, Embeddings in sesquilateral division semigroups. J. Lond. Math. Soc. **31**, 181–191 (1956) (Cited on page 54).

72. P.M. Cohn, *Free Rings and Their Relations*. London Mathematical Society Monographs, vol. 19, 2nd edn. (Academic/Harcourt Brace Jovanovich, London, 1985) (Cited on pages 164 and 166).

73. M. Coornaert, A. Papadopoulos, *Symbolic Dynamics and Hyperbolic Groups*. Lecture Notes in Mathematics, vol. 1539 (Springer, Berlin, 1993) (Cited on page 322).

74. M. Coudrain, M.P. Schützenberger. Une condition de finitude des monoides finiment engendrés. C. R. Acad. Sci. Paris. Serie A **262**, 1149–1151 (1966) (Cited on page 77).

75. M. Crochemore, Sharp characterizations of square-free morphisms. Theor. Comput. Sci. **18**, 221–226 (1982) (Cited on page 75).

76. K. Culik II, J. Karhumäki, J. Kari, A note on synchronized automata and Road Coloring Problem. Int. J. Found. Comput. Sci. **13**, 459–471 (2002) (Cited on page 141).

77. J.D. Currie, Pattern avoidance: themes and variations. Theor. Comput. Sci. **339**(1), 7–18 (2005) (Cited on pages 83 and 84).

78. J.D. Currie, V. Linek, Avoiding patterns in the abelian sense. Can. J. Math. **51**(4), 696–714 (2001) (Cited on page 84).

79. M.M. Day, Amenable semigroups. Ill. J. Math. **1**, 509–544 (1957)

80. P. Dehornoy, Geometric presentations for Thompson's groups. J. Pure Appl. Algebra **203**(1–3), 1–44 (2005) (Cited on page 314). (Cited on page 277).

81. P. de la Harpe, *Topics in Geometric Group Theory*. Chicago Lectures in Mathematics (University of Chicago Press, Chicago, 2000) (Cited on page 298).

82. P. de la Harp, R.I. Grigorchuk, T. Ceccherini-Silberstein, Amenability and paradoxical decompositions for pseudogroups and discrete metric spaces. Tr. Mat. Inst. Steklova **224** (1999), Algebra. Topol. Differ. Uravn. i ikh Prilozh. 68–111; trans. in Proc. Steklov Inst. Math. **224**(1), 57–97 (1999) (Cited on pages 300, 312, 327, 329, and 330).

83. A. de Luca, S. Varricchio, Combinatorial properties of uniformly recurrent words and an application to semigroups. Int. J. Algebra Comput. **1**(2), 227–246 (1991) (Cited on page 174).

84. N. Dershowitz, Orderings for term-rewriting systems. Theor. Comput. Sci. **17**, 279–301 (1982) (Cited on page 6).

85. W. Dicks, Joel Friedman's proof of the strengthened Hanna Neumann conjecture (preprint), http://mat.uab.es/~dicks/SimplifiedFriedman.pdf (Cited on page 324).

86. W. Dicks, Simplified Mineyev (preprint), http://mat.uab.es/~dicks/SimplifiedMineyev.pdf (Cited on page 324).

87. V. Diekert, M. Kufleitner, K. Reinhardt, T. Walter, Regular languages are Church–Rosser congruential, in *Automata, Languages, and Programming*. Lecture Notes in Computer Science, vol. 7392 (Springer, Berlin/New York, 2012), pp. 177–188 (Cited on page 59).

88. Dnestr notebook, Math. Inst. Siberian Acad. Sci., Novosibirsk, 1993 (Cited on page 174).

89. A.P. Do Lago, On the burnside semigroups $x^n = x^{n+m}$. Int. J. Algebra Comput. **6**(2), 179–227 (1996) (Cited on page 154).

90. A.P. Do Lago, I. Simon, Free burnside semigroups. A tribute to Aldo de Luca. Theor. Inform. Appl. **35** (2001) (6), 579–595 (2002) (Cited on page 154).

91. C. Druţu, M. Kapovich, Lectures on geometric group theory (2012, preprint) (Cited on pages 282 and 320).

92. J. Dubnov, V. Ivanov, Sur l'abaissement du degré des polynômes en affineurs. C. R. (Doklady) Acad. Sci. URSS (N.S.) **41**, 95–98 (1943) (Cited on page 175).

93. S. Eilenberg, *Automata, Languages, and Machines, Vol. A*. Pure and Applied Mathematics, vol. 58 (Academic [A subsidiary of Harcourt Brace Jovanovich, Publishers], New York, 1974) (Cited on pages 12, 13, and 65).

94. M. Elder, É. Fusy, A. Rechnitzer, Counting elements and geodesics in Thompson's group F. J. Algebra **324**(1), 102–121 (2010) (Cited on page 325).

95. G. Endimioni, Conditions de finitude pour un groupe d'exposant fini (Finiteness conditions for a group with finite exponent). J. Algebra **155**, 290–297 (1993) (Cited on page 327).

96. D.B.A. Epstein, J.W. Cannon, D.F. Holt, S.V.F. Levy, M.S. Paterson, W.P. Thurston, *Word Processing in Groups* (Jones and Bartlett, Boston, 1992) (Cited on page 321).

97. M. Ershov, Golod–Shafarevich groups: a survey. Int. J. Algebra Comput. **22**(5), 1230001, 68 (2012) (Cited on page 328).

98. M. Ershov, G. Golan, M. Sapir, The Tarski numbers of groups (2014, preprint) (Cited on page 330).

99. D.S. Farley, Proper isometric actions of Thompson's groups on Hilbert space. Int. Math. Res. Not. **45**, 2409–2414 (2003) (Cited on page 324).

100. D.S. Farley, Finiteness and CAT(0) properties of diagram groups. Topology **42**(5), 1065–1082 (2003) (Cited on page 324).

101. A.F. Filippov, An elementary proof of Jordan's theorem. Uspehi Matem. Nauk (N.S.) **5**(39), 173–176 (1950) (Cited on page 216).

102. N. Fine, H. Wilf, Uniqueness theorems for periodic functions. Proc. Am. Math. Soc. **16**, 109–114 (1965) (Cited on page 8).

103. N.P. Fogg, *Substitutions in Dynamics, Arithmetics and Combinatorics*, ed. by V. Berthé, S. Ferenczi, C. Mauduit, A. Siegel. Lecture Notes in Mathematics, vol. 1794 (Springer, Berlin, 2002) (Cited on page 83).

104. J. Friedman, Sheaves on graphs, their homological invariants, and a proof of the Hanna Neumann conjecture (submitted). arXiv:1105.0129 (Cited on page 323).

105. H. Furstenberg, Poincaré recurrence and number theory. Bull. Am. Math. Soc. **5**, 211–234 (1981) (Cited on page 95).

106. F. Galvin, R. Thompson, Unpublished notes (Cited on page 324).

107. I.M. Gelfand, A.A. Kirillov, Sur les corps liés aux algébres enveloppantes des algébres de Lie. Publ. Math. I.H.E.S. **31**, 5–19 (1966) (Cited on page 125).

108. R. Geoghegan (ed.), *Open Problems in Infinite-Dimensional Topology: The Proceedings of the 1979 Topology Conference*, Ohio University, Athens, 1979. Topology Proceedings, vol. 4 (1979) no. 1) (1980), pp. 287–338 (Cited on page 325).

109. R. Geoghegan, F. Guzmán, Associativity and Thompson's group, in *Topological and Asymptotic Aspects of Group Theory*. Contemporary Mathematics, vol. 394 (American Mathematical Society, Providence, 2006), pp. 113–135 (Cited on page 277).

110. E. Ghys, P. de la Harpe (eds.), *Sur les groupes hyperboliques d'aprés Mikhael Gromov*. Progress in Mathematics, vol. 83 (Birkhäuser, Boston, 1990) (Cited on pages 320 and 321).

111. S. Ginsburg, On the length of the smallest uniform experiment which distinguishes the terminal states of a machine. J. Assoc. Comput. Mach. **5**, 266–280 (1958) (Cited on page 138).

112. E.S. Golod, On nil-algebras and finitely approximable p-groups. Izv. Akad. Nauk SSSR. Ser. Mat. **28**, 273–276 (1964) (Cited on pages 166, 176, and 220).

113. E.S. Golod, Some problems of Burnside type, in *Proceedings of International Congress of Mathematicians*, Moscow, 1966 (Izdat. "Mir", Moscow, 1968), pp. 284–289 (Cited on page 328).

114. E.S. Golod, I.R. Shafarevich, On the class field tower. Izv. Akad. Nauk SSSR. Ser. Mat. **28**, 261–272 (1964) (Cited on page 176).

115. W.H. Gottschalk, G.A. Hedlund, *Topological Dynamics*. AMS Colloquium Publications, vol. 36 (American Mathematical Society, Providence, 1955) (Cited on page 39).

116. J.A. Green, D. Rees, On semigroups in which $x^r = x$. Proc. Camb. Philos. Soc. **48**, 35–40 (1952) (Cited on pages 107 and 154).

117. F.P. Greenleaf, *Invariant Means on Topological Groups and Their Applications* (Van Nostrand Reinhold, New York, 1969) (Cited on page 300).

118. M. Greendlinger, Dehn's algorithm for the word problem. Commun. Pure Appl. Math. **13**, 67–83 (1960) (Cited on page 211).

119. R.I. Grigorcuk, On Burnside's problem on periodic groups. Funktsional. Anal. i Prilozhen. **14**(1), 53–54 (1980) (Cited on pages 294 and 295).

120. R.I. Grigorchuk, Symmetrical random walks on discrete groups, in *Multicomponent Random Systems*. Advances in Probability and Related Topics, vol. 6 (Dekker, New York, 1980), pp. 285–3251 (Cited on pages 313 and 314).

121. R.I. Grigorchuk, Symmetric random walks on discrete groups. Uspehi Mat. Nauk **32**(6)(198), 217–218 (1977) (Cited on page 313).

122. R.I. Grigorchuk, On the Milnor problem of group growth. Dokl. Akad. Nauk SSSR **271**(1), 30–33 (1983) (Cited on pages 282, 294, and 295).

123. R.I. Grigorchuk, Degrees of growth of finitely generated groups and the theory of invariant means. Izv. Akad. Nauk SSSR Ser. Mat. **48**(5), 939–985 (1984) (Cited on pages 294, 295, 296, and 299).

124. A.V. Grishin, Examples of T-spaces and T-ideals of characteristic 2 without the finite basis property. Fundam. Prikl. Mat. **5**(1), 101–118 (1999) (Cited on page 191).

125. A.V. Grishin, The variety of associative rings, which satisfy the identity $x^{32} = 0$, is not Specht, in *Formal Power Series and Algebraic Combinatorics* Moscow, 2000 (Springer, Berlin/New York, 2000), pp. 686–691 (Cited on page 191).

126. M. Gromov, Groups of polynomial growth and expanding maps (with an appendix by Jacques Tits). Publ. Math. Inst. Hautes Études Sci. **53**, 53–73 (1981) (Cited on pages 282 and 299).

127. M. Gromov, Hyperbolic groups, in *Essays in Group Theory*. Mathematical Sciences Research Institute Publications, vol. 8 (Springer, New York, 1987), pp. 75–263 (Cited on pages 213, 320, and 321).

128. V.S. Guba, The word problem for the relatively free semigroups satisfying $t^m = t^{m+n}$ with $m \geq 3$. Int. J. Algebra Comput. **3**(3), 335–348 (1993) (Cited on page 154).

129. V.S. Guba, On some properties of periodic words. Math. Notes **72**(3-4), 301–307 (2002) (Cited on page 10).

130. V.S. Guba, The Dehn function of Richard Thompson's group F is quadratic. Invent. Math. **163**(2), 313–342 (2006) (Cited on page 324).

131. V.S. Guba, Notes on the Burnside problem for odd exponents, private communication, Oct 2012 (Cited on page 223).

132. V.S. Guba, S.J. Pride, Low-dimensional (co)homology of free Burnside monoids. J. Pure Appl. Algebra **108**(1), 61–79 (1996) (Cited on page 154).

133. V.S. Guba, M. Sapir, *Diagram Groups*. Memoirs of the AMS, vol. 130 (American Mathematical Society, Providence, 1997) (Cited on pages 265, 266, 272, 273, and 324).

134. V.S. Guba, M. Sapir, The Dehn function and a regular set of normal forms for R. Thompson's group F. J. Aust. Math. Soc. Ser. A **62**(3), 315–328 (1997) (Cited on page 279).

135. V.S. Guba, M. Sapir, On subgroups of R. Thompson's group F and other diagram groups. Mat. Sb. **190**(8), 3–60 (1999) (Cited on pages 278 and 282).

136. V.S. Guba, M. Sapir, Diagram groups and directed 2-complexes: homotopy and homology. J. Pure Appl. Algebra **205**(1), 1–47 (2006) (Cited on page 324).

137. V.S. Guba, M. Sapir, Diagram groups are totally orderable. J. Pure Appl. Algebra **205**(1), 48–73 (2006) (Cited on pages 273 and 324).

138. Y. Guivarc'h, Groupes de Lie à croissance polynomiale. C. R. Acad. Sci. Paris, Sér. A–B **271**, A237–A239 (1970) (Cited on pages 285 and 292).

139. Y. Guivarc'h, Croissance polynomiale et périodes des fonctions harmoniques. Bull. Soc. Math. Fr. **101**, 333–379 (1973) (Cited on pages 285 and 292).

140. C.K. Gupta, A.N. Krasilnikov, A simple example of a non-finitely based system of polynomial identities. Comm. Algebra **30**(10), 4851–4866 (2002) (Cited on page 191).

141. C.K. Gupta, A.N. Krasilnikov, The finite basis question for varieties of groups – some recent results. Special issue in honor of Reinhold Baer (1902–1979). Ill. J. Math. **47**(1–2), 273–283 (2003) (Cited on page 326).

142. M. Hall Jr., *The Theory of Groups* (Macmillan, New York, 1959) (Cited on pages 23 and 221).

143. P. Hall, G. Higman, On the p-length of p-soluble groups and reduction theorems for Burnside's problem. Proc. Lond. Math. Soc. **6**(3), 1–42 (1956) (Cited on page 153).

144. G.H. Hardy, S. Ramanujan, Asymptotic formulae in combinatory analysis. Proc. Lond. Math. Soc. **17**, 75–115 (1918) (Cited on page 128).

145. G.H. Hardy, E.M. Wright, *An Introduction to the Theory of Numbers*, 5th edn. (Clarendon Press, Oxford, 1979) (Cited on pages 127 and 130).

146. T. Harju, D. Nowotka, The equation $x^i = y^j z^k$ in a free semigroup. Semigroup Forum **68**(3), 488–490 (2004) (Cited on page 10).

147. T. Harju, D. Nowotka, On the equation $x^k = z_1^{k_1} z_2^{k_2} \ldots z_n^{k_n}$ in a free semigroup. Theor. Comput. Sci. **330**(1), 117–121 (2005) (Cited on page 10).

148. P.J. Heawood, Map-colour theorems. Q. J. Math. Oxf. **24**, 332–338 (1890) (Cited on page 210).

149. G.A. Hedlund, Endomorphisms and automorphisms of the shift dynamical system. Math. Syst. Theory **3**, 320–375 (1969) (Cited on page 42).

150. S.M. Hermiller, Rewriting systems for Coxeter groups. J. Pure Appl. Algebra **92**(2), 137–148 (1994) (Cited on page 213).

151. N. Herstein, Wedderburn's theorem and a theorem of Jacobson. Am. Math. Mon. **68**, 249–251 (1961) (Cited on page 183).

152. G. Higman, On a conjecture of Nagata. Proc. Camb. Philos. Soc. **52**, 1–4 (1956) (Cited on page 175).

153. G. Higman, *Finitely Presented Infinite Simple Groups*. Notes on Pure Mathematics, vol. 8 (Department of Pure Mathematics, Department of Mathematics, I.A.S. Australian National University, Canberra, 1974) (Cited on page 324).

154. G. Higman, B.H. Neumann, H. Neumann, Embedding theorems for groups. J. Lond. Math. Soc. **24**, 247–254 (1949) (Cited on page 204).

155. W. Imrich, On finitely generated subgroups of free groups. Arch. Math. **28**, 21–24 (1977) (Cited on page 262).

156. S.V. Ivanov, The Burnside problem for all sufficiently large exponents. Int. J. Algebra Comput. **4**(1, 2), 1–300 (1994) (Cited on page 242).

157. S.V. Ivanov, Embedding free Burnside groups in finitely presented groups. Geom. Dedicata **111**, 87–105 (2005) (Cited on page 325).

158. S.V. Ivanov, A.M. Storozhev, On varieties of groups in which all periodic groups are abelian, in *Group Theory, Statistics, and Cryptography*. Contemporary Mathematics, vol. 360 (American Mathematical Society, Providence, 2004), pp. 55–62 (Cited on page 319).

159. M. Jackson, Small inherently nonfinitely based finite semigroups. Semigroup Forum, **64**, 297–324 (2002) (Cited on page 125).

160. M. Jackson, O. Sapir, Finitely based, finite sets of words. Int. J. Algebra Comput. **10**(6), 683–708 (2000) (Cited on page 155).

161. N. Jacobson, *Structure of Rings* (American Mathematical Society, Providence, 1956) (Cited on pages 194 and 195).

162. N. Jacobson, *Lie Algebras* (Dover, New York, 1962) (Cited on page 22).

163. B. Jónsson, A. Tarski, On two properties of free algebras. Math. Scand. **9**, 95–101 (1961) (Cited on page 26).

164. T.J. Kaczynski, Another proof of Wedderburn's theorem. Am. Math. Mon. **71**(6), 652–653 (1964) (Cited on page 183).

165. J. Kaďourek, Uncountably many varieties of semigroups satisfying $x^2y = xy$. Semigroup Forum **60**(1), 135–152 (2000) (Cited on page 155).

166. J. Kaďourek, On bases of identities of finite inverse semigroups with solvable subgroups. Semigroup Forum **67**(3), 317–343 (2003) (Cited on page 156).

167. I. Kaplansky, Rings with a polynomial identity. Bull. Am. Math. Soc. **54**, 575–580 (1948) (Cited on pages 167 and 169).

168. I. Kapovich, A. Miasnikov, P. Schupp, V. Shpilrain, Average-case complexity and decision problems in group theory. Adv. Math. **190**(2), 343–359 (2005) (Cited on page 325).

169. M. Kapovich, Representations of polygons of finite groups. Geom. Topol. **9**, 1915–1951 (2005) (Cited on page 323).

170. M.I. Kargapolov, Yu.I. Merzlyakov, *Fundamentals of Group Theory*, 4th edn. (Fizmatlit "Nauka", Moscow, 1996) (Cited on pages 220, 294, and 326).

171. J. Karhumäki, On cube free ω-words generated by binary morphisms. Discret. Appl. Math. **5**, 279–297 (1983) (Cited on page 83).

172. J. Kari, Synchronizing finite automata on Eulerian digraphs, in *Mathematical Foundations of Computer Science*, Mariánské Lázne, 2001. Theor. Comput. Sci. **295**(1–3), 223–232 (2003) (Cited on pages 142 and 159).

173. A.R. Kemer, Finite basability of identities of associative algebras. Algebra i Logika **26**(5), 597–641 (1987) (Cited on page 191).

174. A.R. Kemer, *Ideals of Identities of Associative Algebras*. Volume 82 of Translations of Mathematical Monographs (American Mathematical Society, Providence, 1991) (Cited on page 191).

175. H. Kesten, Symmetric random walks on groups. Trans. Am. Math. Soc. **92**, 336–354 (1959) (Cited on pages 313 and 314).

176. B.P. Kitchens, *Symbolic Dynamics. One-Sided, Two-Sided and Countable State Markov Shifts*. Universitext (Springer, Berlin, 1998) (Cited on page 159).

177. O. Kharlampovich, M. Sapir, Algorithmic problems in varieties. Int. J. Algebra Comput. **5**, 379–602 (1995) (Cited on pages xv and 152).

178. E.I. Kleiman, On bases of identities of Brandt semigroups. Semigroup Forum **13**, 209–218 (1977) (Cited on page 156).

179. E.I. Kleiman, Bases of identities of varieties of inverse semigroups. Sibirsk. Mat.Zh. **20**, 760–777 (1979) (Cited on page 156).

180. Yu.G. Kleiman, On the basis of products of varieties of groups. Izv. AN SSSR, Ser. Mat. **37**, 95–97 (1973) (Cited on page 243).

181. B. Kleiner, A new proof of Gromov's theorem on groups of polynomial growth. J. Am. Math. Soc. **23**(3), 815–829 (2010) (Cited on page 282).

182. D. Knuth, P. Bendix, Simple word problems in universal algebras, in *Computational Problems in Abstract Algebra*, ed. by J. Leech (Pergamon, New York, 1970), pp. 263–297 (Cited on page 50).

183. G. Köthe, Die Struktur der Ringe, deren Restklassenring nach dem Radikal vollstandig irreduzibelist. Math. Z. **32**, 161–186 (1930) (Cited on page 195).

184. A.T. Kolotov, Algebras and semigroups with quadratic growth functions. Algebra i Logika **19**(6), 659–668 (1980) (Cited on page 157).

185. A.I. Kostrikin, *Around Burnside* (trans. from the Russian and with a preface by James Wiegold). Ergebnisse der Mathematik und ihrer Grenzgebiete (3) (Results in Mathematics and Related Areas (3)), vol. 20 (Springer, Berlin, 1990) (Cited on pages 152 and 153).

186. A.N. Krasil'nikov, On the finiteness of the basis of identities of groups with a nilpotent commutator group. Izv. Akad. Nauk SSSR Ser. Mat. **54**, 1181–1195 (1990); trans. in Math. USSR Izv. **37**, 539–553 (1991) (Cited on page 326).

187. G.R. Krause, T.H. Lenagan, *Growth of Algebras and Gelfand–Kirillov Dimension* (Pitman, London, 1985) (Cited on page 194).

188. R.L. Kruse, Identities satisfied by a finite ring. J. Algebra **26**, 298–318 (1973) (Cited on page 29).

189. C. Kuratowski, Une méthode d'élimination des nombres transfinis des raisonnements mathématiques. Fundam. Math. **3**, 76–108 (1922) (Cited on page 3).

190. E.N. Kuzmin, On the Nagata–Higman theorem, in *Mathematical Structures–Computational Mathematics–Mathematical Modeling, Proceedings Dedicated to the 16th Birthday of Academician L. Iliev*, Sofia, 1975, pp. 101–107 (Cited on page 176).

191. G. Lallement, *Semigroups and Combinatorial Applications* (Wiley, New York/Chichester/Brisbane, 1979) (Cited on page 8).

192. V.N. Latyshev, Finite basis property of identities of certain rings. Usp. Mat. Nauk **32**(4), 259–260 (1977) (Cited on page 186).

193. A.A. Lavrik-Männlin, On some semigroups of intermediate growth. Int. J. Algebra Comput. **11**(5), 565–580 (2001) (Cited on page 131).

194. M.V. Lawson, *Finite Automata* (Chapman & Hall/CRC, Boca Raton, 2004) (Cited on pages 12 and 69).

195. P. Le Chenadec, *Canonical Forms in Finitely Presented Algebras*. Research Notes in Theoretical Computer Science (Pitman/Wiley, London/New York, 1986) (Cited on page 60).

196. E.W.H. Lee, Finite basis problem for semigroups of order five or less: generalization and revisitation. Studia Logica **101**(1), 95–115 (2013) (Cited on page 155).

197. E.W.H. Lee, J.R. Li, W.T. Zhang, Minimal non-finitely based semigroups. Semigroup Forum **85**(3), 577–580 (2012) (Cited on page 155).

198. T.H. Lenagan, A. Smoktunowicz, An infinite dimensional affine nil algebra with finite Gelfand-Kirillov dimension. J. Am. Math. Soc. **20**(4), 989–1001 (2007) (Cited on page 180).

199. J.M. Lever, *The Elizabethan Love Sonnet* (Barnes & Noble, London, 1968) (Cited on page 225).

200. F.W. Levi, On semigroups. Bull. Calcutta Math. Soc. **36**, 141–146 (1944) (Cited on page 56).

201. J. Levitzki, Prime ideals and the lower radical. Am. J. Math. **73**(1), 25–29 (1951) (Cited on pages 77 and 181).

202. J. Lewin, Subrings of finite index in finitely-generated rings. J. Algebra, **5**, 84–88 (1967) (Cited on page 190).

203. D. Lind, B. Marcus, *An Introduction to Symbolic Dynamics and Coding* (Cambridge University Press, Cambridge/New York, 1995) (Cited on pages 42, 148, 151, and 159).

204. Y. Lodha, J. Moore, A geometric solution to the von Neumann-Day problem for finitely presented groups (2013), arXiv:1308.4250 (Cited on pages 325 and 329).

205. A.A. Lopatin, On the nilpotency degree of the algebra with identity $x^n = 0$. J. Algebra **371**, 350–366 (2012) (Cited on page 176).

206. M. Lothaire, *Combinatorics on Words*. Volume 17 of Encyclopedia of Mathematics and Its Applications (Addison-Wesley, Reading, 1983) (Cited on page 83).

207. M. Lothaire. *Algebraic Combinatorics on Words*. Encyclopedia of Mathematics and Its Applications, vol. 90 (Cambridge University Press, Cambridge, 2002) (Cited on pages 41, 42, and 83).

208. I.V. L'vov, Varieties of associative rings. I. Algebra i Logika **12**(3), 269–297 (1973) (Cited on pages 29 and 116).

209. R.C. Lyndon, Identities in two-valued calculi. Trans. Am. Math. Soc. **71**, 457–465 (1951) (Cited on page 32).

210. R.C. Lyndon, P.E. Schupp, *Combinatorial Group Theory* (Springer, Berlin/ New York, 1977) (Cited on pages 200, 209, and 211).

211. R.C. Lyndon, M.P. Schützenberger, The equation $a^M = b^N c^P$ in a free group. Mich. Math. J. **9**, 289–298 (1962) (Cited on page 10).

212. I.G. Lysenok, A system of defining relations for the Grigorchuk group. Mat. Zametki **38**, 503–511 (1985) (Cited on pages 300 and 311).

213. I.G. Lysenok, Infinite Burnside groups of even exponent. Izvestiya: Math. **60**(3), 453–654 (1996) (Cited on page 242).

214. J.H. Maclagan-Wedderburn. A theorem on finite algebras. Trans. Am. Math. Soc. **6**, 349–352 (1905) (Cited on page 183).

215. W. Magnus, Beziehungen zwischen Gruppen und Idealen in einem speziellen Ring. Math. Ann. **111**, 259–280 (1935) (Cited on page 64).

216. W. Magnus, A connection between the Baker-Hausdorff formula and a problem of Burnside. Ann. Math. **52**, 11–26 (1950); Errata Ann. Math. **57**, 606 (1953) (Cited on page 152).

217. A.I. Mal'cev, Nilpotent semigroups. Uch. Zap. Ivanovsk. Pedagog. Inst. **4**, 107–111 (1953) (Cited on page 94).

218. A.I. Mal'cev, *Algebraic Systems* (Springer, Berlin/New York, 1973) (Cited on pages 24 and 25).

219. A. Mann, *How Groups Grow*. London Mathematical Society Lecture Note Series, vol. 395 (Cambridge University Press, Cambridge, 2012) (Cited on page 282).

220. S. Margolis, J. Meakin, E-unitary inverse monoids and the Cayley graph of a group presentation. J. Pure Appl. Algebra **58**, 45–76 (1989) (Cited on page 136).

221. S. Margolis, J. Meakin, M. Sapir, Algorithmic problems in groups, semigroups and inverse semigroups, in *Semigroups, Formal Languages and Groups*, York, 1993, pp. 147–214 (Cited on pages 137 and 261).

222. G.I. Mashevitsky. On bases of completely simple semigroup identities. Semigroup Forum **30**(1), 67–76 (1984) (Cited on page 155).

223. A Mathoverflow.net discussion # 60897 of invariant means on the group of integers, http://mathoverflow.net/questions/60897/invariant-means-on-the-integers (Cited on page 307).

224. N.H. McCoy, Prime ideals in general rings. Am. J. Math. **71**(4), 823–833 (1949) (Cited on page 181).

225. R. McKenzie, Tarski's finite basis problem is undecidable. Int. J. Algebra Comput. **6**, 49–104 (1996) (Cited on pages 30 and 31).

226. R. McNaughton, S. Papert, *Counter-Free Automata*. With an appendix by William Henneman. M.I.T. Research Monograph, vol. 65 (M.I.T., Cambridge/Mass.-London, 1971) (Cited on page 69).

227. G.F. McNulty, *Talks at the Twenty First Victorian Algebra Conference with Workshop on Universal Algebraic Techniques in Semigroup Theory and Algebraic Logic*, Melbourne, Oct 2003 (Cited on page 83).

228. G.F. McNulty, C.R. Shallon, Inherently non-finitely based finite algebras, in *Universal Algebra and Lattice Theory* Puebla, 1982. Volume 1004 of Lecture Notes in Mathematics (Springer, Berlin/New York, 1983), pp. 206–231 (Cited on page 31).

229. I.L. Mel'nichuk, Existence of infinite finitely generated free semigroups in certain varieties of semigroups, in *Algebraic Systems with One Action and Relation* (Leningrad. Gos. Ped. Inst., Leningrad, 1985), pp. 74–83 (Cited on page 84).

230. Yu.I. Merzlyakov, Infinite finitely generated periodic groups. Dokl. Akad. Nauk SSSR **268**(4), 803–805 (1983) (Cited on pages 220 and 294).

231. F. Mignosi, P. Séébold, If a D0L language is k-power free then it is circular, in *Proceedings of 20th International Conference on Automata, Languages, and Programming (ICALP)*, Lund. Lecture Notes in Computer Science, vol. 700 (Springer, Berlin/New York, 1993), pp. 507–518 (Cited on page 83).

232. K.V. Mikhajlovskii, A.Yu. Olshanskii, Some constructions relating to hyperbolic groups, in *Geometry and Cohomology in Group Theory*, Durham, 1994. London Mathematical Society Lecture Note Series, vol. 252 (Cambridge University Press, Cambridge, 1998), pp. 263–290 (Cited on page 319).

233. J. Milnor, Problem 5603. Amer. Math. Mon. **75**, 685–686 (1968) (Cited on page 282).

234. I. Mineyev, submultiplicativity and the Hanna Neumann conjecture. Ann. Math. (2) **175**(1), 393–414 (2012) (Cited on page 324).

235. N. Monod, Groups of piecewise projective homeomorphisms. Proc. Natl. Acad. Sci. U.S.A **110**(12), 4524–4527 (2013) (Cited on page 329).

236. E.F. Moore, Gedanken experiments on sequential machines, in *Automata Studies*, ed. by C.E. Shannon, J. McCarthy (Princeton University Press, Princeton, 1956), pp. 129–153 (Cited on page 138).

237. J. Moore, Fast growth in the Følner function for Thompson's group F. Groups Geom. Dyn. **7**(3), 633–651 (2013) (Cited on page 325).

238. M. Morse, G.A. Hedlund, Symbolic dynamics. Am. J. Math. **60**, 815–866 (1938) (Cited on pages 39 and 72).

239. M. Morse, G.A. Hedlund, Symbolic dynamics II. Sturmian trajectories. Am. J. Math. **62**, 1–42 (1940) (Cited on page 41).

240. M. Morse, G.A. Hedlund, Unending chess, symbolic dynamics and a problem in semigroups. Duke Math. J. **11**, 1–7 (1944) (Cited on page 89).

241. W.D. Munn, Free inverse semigroups. Proc. Lond. Math. Soc. **30**, 385–404 (1974) (Cited on page 137).

242. V.L. Murskii, The existence in the three-valued logic of a closed class with a finite basis having no finite complete system of identities. Dokl. Akad. Nauk SSSR **163**, 815–818 (1965) (Cited on page 32).

243. V.L. Murskii, The number of k-element algebras with a binary operation which do not have a finite basis of identities. Probl. Kibernet. **35**(5–27), 208 (1979) (Cited on page 31).

244. M.B. Nathanson, Number theory and semigroups of intermediate growth. Am. Math. Mon. **106**, 666–669 (1999) (Cited on page 129).

245. B.H. Neumann, T. Taylor, Subsemigroups of nilpotent groups. Proc. R. Soc. Lond. Ser. A 274, **1**, 1–4 (1963) (Cited on page 158).

246. H. Neumann, On the intersection of finitely generated free groups. Publ. Math. Debr. **4**, 186–189 (1956) (Cited on pages 262 and 323).

247. H. Neumann, On the intersection of finitely generated free groups. Addendum. Publ. Math. Debr. **5**, 128 (1957) (Cited on page 262).

248. H. Neumann, *Varieties of Groups* (Springer, New York, 1967) (Cited on pages 29, 243, and 326).

249. W.D. Neumann, On intersections of finitely generated subgroups of free groups, in *Groups-Canberra 1989*. Lecture Notes in Mathematics, vol. 1456 (Springer, Berlin/New York, 1990), pp. 161–170 (Cited on pages 262 and 323).

250. M.H.A. Newman, On theories with a combinatorial definition of "equivalence". Ann. Math. **43**(2), 223–243 (1942) (Cited on page 48).

251. P.S. Novikov, S.I. Adian, On infinite periodic groups. I, II, III. Izv. Akad. Nauk SSSR. Ser. Mat. **32**, 212–244; 251–524; 709–731 (1968) (Cited on pages 222, 223, 224, and 319).

252. S. Oates, M.B. Powell, Identical relations in finite groups. J. Algebra **1**, 11–39 (1964) (Cited on page 29).

253. J. Okniński, Linear semigroups with identities. in *Semigroups: Algebraic Theory and Applications to Formal Languages and Codes*, Luino 1992 (World Scientific, Singapore, 1993), pp. 201–211 (Cited on page 131).

254. J. Okniński, *Semigroups of Matrices* (World Scientific, Singapore, 1998) (Cited on page 131).

255. J. Okniński, A. Salwa, Generalised tits alternative for linear semigroups. J. Pure Appl. Algebra **103**(2), 211–220 (1995) (Cited on page 131).

256. A.Yu. Olshanskii, On the problem of a finite basis of identities in groups. Izv. Akad. Nauk SSSR, Ser. Mat. **34**, 376–384 (1970) (Cited on pages 197 and 243).

257. A.Yu. Olshanskii, On the question of the existence of an invariant mean on a group. Uspekhi Mat. Nauk **35** no. 4(214), 199–200 (1980) (Cited on page 314).

258. A.Yu. Olshanskii, The Novikov-Adyan theorem. Mat. Sb. (N.S.) **118**(160)(2), 203–235 (1982) (Cited on pages 222, 223, 224, 225, 228, 229, 230, 231, 232, 233, 234, 235, 236, 237, 240, and 242).

259. A.Yu. Olshanskii, Varieties in which all finite groups are abelian. Mat. Sb. (N.S.) **126**(168)(1), 59–82 (1985) (Cited on page 319).

260. A.Yu. Olshanskii, *The Geometry of Defining Relations in Groups*. Nauka, Moscow, 1989 (trans. from Russian by Yu. A. Bahturin). Mathematics and Its Applications (Soviet Series), vol. 70 (Kluwer, Dordrecht, 1991) (Cited on pages 200, 211, 217, 221, 223, 225, 229, 240, 319, and 322).

261. A.Yu. Olshanskii, Hyperbolicity of groups with subquadratic isoperimetric inequalities. Int. J. Algebra Comput. **1**, 282–290 (1991) (Cited on page 321).

262. A.Yu. Olshanskii, D.V. Osin, Large groups and their periodic quotients. Proc. Am. Math. Soc. **136**(3), 753–759 (2008) (Cited on page 221).

263. A. Olshanskii, D. Osin, M. Sapir, Lacunary hyperbolic groups (with appendix by M. Kapovich and B. Kleiner). Geom. Topol. **13**, 2051–2140 (2009) (Cited on page 221).

264. A. Olshanskii, M. Sapir, Non-amenable finitely presented torsion-by-cyclic groups. Publ. Math. Inst. Hautes Études Sci. **96**(2002), 43–169 (2003) (Cited on pages 325 and 329).

265. D. Osin, Small cancellations over relatively hyperbolic groups and embedding theorems. Ann. Math. (2) **172**(1), 1–39 (2010) (Cited on page 319).

266. P. Perkins, Bases for equational theories of semigroups. J. Algebra **11**, 298–314 (1969) (Cited on page 155).

267. P. Perkins, Basic questions for general algebras. Algebra Universalis **19**(1), 16–23 (1984) (Cited on page 31).

268. K. Petersen, *Ergodic Theory* (corrected reprint of the 1983 original). Cambridge Studies in Advanced Mathematics, vol. 2 (Cambridge University Press, Cambridge, 1989) (Cited on page 149).

269. M. Petrich, *Inverse Semigroups*. Pure and Applied Mathematics (New York). Wiley-Interscience Publication (Wiley, New York, 1984) (Cited on pages 133 and 156).

270. A.N. Plyushchenko, On the word problem in free Burnside semigroups with the identity $x^2 = x^3$. Izv. Vyssh. Uchebn. Zaved. Mat. **11**, 89–93 (2011); trans. in Russian Math. (Iz. VUZ) **55**(11), 76–79 (2011) (Cited on page 154).

271. M.E. Prouhet, Mémoire sur quelques relations entre les puissances des nombres. C. R. Acad. Sci. Paris. Serie A **33**, 31 (1851) (Cited on page 71).

272. Yu.P. Razmyslov, *Identities of Algebras and their Representations* (trans. from the 1989 Russian original by A.M. Shtern). Translations of Mathematical Monographs, vol. 138 (American Mathematical Society, Providence, 1994) (Cited on page 176).

273. P.F. Reichmeider, *The Equivalence of Some Combinatorial Matching Theorems* (Polygonal Publishing House, Washington, 1984) (Cited on page 308).

274. L.H. Rowen, *Polynomial Identities in Ring Theory*. Pure and Applied Mathematics, vol. 84 (Academic, [Harcourt Brace Jovanovich, Publishers], New York/London, 1980) (Cited on pages 194 and 195).

275. J. Sakarovitch, *Elements of Automata Theory* (trans. from the 2003 French original by Reuben Thomas) (Cambridge University Press, Cambridge, 2009) (Cited on pages 12, 13, and 65).

276. I.N. Sanov, Solution of Burnside's problem for exponent 4. Leningr. State Univ. Ann. [Uchenye Zapiski] Math. Ser. **10**, 166–170 (1940) (Cited on page 222).

277. M. Sapir, Problems of Burnside type and the finite basis property in varieties of semigroups. Izv. Akad. Nauk. SSSR. Ser. Mat. **51**(2), 319–340 (1987); trans. in Math USSR-Izv **30**(2), 295–314 (1988) (Cited on pages 79, 93, 108, and 116).

278. M. Sapir, Inherently non-finitely based finite semigroups. Mat. Sb. **133**(2), 154–166 (1987) (Cited on pages 118 and 125).

279. M. Sapir, The restricted Burnside problem for varieties of semigroups. Izv. Akad. Nauk SSSR Ser. Mat. **55**(3), 670–679 (1991); trans. in Math. USSR-Izv. **38**(3), 659–667 (1992) (Cited on page 153).

280. M. Sapir, Identities of finite inverse semigroups. Int. J. Algebra Comput. **3**(1), 115–124 (1993) (Cited on page 133).

281. M. Sapir, Asymptotic invariants, complexity of groups and related problems. Bull. Math. Sci. **1**(2), 277–364 (1993) (Cited on pages 319 and 321).

282. M. Sapir, Some group theory problems. Int. J. Algebra Comput. **17**(5–6), 1189–1214 (2007) (Cited on pages 282 and 323).

283. O. Sapir, Finitely based words. Int. J. Algebra Comput. **10**(4), 457–480 (2000) (Cited on page 155).

284. O. Sapir, Finitely based words with at most two nonlinear letters (2012, preprint) (Cited on page 155).

285. O.Yu. Schmidt, *Abstract Theory of Groups* (trans. from the Russian by Fred Holling and J.B. Roberts), trans. ed. by J.B. Roberts (W.H. Freeman, San Francisco/London, 1966) (Cited on page 107).

286. O. Schreier, Die Untergruppen der freien Gruppen. Abh. Math. Sem. Univ. Hambg. **5**, 161–188 (1926) (Cited on page 67).

287. A. Schrijver, *Combinatorial Optimization: Polyhedra and Efficiency. Vol. A. Paths, Flows, Matchings. Chapters 1–38.* Algorithms and Combinatorics, vol. 24A (Springer, Berlin, 2003) (Cited on page 308).

288. P. Schupp, On Dehn's algorithm and the conjugacy problem. Math. Ann. **178**, 119–130 (1968) (Cited on page 214).

289. J.-P. Serre, *Trees* (Springer, Berlin/New York, 1980) (Cited on page 209).

290. C. Shallon, Nonfinitely based binary algebra derived from lattices, Ph.D. thesis, University of California, Los Angeles, 1979 (Cited on page 32).

291. V.V. Shchigolev, Examples of infinitely based T-ideals. Fundam. Prikl. Mat. **5**(1), 307–312 (1999) (Cited on page 191).

292. V.V. Shchigolev, Examples of infinitely based T-spaces. Mat. Sb. **191**(3), 143–160 (2000); trans. in Sb. Math. **191**(3–4), 459–476 (2000) (Cited on page 191).

293. L.N. Shevrin, On locally finite semigroups. Dokl. Akad. Nauk SSSR **162**, 770–773 (1965); trans. in Sov. Math., Dokl. **6**, 769–772 (1965) (Cited on page 107).

294. L.N. Shevrin, On the theory of epigroups. I. Mat. Sb., **185**(8), 129–160 (1994); trans. in Russ. Acad. Sci. Sb. Math. **82**(1), 485–512 (1995) (Cited on page 113).

295. L.N. Shevrin, On the theory of epigroups. II. Mat. Sb. **185**(9), 153–176 (1994); trans. in Russ. Acad. Sci. Sb. Math. **83**(2), 133–154 (1995) (Cited on page 113).

296. L.N. Shevrin, M.V. Volkov, Identities of semigroups. Izv. Vyssh. Uchebn. Zaved. Mat. **11**, 3–47 (1985) (Cited on page 154).

297. A.I. Shirshov, On some nonassociative nil-rings and algebraic algebras. Mat. Sb. **41**(3), 381–394 (1957) (Cited on pages 167 and 170).

298. A.I. Shirshov, On rings with identity relations. Mat. Sb. **43**(2), 277–283 (1957) (Cited on page 170).

299. A.I. Shirshov, in *Selected Works of A.I. Shirshov* (trans. from the Russian by Murray Bremner and Mikhail V. Kotchetov), ed. by L.A. Bokut, V. Latyshev, I. Shestakov, E. Zel'manov. Contemporary Mathematicians (Birkhäuser, Basel, 2009) (Cited on page 167).

300. L.M. Shneerson, Relatively free semigroups of intermediate growth. J. Algebra **235**, 484–546 (2001) (Cited on page 158).

301. L.M. Shneerson, Growth, unavoidable words, and M. Sapir's conjecture for semigroup varieties. J. Algebra **271**(2), 482–517 (2004) (Cited on page 158).

302. L.M. Shneerson, On semigroups of intermediate growth. Commun. Algebra **32**(5), 1793–1803 (2004) (Cited on page 157).

303. L.M. Shneerson, Polynomial growth in semigroup varieties. J. Algebra **320**(6), 2218–2279 (2008) (Cited on page 158).

304. A.M. Shur, Growth properties of power-free languages, in *Developments in Language Theory*. Lecture Notes in Computer Science, vol. 6795 (Springer, Heidelberg, 2011), pp. 28–43 (Cited on page 83).

305. A. Smoktunowicz, Some results in noncommutative ring theory, in *International Congress of Mathematicians*, vol. II (European Mathematical Society, Zürich, 2006), pp. 259–269 (Cited on page 195).

306. W. Specht, Gesetze in Ringen. I. Math. Z. **52**, 557–589 (1950) (Cited on page 191).

307. C.C. Squier, Word problems and a homological finiteness conditions for monoids. J. Pure Appl. Algebra **49**, 201–217 (1987) (Cited on page 262).

308. R.P. Stanley, *Enumerative Combinatorics* (Vol. 2. With a foreword by Gian-Carlo Rota and appendix 1 by Sergey Fomin). Cambridge Studies in Advanced Mathematics, vol. 62 (Cambridge University Press, Cambridge, 1999) (Cited on pages 128 and 297).

309. V.I. Sushchansky, Periodic p-groups of permutations and the unrestricted Burnside problem. Dokl. Akad. Nauk SSSR **247**(3), 557–561 (1979) (Cited on page 221).

310. A.S. Švarč, A volume invariant of coverings. Dokl. Akad. Nauk SSSR **105**, 32–34 (1955) (Cited on page 37).

311. S. Świerczkowski, On a free group of rotations of the Euclidean space. Indag. Math. **20**, 376–378 (1958) (Cited on page 301).

312. R. Szwarc, A short proof of the Grigorchuk-Cohen cogrowth theorem. Proc. Am. Math. Soc. **106**(3), 663–665 (1989) (Cited on page 313).

313. J. Tits, Free subgroups in linear groups. J. Algebra **20**, 250–270 (1972) (Cited on page 326).

314. A. Thue, Über unendliche Zeichenreihen. Norske Vid. Selsk. Skr. , I. Mat. Nat. Kl. Christiana, **7**, 1–22 (1906) (Cited on pages 72 and 74).

315. W.P. Thurston, Conway's tiling groups. Am. Math. Mon. **97**(8), 757–773 (1990) (Cited on page 202).

316. A.N. Trahtman, Finiteness of a basis of identities of five-element semigroups, in *Semigroups and Their Homomorphisms* (Ross. Gos. Ped. Univ., Leningrad, 1991), pp. 76–97 (Cited on page 155).

317. A.N. Trahtman, The road coloring problem. Isr. J. Math. **172**(1), 51–60 (2009) (Cited on page 141).

318. V.I. Trofimov, The growth functions of finitely generated semigroups. Semigroup Forum **21**(4), 351–360 (1980) (Cited on page 157).

319. V.I. Trofimov, Growth functions of algebraic systems. Candidate's Thesis, Physicomathematical Sciences, Sverdlovsk, 1982 (Cited on page 157).

320. V.A. Ufnarovskii, On the use of graphs for calculating the basis, growth, and Hilbert series of associative algebras. Mat. Sb. **180**(11), 1548–1560 (1989) (Cited on pages 157 and 194).

321. V.A. Ufnarovskij, Combinatorial and asymptotic methods in algebra, in *Algebra, VI*. Volume 57 of Encyclopaedia Mathematical Sciences (Springer, Berlin, 1995), pp. 1–196 (Cited on pages 157 and 194).

322. L. van den Dries, A.J. Wilkie, Gromov's theorem on groups of polynomial growth and elementary logic. J. Algebra **89**(2), 349–374 (1984) (Cited on page 282).

323. M.R. Vaughan-Lee, Uncountably many varieties of groups. Bull. Lond. Math. Soc. **2**, 280–286 (1970) (Cited on page 243).

324. M.R. Vaughan-Lee, *The Restricted Burnside Problem*, 2nd edn. Volume 8 of London Mathematical Society Monographs. New Series. (The Clarendon Press/Oxford University Press, Oxford/New York, 1993) (Cited on page 221).

325. M.V. Volkov, The finite basis problem for finite semigroups: a survey, in *Semigroups*, Braga, 1999 (World Scientific, River Edge, 2000), pp. 244–279 (Cited on page 154).

326. M.V. Volkov, The finite basis problem for finite semigroups. Sci. Math. Jpn. **53**(1), 171–199 (2001) (Cited on page 154).

327. M.V. Volkov, Synchronizing automata and the Černý conjecture, in *Language and Automata Theory and Applications*, ed. by C. Martín-Vide, F. Otto, H. Fernau. Volume 5196 of Lecture Notes in Computer Science (Springer, Berlin/Heidelberg, 2008), pp. 11–27 (Cited on pages 139 and 159).

328. J. von Neumann, Zur allgemeinen Theorie des Masses. Fund. Math. **13**, 73–116 (1929); Collected Works, vol. I (Pergamon Press, New York/Oxford/London/Paris, 1961), pp. 599–643 (Cited on pages 304 and 314).

329. S. Wagon, *The Banach–Tarski Paradox*. Corrected reprint of the 1985 original (Cambridge University Press, Cambridge, 1993) (Cited on pages 300, 304, 306, 312, and 314).

330. D.T. Wise, Cubulating small cancellation groups. Geom. Funct. Anal. **14**(1), 150–214 (2004) (Cited on page 322).

331. W. Woess, Cogrowth of groups and simple random walks. Arch. Math. (Basel) **41**(4), 363–370 (1983) (Cited on page 313).

332. J.A. Wolf, Growth of finitely generated solvable groups and curvature of riemannian manifolds. J. Differ. Geom. **2**, 421–446 (1968) (Cited on pages 282, 285, and 311).

333. E.I. Zel'manov, The solution of the restricted Burnside problem for groups of odd exponent. Izv. Akad. Nauk. SSSR. Ser. Mat. **54**(1), 42–59 (1990); trans. in Math. USSR-Izv. **36**(1), 41–60 (1991) (Cited on pages 153 and 299).

334. E.I. Zel'manov, The solution of the restricted Burnside problem for 2-groups. Mat. Sb. (N.S.), **182**(4), 568–592 (1991) (Cited on pages 153 and 299).

335. E.I. Zel'manov, On additional laws in the Burnside problem for periodic groups. Int. J. Algebra Comput. **3**(4), 583–600 (1993) (Cited on pages 326 and 327).

336. E.I. Zel'manov, Some open problems in the theory of infinite dimensional algebras. J. Korean Math. Soc. **44**(5), 1185–1195 (2007) (Cited on page 194).

337. A.I. Zimin, Blocking sets of terms, in *Fifteenth All-Union Algebra Conference, Krasnoyarsk, Abstract of Reports, Part 1*, 1979, p. 63 (Cited on pages 77 and 84).

338. A.I. Zimin, Blocking sets of terms. Mat. Sb. **119**(3), 363–375 (1982) (Cited on page 77).

Name Index

Abel, 84
Abért, xvi, 158, 198, 255
Adian, xiv–xvi, 28, 197, 198, 213, 217,
    222, 223, 243, 314, 315, 319, 322,
    329
Adler, 86, 139, 141
Agol, 322
Aleshin, 220
Amitsur, 169
Arshon, 71, 72

Baer, xv, xvi, 161, 180–183, 188, 189, 195
Baker, 1, 32, 84
Banach, 198, 300–302, 304
Bartholdi, xvi, 86, 131, 282, 295, 299, 300
Bass, xiv, 197, 198, 282, 285, 292
Baumslag, 207, 289
Bean, 71, 76, 77, 79, 103
Belov-Kanel, xiv, xvi, 157, 161, 172, 174,
    176, 191, 194, 195
Belyaev, 131
Bendix, 49, 50, 53, 60
Bergman, xvi, 54, 161, 164
Bernoulli, 148
Berstel, 83
Bestvina, 323
Garrett Birkhoff, xv, 25
George Birkhoff, xv, 1, 22, 39, 149
Book, 54
Borisenko, 157, 194
Brandt, 19, 56, 60, 109, 112, 113, 115,
    133–136, 156, 185
Braun, 195
Bridson, 153
Brin, 278
Brown, 85, 106
Bryant, 191, 197, 198, 242, 243

Brzozowski, 154
Burillo, 325
Burnside, xiii–xv, 1, 28, 30, 31, 77, 152,
    154, 166, 194, 197, 198, 220, 221,
    223, 242, 299, 314, 318, 328

Cannon, 321
Capri, 83
Cayley, 23, 66, 67, 69, 95, 136–138, 140,
    143, 294, 320
Ceccherini-Silberstein, 329
Chebyshev, 85, 127
Černý, 159
Chomsky, 1, 65
Church, xv, 45, 46, 48–50, 54, 59–61, 63,
    65, 130–132, 154, 209, 213, 244, 259,
    264, 276, 279–282, 284, 324, 325
Clark, 84
Cohn, 54, 164
Coornaert, 322
Coudrain, 77
Crochemore, 75
Cross, 29, 30, 186, 190, 191
Culik, 141
Currie, 83, 84

Darboux, 317
Day, xiv, 197, 313, 314, 325, 327
de Bruijn, 44
Dehn, xiv, 209, 213, 242, 314, 315, 318,
    320, 321, 324
Dekker, 198, 312
de la Harpe, 298, 329
de Luca, 174
Dershowitz, 6
Descartes, 2
Dicks, 324

© Springer International Publishing Switzerland 2014
M.V. Sapir, *Combinatorial Algebra: Syntax and Semantics*,
Springer Monographs in Mathematics, DOI 10.1007/978-3-319-08031-4

Subject Index

0-cell, 217
0-direct sum, 133
0-edge, 217
0-relation, 217
2-complex, 256
  cells of, 256
  connected, 256
  fundamental group of, 256
  spanning tree of, 256

absolutely negligible subset of a subshift,
  148
action of a semigroup, 67
  free, 304
adjacency matrix, 65
algebra, 15
  inherently non-finitely based, 30
  absolutely free, 24
  critical, 29
  finitely based, 28
  free in a class, 24
  locally finite, 27
  relatively free, 24
    rank of, 24
amalgam of two groups, 207
associative algebra, 22
  absolutely free, 163
  n-dimensional, 168
  finite dimensional, 166
associativity, 18
automaton, 11
  bi-rooted, 137
  complete, 12
  deterministic, 12
  equalizer, 12
  finite, 11
  inverse, 15

  Mealy, 13
  rooted, 12
  strongly deterministic, 13
  synchronizing, 138
  underlying graph of, 11
automorphism, 18

Baer radical, 181
ball in a metric space, 39
band of bonds, 227
band of cells, 204
  end edge of, 204
  sides of, 204
  start edge of, 204
band of semigroups, 87
basis, 21
binary relation, 2
  anti-symmetric, 2
  equivalence, 2
    generated by a set of pairs, 3
    trivial, 2
  partial order, 2
  reflexive, 2
  symmetric, 2
  total order, 2
  transitive, 2
binomial coefficient, 2
block, 81
bond, 218
boundary of a van Kampen diagram,
  199

canonical representative, 48
Cartesian product, 2, 16
Catalan number, 297
Cayley graph of a semigroup, 66
  labeled, 66

© Springer International Publishing Switzerland 2014
M.V. Sapir, *Combinatorial Algebra: Syntax and Semantics*,
Springer Monographs in Mathematics, DOI 10.1007/978-3-319-08031-4

Printed in the United States
By Bookmasters